Lecture Notes in Applied and Computational Mechanics

Volume 23

Series Editors

Prof. Dr.-Ing. Friedrich Pfeiffer
Prof. Dr.-Ing. Peter Wriggers

Lecture Notes in Applied and Computational Mechanics

Edited by F. Pfeiffer and P. Wriggers

Mechanical Modelling and Computational Issues in Civil Engineering

Michel Frémond
Franco Maceri (Eds.)

 Springer

Professor Dr. Michel Frémond
Laboratoire Lagrange
Laboratoire Central des Ponts et Chaussées
75732 Paris cedex 15
France
michel.fremond@lcpc.fr

Professor Dr. Franco Maceri
Dipartimento di Ingegneria Civile
Università di Roma "Tor Vergata"
Via del Politecnico, 1
00133 Roma
Italy
Franco.Maceri@uniroma2.it

With 236 Figures

ISSN 1613-7736

ISBN-10   3-540-25567-2 Springer Berlin Heidelberg New York

ISBN-13   978-3-540-25567-3 Springer Berlin Heidelberg New York

Library of Congress Control Number: 2005923656

Springer is a part of Springer Science+Business Media

springeronline.com

Typesetting: Data conversion by the editors.
Final processing by PTP-Berlin Protago-TeX-Production GmbH, Germany
Cover-Design: design & production GmbH, Heidelberg
Printed on acid-free paper   62/3020/Yu - 5 4 3 2 1 0

Preface

This book collects some papers of Italian and French engineers, mechanicians and mathematicians, all associated within a European research network, the Lagrange Laboratory. Many topics are covered, such as monumental dams, soil mechanics and geotechnics, granular media, contact and friction problems, damage and fracture, new structural materials, vibration damping. Modelling and computational aspects are both dealt with. The Lagrange Laboratory met plenarily twice, at Le Mont-Saint-Michel in 2001 and at Ravello in 2002, and many individual exchanges took place meanwhile as a result of the activities of a true community, sharing scientific culture and a common understanding of modern civil engineering. In our opinion, this book offers a good example of cooperation between Italian and French scientists, and we believe that it will be of great interest for the reader.

Rome,
December 20, 2004

Michel Frémond
Franco Maceri

Contents

A Non-linear Hardening Model Based on Two Coupled Internal Hardening Variables: Formulation and Implementation

Nelly Point, Silvano Erlicher

Comparison between Static and Dynamic Criteria of Material Stability

Antonio Grimaldi, Raimondo Luciano

Material Damage Description via Structured Deformations . . . 235

Marc François, Gianni Royer-Carfagni

Singular equilibrated stress fields for no-tension panels 255

Massimiliano Lucchesi, Miroslav Silhavy, Nicola Zani

Damage of Materials: Damaging Effects of Macroscopic Vanishing Motions . . . 267

Elena Bonetti, Michel Frémond

Monumental Dams

Ruggiero Jappelli

Dipartimento di Ingegneria Civile
Università di Roma "Tor Vergata",
via del Politecnico, 1
00133 Roma, Italy

Abstract. Like silent witnesses to the past, large dams built to create manmade reservoirs often deserve the privilege of monument status for their age, function, performance, grandeur and even solemnity. Due to their amazing architectural characteristics and to both their appurtenant temporary and permanent works for diversion, use, and release of water, well-designed dams and reservoirs integrate themselves into the environment, positively changing the features of the surrounding landscape and the liveability of the area. Although designed for a long duration, dams, like other monuments, require constant and careful inspection and both ordinary and extra-ordinary maintenance in order to safeguard their safety. In some cases, after more than 50 years of distinguished service, it was necessary to abandon plants that no longer conform to modern standards. In other cases, the abandonment is expected. Amid the interesting and varied set of existing large dams in Italy (more than 500), one can find examples of both the causes and the effects of aging, which require a watchful surveillance and courageous measures for restoration, improvement and/or refurbishment. The appeal of Italian dams is often increased by the proximity to sites of important monuments of the past.

1   Monuments

Michel Frémond and Franco Maceri have given me the honour of opening this book with an unconventional attempt to approach two apparently very distant topics to which I have dedicated more than half a century of work: *monuments* and *dams*. Therefore, perhaps too boldly, I have entitled this discourse: "Monumental Dams" ("Les barrages monumentaux"), which I hope is going to be an easily digestible presentation, as if it were but an appetizer or an *ouverture*, preceding deeper treatments of other subjects.

I remember one stormy winter's day, when, from afar, I first saw the Doric Greek temple at Segesta that towers all alone in the deserted and barren landscape of western Sicily, elegant in its simplicity of structure and slenderness of line. I remember to this day how the sight moved me. Drawing ever closer, I began excitedly snapping the photos that I now present to you (Fig. 1), accompanied by the following romantic description by Rose Macaulay: "... Segesta towered lonely on its wild mountain over a desert..." [87] and another written by an enthusiastic French traveller: "...C'est au détour du chemin, que, soudainement éclose au sommet d'une colline, surgit

Fig. 1. The Segesta temple (photos by R. Jappelli, 1955)

une suite fermée de colonnes, composant un accord aussi pur que le cri qu'il suscite en toi: apparition rapide d'un temple perdu dans la solitude abrupte que brûle le soleil du Midi..." [8].

Monuments are important buildings, which, either by some intrinsic, historic or artistic trait, or by an acquired value, become inimitable witnesses

Fig. 2. A definition of "Monument" by Umberto Eco [38]

Fig. 3. Examples of ancient and modern "monumental" structures: a) Saint Angelo's Castle, Rome; b) The headquarter of the Post Office, Naples

to the past; or, according to a prosaic definition given by Umberto Eco, in one of his pieces entitled "La Bustina di Minerva" (Fig. 2).

If Segesta's temple is fascinating mainly because of its slenderness of form, other monuments are equally imposing due to the compactness of their great mass (Fig. 3). However, both types possess a common "monumental quality" which I intend to include architectural beauty (grandiose or slender according to the case) along with sturdiness, historical background, survival, and even a sort of solemnity, which arises from a three-dimensional appearance in solitary sites (Fig. 4).

Monuments are always marked by their close relationship with man and both happy and sad events. Therefore, in order to fully appreciate their

Fig. 4. Remains of Monuments in Piranesi's visions

grandeur and importance, accessibility must be guaranteed so that mankind can benefit from these sites.

2   Dams

A dam is a great feat of workmanship, built with various materials across a selected section of a river in order to regulate its flow and to create an artificial reservoir (Fig. 5)

The main function of such a reservoir is to *regulate* the accumulation of rainfall, which, in the catchments, is irregularly distributed in space and time. This regulation consists of collecting water in the reservoir during the rainy seasons and releasing it for different uses during the dry seasons. In economic

Fig. 5. The ROOSVELT dam on the Salt River, Colorado, U.S.A. [136]

terms, this function can be compared to savings, which allow man to face difficult times by turning to funds accumulated in better times.

According to the dictionary of the International Committee on Large Dams [53], a "large dam" is one that reaches a height of more than 15 m and that meets at least one of the following conditions:

- the crest length is not less than 500 m;
- the capacity of the reservoir formed by the dam is not less than one million cubic metres;
- its spillway is proportionate to a flood not less than 2000 m^3/s
- the geotechnical conditions at the site are particularly difficult;
- the dam is of unusual design.

Fig. 6. The TARBELA dam on the Indo River, Pakistan [59]

The dams of Italy, like those in the rest of Europe, cannot compare in dimension to those recognized as the largest dams in the world[1] but that does not mean that they are less interesting. The MONTE COTUGNO dam near Potenza, Italy, on the Sinni River (Fig. 7) between Basilicata and Calabria, is a homogeneous embankment dam with a facing of bituminous concrete and a volume of $12 \cdot 10^6 \ m^3$, the largest in Europe [17]. The reservoir's capacity is $500 \cdot 10^6 \ m^3$ and a huge aqueduct conveys its water to the far ends of Puglia. The ANCIPA dam near Enna, Italy, is a hollow gravity structure with buttresses and reaches a height of $110 \ m$. It is among the top of its class

---

[1] ROOSVELT, U.S.A. (Fig. 5): TARBELA, Pakistan: $V = 121 \cdot 10^6 \ m^3$ (Fig. 6); FORT PECK, U.S.A.: $V = 100 \cdot 10^6 \ m^3$, reservoir $V = 25 \cdot 10^9 \ m^3$; ROGUN, Russia: $H = 320 \ m$

Fig. 7. The MONTE COTUGNO embankment dam on the Sinni River near Potenza, Italy [17]

in the world (Fig. 8 [2]). The PLACE MOULIN, an arch gravity dam (Fig. 9) across the Buthier River, is a gigantic structure in Val d'Aosta.

Dam design is strictly dictated by several factors: the hydrological characteristics of the catchments; the morphology of the selected site and the regime of the river; the mechanical properties of the ground and the construction materials prevailing at the site; the climate; and, finally, the time schedule given for completion.

For each structure, the final specifications concerning height, foundation, and waterproofing are quite various. Figures 10 and 11 represent only a few of the typically employed solutions in terms of workmanship and materials.

Fig. 8. The ANCIPA near Enna, Italy, is a gravity dam with hollow buttresses [64]

Fig. 9. The PLACE MOULIN is an arch gravity dam on the Buthier River in Val d'Aosta, Italy [64]

In the past, dam classifications have been proposed according to different building features, by type of material (concrete, earth) or by the design solution for the dam body (gravity, hollow, embankment,...). Today these

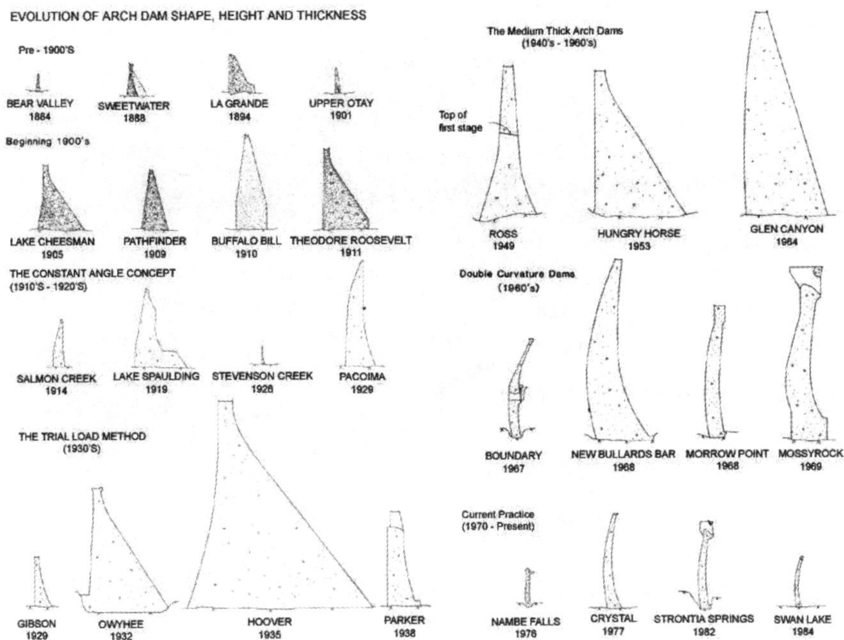

Fig. 10. The evolution of masonry dams in the U.S.A. [136]

classification criteria are found to be insufficient or defective because of intermediate solutions that are now available. In fact, on one hand, earth can be reinforced with different materials and acquires qualities resembling concrete; on the other, concrete can be placed with methods used for earth embankment construction. One can also find composite dams, consisting partly of concrete and partly of earth or rock materials.

There are presently more than 40,000 large dams in the world; according to 1997 statistics, Italy boasts among the most interesting and varied large dams, which amount to a total of 551 [33].

Whoever looks at one of these remarkable works for the first time is greatly impressed, as if he were before an historic monument. This is due to the uniqueness, solemnity and *grandeur* of the fabric, especially if the structure is old. Dams, like monuments, have a close relationship with man, linked to his primordial need to control nature and to his increasing demand of water. Both dams and monuments possess the intrinsic character that can be defined by the word "monumentality".

"Dams usually impress people as feats of engineering, fulfilling important functions as drinking water reservoirs and in flood control. In many cases they are major tourist attractions as well... Most people surely are less inclined to view a dam as a cultural monument which testifies to the past and so reflects

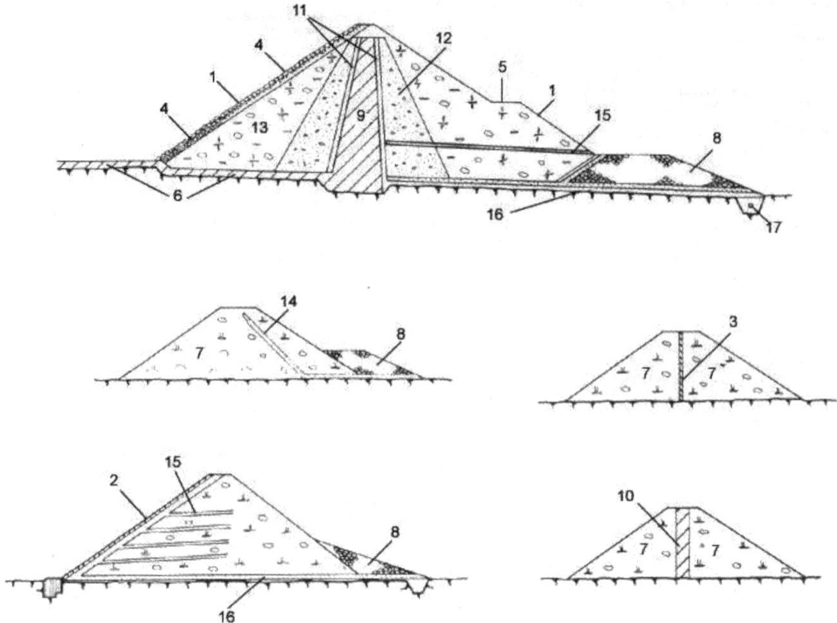

Fig. 11. Types of embankment dams. 1. slope; 2. facing; 3. diaphragm; 4. slope protection; 5. berm; 6. upstream blanket; 7. shoulder or shell; 8. toe weight; 9. core; 10. core wall; 11. filter; 12. transition zone; 13. pervious zone; 14. chimney drain; 15. drainage layers; 16. blanket drain; 17. toe drain [53]

the particular economic and historic situation of a country. But exactly that is characteristic of many of the dams in Saxony" (Kurt Biedenkopf,[2] 2001).

As stated in the 1998 INCOLD Symposium on the Rehabilitation of Dams, regarding the 226 m high BHAKRA concrete gravity dam constructed in India in the 1950s: "The world's most formidable, Bhakra dam, unique in many ways with intricate construction features and ranked as the second highest concrete gravity dam in the world, has a very prestigious place in the "World Register of Large Dams". It is a silent majestic monument built across the river Sutley at the village of Bhakra in India, providing irrigation to millions of hectares of parched land and lighting millions of houses by rendering service round-the-clock" [106].

Today, in Italy, as perhaps in much of the world, the primary task of the engineer is not so much to design new dams, but rather to maintain the functionality of existing structures, a far more difficult task.

---

[2] Minister President of Saxony

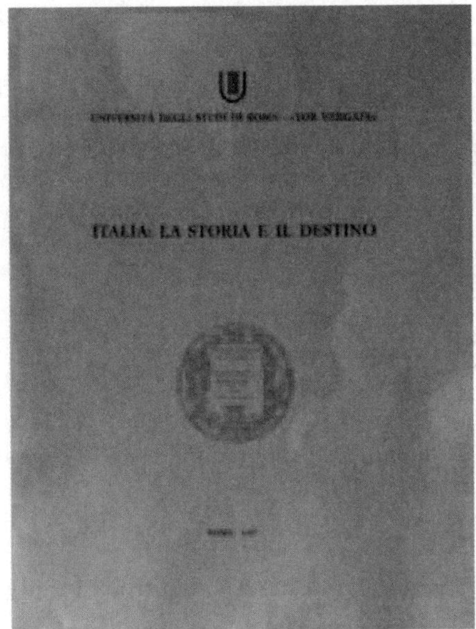

Fig. 12. In memory of Sabatino Moscati [99]

Fig. 13. Sites of the most important Pre-Roman (white) and Roman (black) dams in the Mediterranean basin [121,80]

3 Archaeology

In order to understand the sense of the relation of dams to Archaeology it would help to read a few texts.

The romantic sight of ruins, their irresistible attraction and their impact upon the soul of both poets and travellers alike are described with extraordinary depth and richness of imagery and prose in the classic book by Rose Macaulay [87].

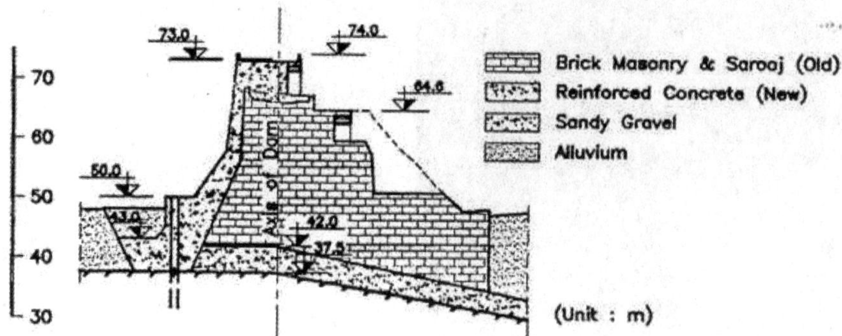

Fig. 14. A section of the ancient FARIMAN dam in Iran (approximately 1000 years old), recently waterproofed, reinforced and raised [110]

The close relationship between Archaeology and man is described briefly and simply by Vincenzo Tusa [135].

Sabatino Moscati gives a glimpse into the unique bond between Archaeology and Italy's history at a conference he held near the conclusion of his academic career at the University of Rome Tor Vergata in 1997 [99] (Fig. 12).

The relationship between archaeological remains and the ground they rest upon is well known; less-known, but certainly not new, is their relationship with *water*. One such example would be the ruins of ancient Sibari, where in the *Parco del Cavallo* site, a large system of well points has been in operation for dozens of years to lower the ground water level and allow the exploration of the archaeological site for the benefit of visitors [70].

In the world, there are a number of so-called "archaeological dams"; so many, in fact, that Jean Kerisel reports a map which indicates sites of ancient dams in the Mediterranean area (Fig. 13), including FARIMAN (Fig. 14), MARIB, JAWA and SADD-El-KAFARA [121].

The MARIB dam on the Danah River in Yemen was 20 m high and 700 m long. The earth embankment, with its very steep slopes, was constructed in layers parallel to the facing. At the abutment, two large spillways are still visible. This great work was partially destroyed on more than one occasion by recurring floods every 50 years or so. It was reconstructed and perhaps even raised. It was finally destroyed after 1300 years of service and the 50,000 local inhabitants it served were transferred to other areas.

The JAWA dam in Jordan is formed by an earth fill protected by masonry revetment. J. Kerisel mentions it as being the oldest dam in the Mediterranean basin [80].

The same Author reports that the SADD-EL-KAFARA dam on the Wadi Garawi River in Egypt (Fig. 15) is the oldest known large dam. It was built under the IV Dynasty (2600 BC) and was discovered roughly 100 years ago

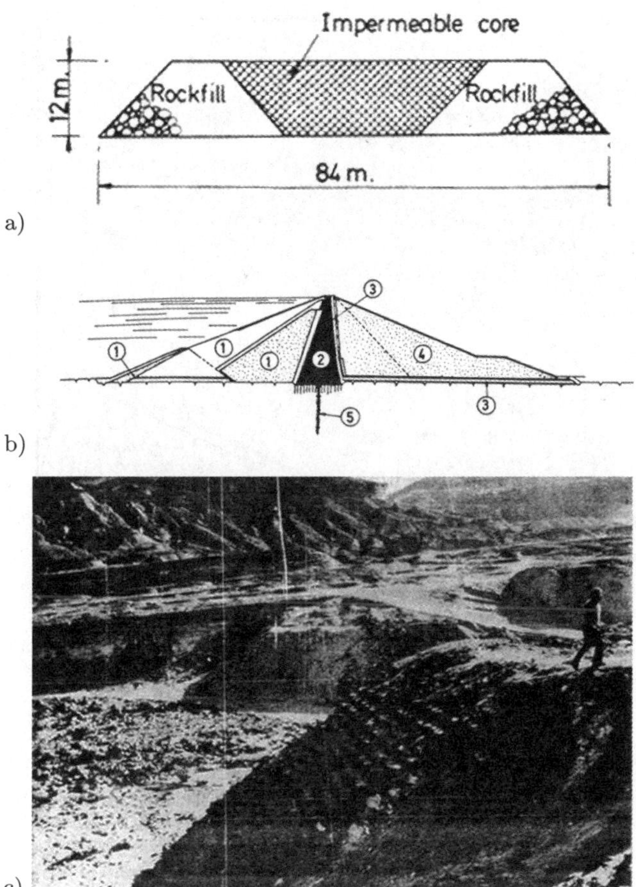

Fig. 15. a) A section of the SADD-EL-KAFARA dam in Egypt (IV Dynasty, 2600 b.c.) on a tributary of the Nilo, 39 km south of Cairo; b) A comparison with a modern embankment dam [80]; c) remains of the dam after erosion of its core caused by the river's stream; the upstream slope is protected by hand-placed dry masonry [45]

among the ruins at Garaw. It was 14 m in height and had a length at the crest of 113 m. It was designed to regulate floods, which were rare, but violent. This great construction was probably built without having deviated the course of the river and was then destroyed immediately upon its completion. The consequences were so severe that the Egyptians did not build any other dam for at least the next eight centuries. Nevertheless, the builders of the SADD-EL-KAFARA deserve the utmost respect for the sheer genius of the project. Figure 15a shows a typical section, formed by a gravel and silt core between rock shells, protected by square-block masonry. In Fig. 15b, Kerisel compares

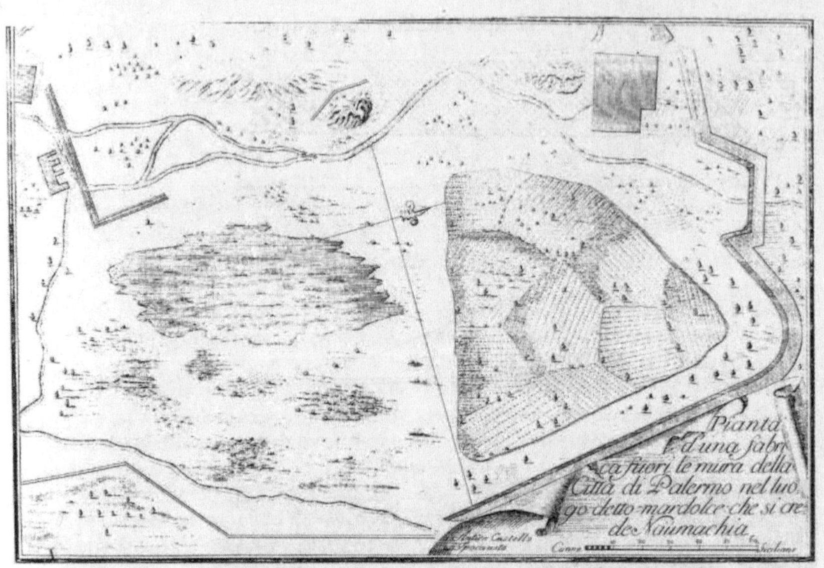

Palermo. La Favara nel XVIII secolo (*Pigonati 1767*).

Fig. 16. The pond in which the Fawara or Maredolce Castle was reflected and the weir built by the Arabs in Palermo, Italy

this rudimental construction to a modern earth dam with a central core. Figure 15c looks inside the ruins of the dam whose core has been eroded by the stream.

There are no such "archaeological" large dams in Italy. However, a little-known fact is that the ruins of the old Arab castle called "Maredolce" in the middle of a poor neighbourhood of Palermo stands on an ancient weir built by the Arabs to create a little pond, known as "Fawara", in which the image of the castle was reflected (Fig. 16–19) [15].

Architectural scholars [32] report that "very little is now left of the monuments from the Arab age due to the Norman conquest of the island. The Normans were zealous keepers of the Christian religion and thus, all the structures of the "unfaithful" were destroyed by them. The only Arab architecture that has survived, enclosed in the later Norman one, is a small structure on the site of the hot springs near the capital at Cefalà Diana and, in Palermo, part of the mosque of Saint John of the Hermits and some remains of the Emiris Palace and of the Fawara castle (cf. Pietro Laura's, "About the Arabs and their sojourn in Palermo, Sicily, 1832").

S. Braida reports in 1988: "This pleasant view and this enchanting park where the world smiles, recalling the memories of the past, was probably first used by the hermits of Palermo and then later renovated by the Normans... Within this city, sprung the greatest fount of all time, which was surrounded

Fig. 17. The bottom outlet of the Maredolce pond (photo by C. Gambino, 2002)

Fig. 18. The remains of the Fawara or Maredolce Castle in Palermo (photo by C. Gambino, 2002)

by a wall and formed an enclosed garden, which was called "Albehira" by the Arabs. Fish of many types were closed therein and the lake was adorned with little ornate boats decorated with gold and silver in which the King and his wife often entertained for their own personal amusement. It has been passed down from the chronicles of those times, that King Roggerious, who in peaceful times could not remain idle, commanded that a marvellous structure

Fig. 19. An imaginary view of the Maredolce complex around the end of the XVII century (from a sketch by V. Auria, reproduced by Braida [15])

Fig. 20. The necropolis of Mount Canalotti, near the site of the new DISUERI dam in Sicily on the Gela River (photo by C. Gambino, 2002)

be built in Palermo, full of every delightful thing and every possible precious ornament...

...Quondam autem montes et nemora quae sunt circa Panormum muro fecit lapideo circumcludi et parcum deliciosum satis et amenum diversis arboribus insitum et plantatum construi iussit et in eo damas, capreolos, por-

Fig. 21. The small necropolis at Garrasia within the quarry of the evaporitic limestone used for the construction of the new DISUERI dam in Sicily [35]

Fig. 22. The remains of the Roman "Statio" discovered in the area of the PIETRAROSSA reservoir on the Margherito River in Sicily (photo by C. Gambino, 2002)

cos silvestres iussit includi. Fecit et in hoc parco palatium ad quod aquam de fonte lucidissimo per conductus subterraneos... pulchrum fecit bivarium in quo pisces diversarum generum de variis regionibus adductos..." [15].

Fig. 23. The MOEHNE dam, in the Reno basin, after important rehabilitation works of the outlet gates [57]

The construction of large reservoirs can interfere with archaeological remains. This problem has been dramatically raised every time that previously unknown archaeological sites have been discovered, as in Turkey, right in the midst of an artificial reservoir under construction [5].

On the other hand, the very nearness of a reservoir to an important existing archaeological site could be the occasion to promote the establishment of a national park. This could be the case at DISUERI, in Sicily, where the cliffs of Mount Canalotti, right beside the new dam site, are spotted with rupestrian caverns which have been identified by archaeologists as "burial tombs in little caves" (Fig. 20). The existence of this necropolis was related to the inhabited area discovered on the terraced slopes of Mount Maio.

The innumerable manufactured ceramic and bronze items composing both the treasure in these burial sites and the common household objects for everyday use suggest a time period for this inhabited centre between the twelfth and ninth centuries before Christ. However, it has also been ascertained that this civilization survived at least until the seventh century BC and that there was an exodus, which may have been caused by the advance of the Rodio-Cretans, the founders of Gela, in the territories of central Sicily. This area was previously inhabited by indigenous communities, who were a strong obstacle to the expansionist plans of the colonizing Greeks [111]. It was exactly the area of Disueri that the well-known archaeologist Paolo Orsi (to whom the Archaeological Museum of Syracuse was dedicated) was able to explore, thanks to the local police who protected him against bandits.

Yet another tiny necropolis was found in the same spot next to the dam site, during the quarrying of the soft evaporitic limestone used in the dam shells (Fig. 21) [35].

At PETRAROSSA, where the construction of the dam was nearly complete, the work has been suspended for many years due to the evidence of

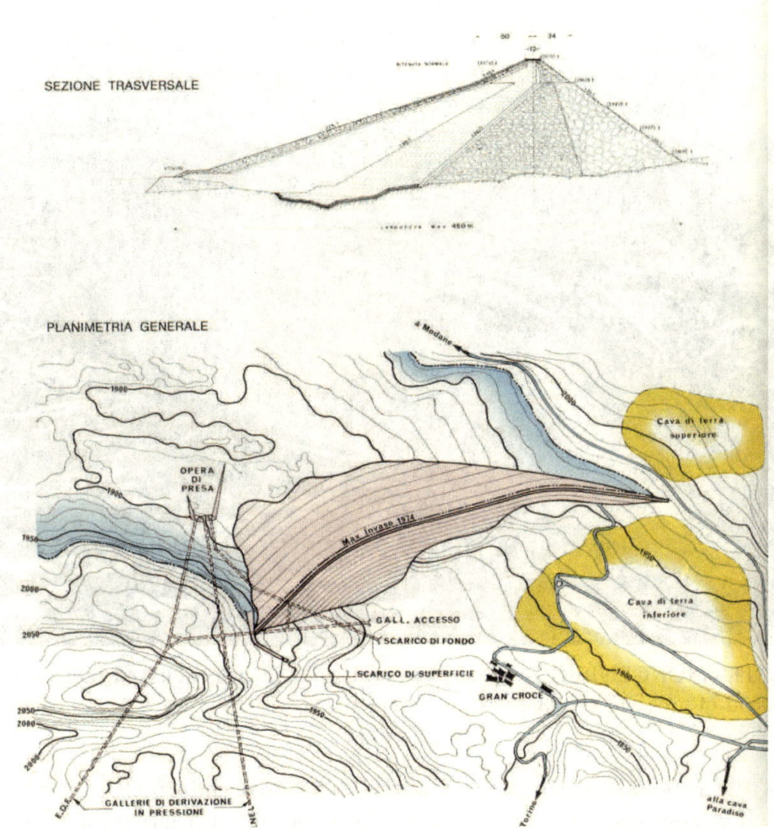

Fig. 24. The MONCENISIO dam is subjected to periodical inspections by an Italian-French committee [63]

important archaeological findings in the middle of the future reservoir on the Margherito River. However, the archaeological research which today has been finally initiated to fully ascertain the true importance of the remains will continue for a long time (Fig. 22). A thorough study will be necessary to conclusively decide, after considering all the information, whether it would be possible to complete the hydraulic work in progress, giving due allowance to the ambitious desire to turn this archaeological heritage into a pleasant historical site. It seems that the importance of this site rests upon the fact that the remains (perhaps a Roman Statio) are from a period even older than the most well-known and grandiose Roman Villa Casale of Piazza Armerina [47].

This is an example of a distressing dilemma between those who would want to preserve *everything* ancient at *any* cost and those who are willing to sacrifice each and every finding to the less noble but equally important

Fig. 25. The MONCENISIO reservoir, at an altitude of two thousand metres, on the occasion of a drawdown [39]

cause of the rational distribution of water to an area enduringly struck by drought. These situations of harsh conflict between two opposite public interests must be resolved case-by-case, and the decision based not only upon the good reasons presented by both parties, but upon reasonableness and sound engineering judgement [116].

4   History

Dams, like monuments, can reach a considerable age. They are subject to weathering and deterioration and it is man's responsibility to preserve them. Therefore, they should undergo constant and careful inspection, not only for reasons of safety but also for maintenance and conservation.

Let's listen to an authoritative voice from India: "...For harnessing water resource of the rivers for catering to various human needs like drinking water supply, irrigation, hydropower generation, industrial needs etc. man has been constructing dams from time immemorial. The ancient civilisation of India, Egypt, Mesopotamia and China shows that making water available for domestic use, harnessing it for irrigation and controlling it for flood protection against floods were basic pre-occupations in these societies. In India, there is evidence of canals, dams and dugwells from pre-historic times. As

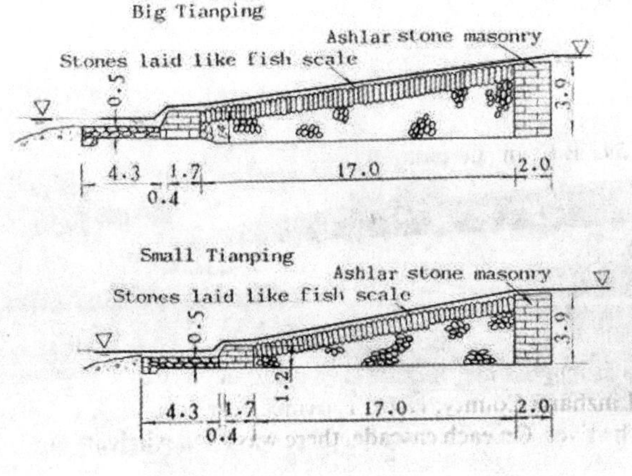

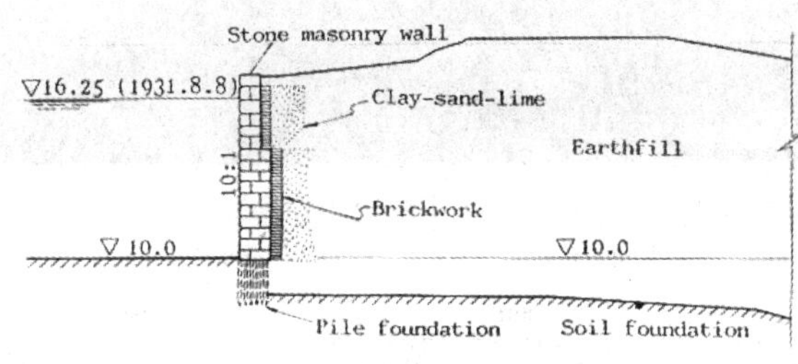

Fig. 26. a) The dry masonry TIANPING dam, built in 219 b.c. on the Xiangjan River in China is still in good condition [23]; b) The GAOJIAYAN dry masonry dam with earthfill shells on the Huaihe River in China is in service since the days of the Ming Dynasty in the VII century [23]

dam engineering developed with the advancement of technology, new types of dams employing different construction materials came into operation, adopting modern design criteria and construction methods that stressed the safety of these structures, since it is considered imperative to derive economic benefits from these dams for as long as possible. This warrants that special attention be given to dams built in earlier times not conforming to modern design and construction practices which may have to be rehabilitated through universally accepted current standards and measures to make these structures quite safe... " (C.V.J. Varma, Chairman, [60])

Every dam has a story that is linked to the history of civilization and water and this story coincides with its rate of aging and progressive deterioration, requiring the maintenance and repair procedures necessary for its preservation (Fig. 23) besides strict and regular inspections (Figs. 24, 25). Evidence

Fig. 27. The AMIR dam in Iran, in service for 1000 years [61]

of the past survives within these reservoirs. Often the materials used in dam construction are similar to those used in nearby historical monuments.

Dams with a long history are reported from China (Fig. 26) and Iran, where the AMIR dam is still in use today (Fig. 27). In Sicily, there are still traces of small dams, such as at the San Cusumano site on the Megarese harbor (Fig. 28) and at Lentini, where stories are told of a certain *"Piverium"*[3] (perhaps a vivarium for fishes).

In the south of Spain, the Romans built the CORNALVO and PROSERPINA dams which are still in service after 1300 years for hydro-potable purposes. Their masonry curtains were filled with stones or clay and faced with mortar.

The GROTTICELLI dam (Fig. 29) is easily visible in the bed of the Gela River, about 1 *km* south of the Ponte Olivo site. It was designed by the architect Carlo Cadorna in 1563. The dry masonry work is composed of large evaporitic limestone blocks; interstices are filled with gravel and silt.

The SAINT FERREOL dam, built in the years 1666-1675, thanks to Louis XIV, was built as a reservoir to regulate the water of the Languedoc canal in

---

[3] The word derives presumably from the Latin *vivarium*, and in turn from the Sicilian *biveri*, that is *biviere*, with the meaning of marshy lake; according to A. Traina the *bivieri* is a "small reservoir of water created by a dike for the purpose of growing fishes (vivaio)"

Fig. 28. A section of the old weir in the St. Cusumano Valley in Sicily with mouth into the megarese harbor between the Cape Xiphonio and the island of Tapso (by G. Di Grazia, in [1])

Fig. 29. The GROTTICELLI masonry weir on the Gela River in Sicily, built in 1563 and designed by Architect Carlo Cadorna (photo by C. Gambino)

France, which connected the Mediterranean Sea with the Atlantic Ocean. It is an embankment dam ($H = 36\ m$) with a masonry core. It maintained the record as the highest dam in the world for 165 years.

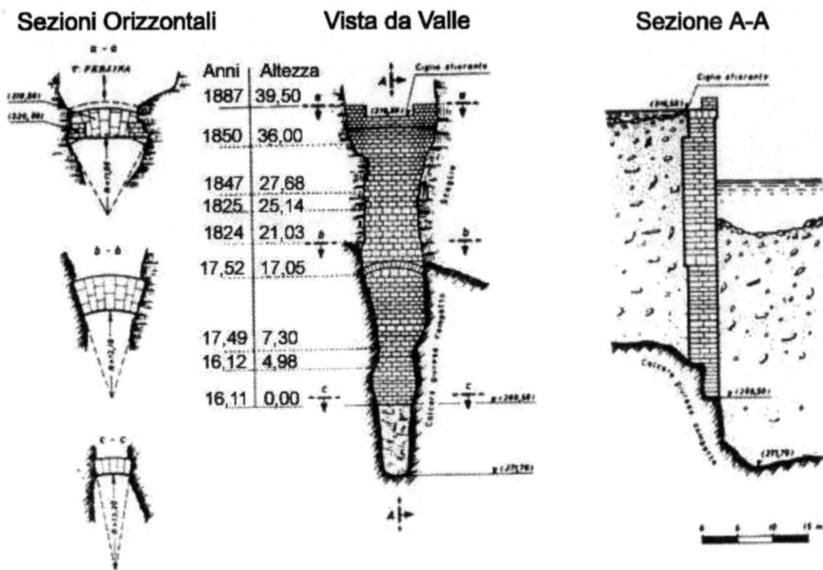

Fig. 30. The PONTE ALTO arch dam in Trento, Italy, firstly built in 1611 to regulate the Fersina river, tributary of the Adige river; the dam was often raised as a consequence of a progressive silting up [41]

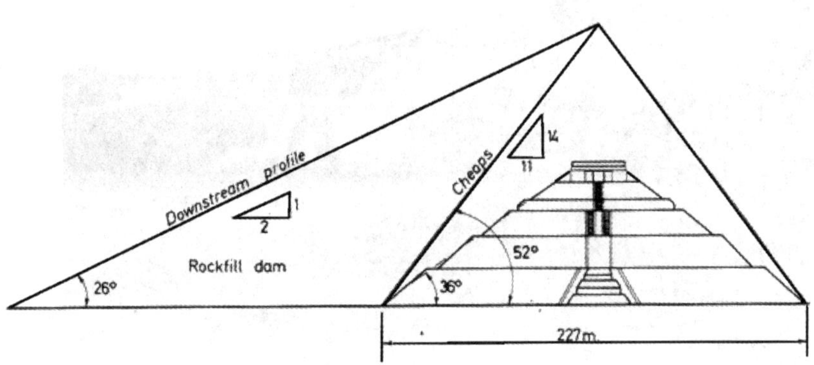

Fig. 31. Comparison of the Cheope Pyramid in Egypt and the Del Sole Pyramid in Mexico to a modern rockfill dam [80]

Throughout Italy, traces of the remains of ancient manmade dams and reservoirs can be found. Near Siena, along the road from Giuncarico to Ribolla, one can still see, though largely covered by vegetation, the impressive remains of a barrier wall some 24 m high, which once dammed the Bruna

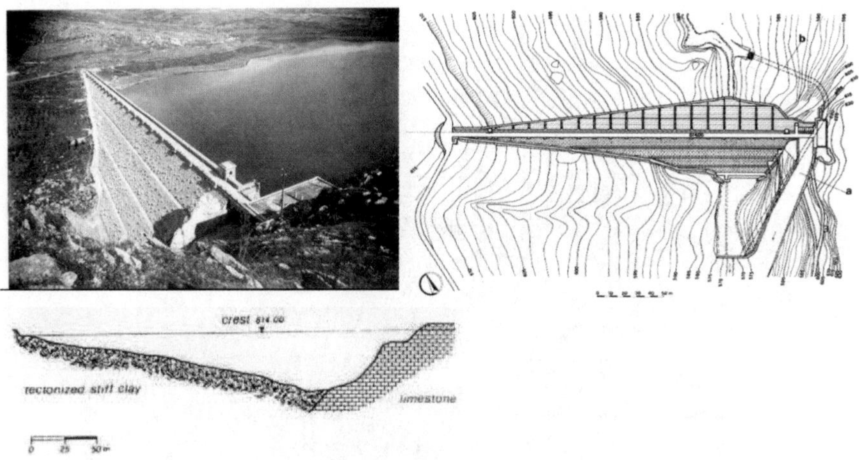

Fig. 32. The PIANA DEI GRECI dam on the Hône River near Palermo, built in 1921-22 of dry masonry with a concrete facing on the upstream slope [6]

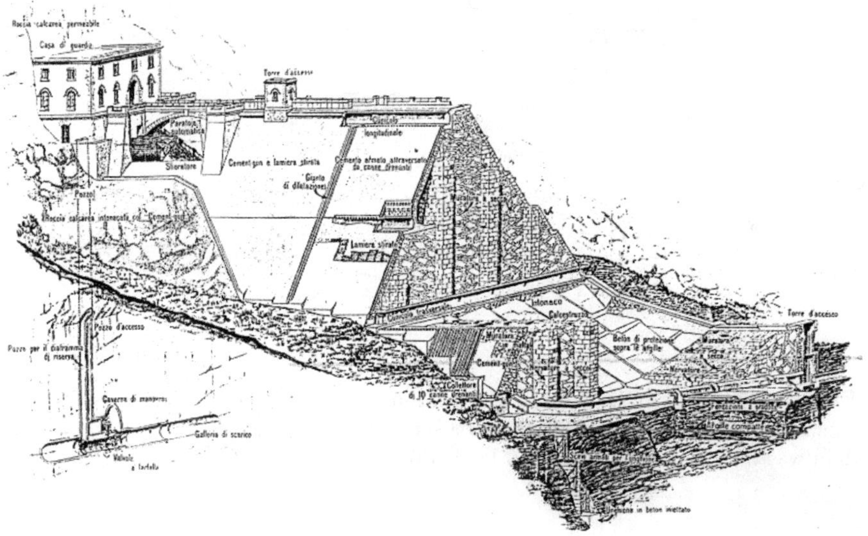

Fig. 33. The PIANA DEI GRECI dam: an amazing original sketch demonstrating the structural features [89]

River. It has become known as the MURACCI, which means "ugly or darned old walls".

The Sienese news chronicle of its day, entitled, "Dè ricordi Maggiori" and written by Siena local Pierantonio Allegretti, indicates that Guidoccio d'Andrea, the "Muracci" dam designer, was charged with the project in 1468.

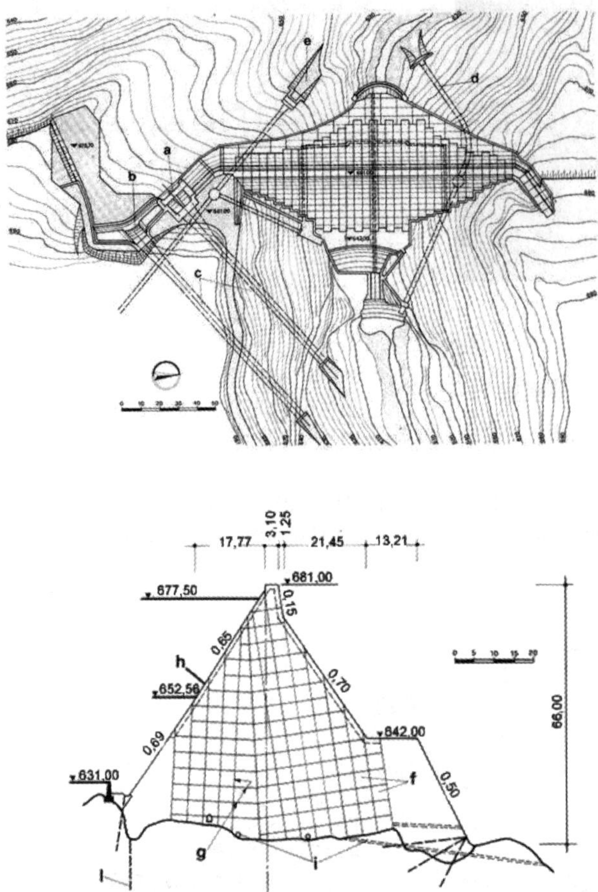

Fig. 34. The FANACO dam on the Platani River, Agrigento, Sicily, is made of concrete blocks with gravel joints and a metallic facing on the upstream slope. The dam was built between 1952-1956 for hydroelectric and hydropotable purposes [6]

The actual work was entrusted to two *comacini* masters and was completed in 1481. It cost more than 15,000 gold florins.

According to Murat. Script. Italic T. XXIII in the city archives of Siena, there is no further mention of Lakes Pietra or Bruna, except in 1476 when the rulers of the city ordered various masters of the arts to visit Lake Bruna on July 25th of that year. Then, one last time in December of 1492, extremely urgent letters were written to architect Francesco di Giorgio di Martino who was in Naples at the time, asking him to return quickly to Siena because the wall at the Bruna was threatening to collapse. As a matter of fact, just a few months later, "The great wall indeed collapsed, due to the immense

Fig. 35. The QUAIRA DELLA MINIERA dam near Bolzano, Italy [63]

pressure from the huge volume of water it contained (where 120,000 pounds of fish were destined to be transported from Lake Perugia!)". Siena journalist Allegretti reported, "As the dawn broke on the 1st of January, 1493, so the Maremma Lake, in which fishing had yet to begin, broke down its wall and flooded many towns. Men and animals died because of the miscalculation of the one who built it."

The PONTE ALTO dam on the Fersina River in Trento, a tributary of the Adige River, dates back to 1611 when it was rebuilt in a narrow gorge from a previous barrier constructed in 1597 that was destroyed by a flood. It initially stood only 5 m in height but was later raised on various occasions (Fig. 30).

At the time of its final elevation in 1883, the arch dam MADRUZZA ($H = 40\ m$) was being built just downstream of it. The basins created by these two structures still serve to adjust the slopes of the river bed even though by now they have become completely filled in with sediments [109].

5   Architecture

Dam bodies and their ancillary works exhibit different characteristics and architectural forms depending upon the requisites. These include support-ing, waterproofing, draining, filtering, separating, protecting, diverting, in-taking, outletting, dissipating, sedimenting and releasing functions, which

Fig. 36. The ZOCCOLO dam and reservoir from the left bank with the spillway and the guardhouse in the foreground [63]

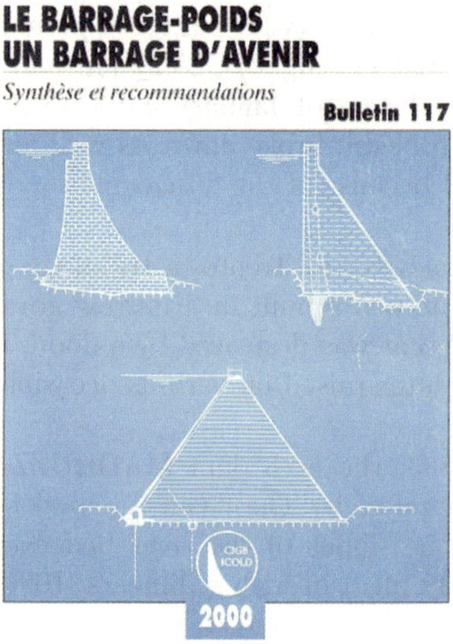

Fig. 37. The re-evaluation of gravity dams

are entrusted to single structural elements in order to ensure the mechanical compatibility of the artifact with water and ground.

The architecture of a dam must, therefore, conform to strict regulations that depend upon the conformation of the site, the properties of the ground,

Fig. 38. The left abutment of the CASTELLO embankment dam near Agrigento (Sicily) rests on a large, stabilized landslide

and the construction materials, which depend in turn upon the physical environment and its relevant transformations. Some factors to be considered are:

- The final choice of both the dam and its appurtenant works must conform to the mechanical characteristics of the ground with respect to permeability, strength and deformability;
- The dam body must be conceived and built with materials available within very close range of the specified construction site;
- Both ground and materials should never be considered "impermeable", that is only an ideal qualification;
- The flow of the waters which interact with the dam should be conveyed downstream by means of a drainage system protected by appropriate filters;
- The primary and interdependent static and waterproofing functions, are better entrusted to different structural members;
- It is not wise to resort to lightweight structural members for the sake of stylishness, as they have not always exhibited good performance against high horizontal forces;
- In embankment dams, the spillways are external to the dam body. They are also of greater dimension than those employed in concrete dams in order to limit the risk of possible overtopping, a task which embankments, with some exceptions, are not qualified for.

These constrains have sparked the conception of brilliant and articulate solutions in the design of both the external and internal architectural elements of dams and their ancillary works.

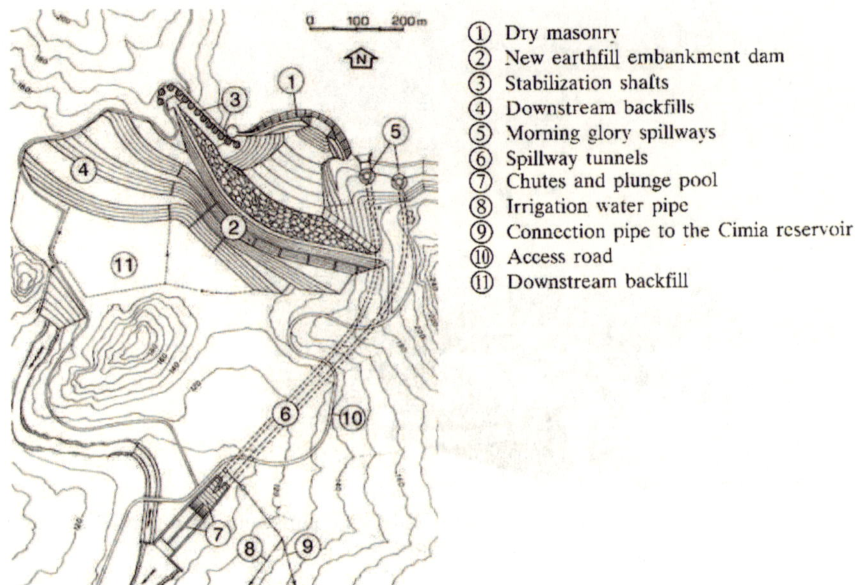

Fig. 39. On the right bank, the embankment of the new DISUERI dam on the Gela River in Sicily rests on a landslide stabilized with a crown of large diameter shafts [35]

Fig. 40. A. The old DISUERI dry masonry dam (1948); B. The new DISUERI embankment dam (1994) with higher normal water level elevation

The imaginative J. Kerisel compares a large modern dam to the pyramids of Cheops in Egypt and "of the Sun" in Mexico in order to point out man's marvellous ability to build great structures since ancient times and to demonstrate the *grandeur* of the embankment dams and the durability of the materials (i.e. earth, dry masonry, rockfill) (Fig. 31) [80].

PIANA DEI GRECI (Fig. 32) is a dry masonry dam on the Hône River near Palermo. It was built between 1921-23, upstream from a narrow gorge

Fig. 41. The breach artificially opened in the body of the old DISUERI dam to allow the free passageway of water [35]

Fig. 42. The soft calcarenite placed in a random fashion in the masonry of the San Corrado church at Noto which failed without premonitory evidence in March of 1996 (photo by V. Jappelli, 1996)

cut into a limestone formation. With the exception of the left abutment, which was placed against the rock, the structure was founded on intensely tectonized, but stable eocenic clay. Hand-placed dry masonry was selected for the dam body and its waterbarrier function was entrusted to a concrete slab facing. After the earthquake of 1968, a metal facing was added to the original waterproofing structure [138].

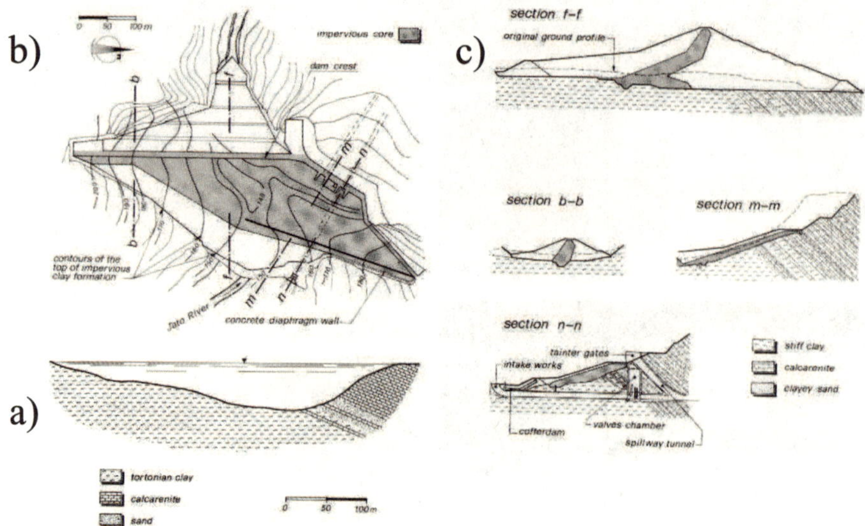

Fig. 43. The internal architecture of the POMA dam on the Iato River near Palermo, Sicily: a) longitudinal section of the ground; b) planimetry of the structure and contour levels of the roof of the clay bed; c) maximum cross section and details of the embankment dam [13]

Fig. 44. The Marcello Theatre in Rome [11]

The internal architecture of the dam (Fig. 33) is equipped with an amazing system of narrow manhole adits and drainage pipes. The system allows for thorough inspection of all the essential parts of the structure in order to reveal any trace of a leak.

The original documents [88,89] show that the designer was not particularly concerned about the foundation, which he merely judged "solid". He did, however, dedicate a lot of attention to the choice of building materials. As

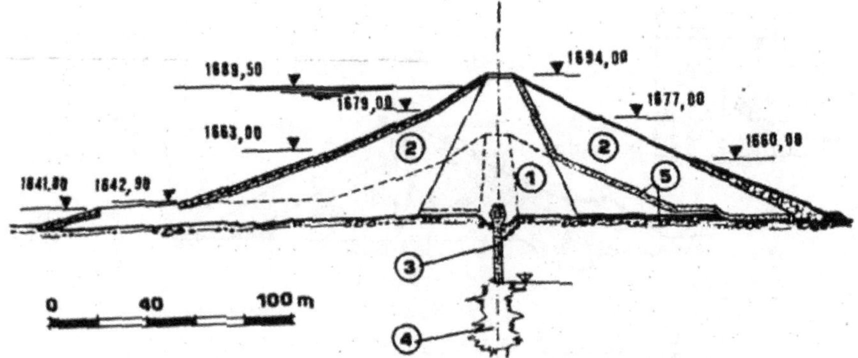

Fig. 45. The maximum cross section of the VERNAGO embankment dam near Bolzano, Italy, built on the Senales River upon a thick fan of detritus. The embankment as built in the first phase (1957) is represented by the broken line; after the observation of the behaviour of the first embankment at the time the crest level was raised (1962), supplementary measures were necessary to improve the water barrier in foundation. 1) central core; 2) shells; 3) concrete cut-off; 4) grout courtain; 5) filters [31]

a matter of fact, he rejected the limestone outcropping near the dam site - material that was later used to make marble slates - and decided, instead, to use a more "solid" limestone, which was quarried from a site a few miles away.

Through the use of this material, it was the designer's intention, to avoid, or at least reduce, any tendency of the dam to settle. Therefore, he was surprised at the completion of the dam, to observe a settlement of almost one metre. He had not considered that the hand-placed blocks could settle almost to the same degree as a rockfill and that the compressibility of the dam body could only be reduced by means of vibratory rollers [26].

PIANA DEI GRECI stands out in the history of dams in Italy, one reason being the low ratio (1.56) between its base-width and its height. Despite its hazardous design, the static performance of the structure has proven excellent, possibly on account of a three-dimensional effect, as demonstrated by recent analyses. The PIANA DEI GRECI dam is 80 years old and is still fully efficient [13,79].

The FANACO dam is sited on the Platani River (Fig. 34) at the foot of Mount Cammarata near Agrigento, Sicily, where the countryside reminds one of a passage by Quasimodo, "from the plains of Acquaviva, where the Platani rolls shells under water between the feet of olive-skinned boys" (S. Quasimodo, New Poetry, 1936-42). This may also be considered a historic dam because of its original design conceived by Claudio Marcello. The well-known Italian dam designer, with his invention of concrete blocks "lubricated" by gravel joints, gave the dam body the characteristics of an embankment, capa-

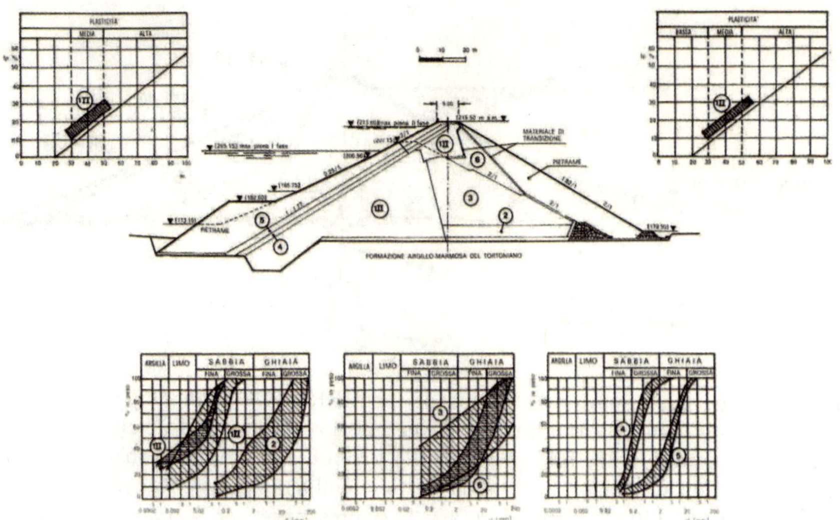

Fig. 46. The maximum cross section and the building materials of the DON STURZO dam on the Gornalunga River in Sicily as predisposed to be raised in 1970 [75]

Fig. 47. The spillway of a large Brazilian dam under construction

ble of adapting to a ground which could not have supported a rigid concrete dam [95].

The QUAIRA DELLA MINIERA near Bolzano (Fig. 35) is an elegant rectilinear arch-gravity dam with a height of 85 m. It was built in the 1960s in a glacial valley upon crystalline shales. The spillway sill is located in the centre of the crest. Its performance has been fully satisfactory and it has never required appreciable repairs.

Fig. 48. The morning glory spillways under construction in the eighties at the site of CAMPOLATTARO embankment dam on the Tammaro River, Benevento, Italy

The ZOCCOLO dam (Fig. 36) in the Val d'Ultimo is an embankment dam with a thin bituminous concrete water barrier on the upstream slope. It has very simple architectural lines in the embankment and the spillway is a long open channel. The performance of this dam and of its foundations will be discussed further on.

These four examples of large modern dams demonstrate the long term good performance of heavy duty structures in which building materials are utilized in massive quantities.

The latest technique in dam building is aimed precisely at the re-evaluation of gravity dams, including embankment dams, as compared to slender ones [56] (Fig. 37). In contrast to slender dams, having an arched geometry with the convexity facing upstream to be able to get the most of the arch effect, the site of a gravity dam depends on completely different factors. These include the need to comply with the stability of the slopes (Fig. 38), by widening, if necessary, the embankment with backfills, berms, earth platforms, flarebacks and fills. In order to turn the flank of a landslide, the embankment may also direct the concavity upstream (Fig. 39).

Fig. 49. The CIGNANA dam (1928) in Val d'Aosta, Italy, with a close-up view of the handplaced masonry facing on the downstream slope

Fig. 50. The small BRUSSON dam in Val d'Aosta, Italy (1928) (photo by S. Di Maio, 1996)

The old DISUERI dam on the Gela River (Fig. 40), recently abandoned after 50 years of service (see Sect. 10), offers an interesting example of internal architecture, that could be inspected when a wide breach was opened in the dam to allow the free flow of water into the new reservoir. The embankment nominally composed of hand-placed dry masonry with square limestone blocks was actually found to be composed, in its innermost part, of loose materials so weak they could be removed with a shovel (Fig. 41). As with some monuments (Fig. 42), the existence of loose materials in fundamental parts of a masonry hand-laid structure can be explained by the difficulty of supplying

Fig. 51. The metallic footbridge (design by F. De Miranda) to access the intake tower of the VILLAROSA reservoir on the Morello River near Enna, Sicily [6]

rock materials at the site with the necessary strength to form an adequate masonry of squared blocks. Even in this case, the waterproofing function was entrusted to a concrete slab facing resting on the upstream slope.

In a similar way, the nearby fortifications, built by the Greeks at Capo Soprano near Gela to reinforce their original settlement, consisted of two strong walls of huge, well-squared-off blocks of solid rock. The material used as filler between the external walls was loose uncompacted soft rock. The walls were elevated in later years by rows of bricks made of clay baked in the sun ("uncooked").

The distribution of the materials in the body of an embankment stringently depends upon the conformation of the dam site. At the site of the POMA dam on the Iato River (PA) the foundation soil consists of stiff clays of Tortonian age, which are overlain on both abutments by poorly cemented, fractured, permeable calcarenite banks, passing gradually, near the contact with the clays, to compressible cohesionless sands including disarranged calcarenite layers (Fig. 43a) [13].

The top surface of the clayey formation was completely uncovered during the preparation of the foundation. The contours of this surface are shown in Fig. 43b. The geometrical configuration of the clay-calcarenite contact was the controlling factor in design decisions on the dam siting and on the choice of dam type. The top surface of the clay is concave and extends uphill on both abutments up and above the maximum reservoir elevation; in upstream-downstream direction, the clay surface exhibits a gentle ridge located approximately in the narrowest section of the gorge, while downstream

Fig. 52. An intake tower with wooden footbridge in use in the 1700s in the mining region of Oberharz, Germany [120]

the clay-calcarenite contact dips and the clay leaves the place to the overhanging permeable calcarenite.

Site and embankment type selection were strictly interconnected. Taking advantage of the shape of the gorge, the valley was closed, near the narrowest section, with a 57 m high embankment, which on the right valley side was gradually conformed to the geomorphology by flanking and wrinkling the abutment (Fig. 43c, sect. n-n). The core is completely keyed into the clayey bed; its gradually variable shape was selected with the aim of avoiding undesirable sharp bends along the longitudinal axis. In fact, from the left abutment, where a pre-existing landslide was completely removed, the water barrier in the maximum cross section is positioned almost centrally (Fig. 43c, sect. b-b), but moving toward the bottom of the valley the core slopes gradually upstream and the cut-off widens appreciably (Fig. 43c, sect. f-f). On the right side of the valley, the internal water barrier was transformed into an earth blanket, protected by an external shell which conforms to the geological situation (Fig. 43c, sect. m-m).

In the gradually variable sections, the geometry of the core was sketched with the condition of using as much of the suitable closely available material as possible, whilst preserving the stability of the upstream shells. The downstream shell at and near the dam's maximum cross section rests on soft calcarenites, which were used extensively in the dam body.

Fig. 53. The guardhouse with the station for operating the outlet valves and for the acquisition and digest of instrument results installed at the ROSAMARINA dam on the San Leonardo River near Palermo, Italy (photo by C. Gambino, 2002)

Fig. 54. Windows for the instruments on the downstream slope of a rockfill dam [34]

The tunnels of the outlet works were allocated in the calcarenite outcropping in the right abutment with their mouth crossing the core water barrier. The final plans for the dam could be locally conformed to the uncommon geomorphological situation of the site during the construction stage with a careful observational approach, requiring the complete uncovering of the contact surface between clay and calcarenite. The inclined core, which elsewhere has been drawn perhaps only in search of an apparent formal design elegance, is here fully justified. The choice of this rather unusual dam layout avoided

Fig. 55. San Colombano monastery, Rovereto, Italy: even the monuments conform to natural structural features with ingenious architectural designs (I Viaggi di Repubblica, 238, 22-8-02)

leakage problems at the right abutment, and, at the same time, permitted considerable savings. The alternate solution, with the dam located upstream entirely on the clay formation, would have been far more expensive, due to the valley geomorphology, the geotechnical properties of the various foundation soils and rocks, and the characteristics of the available construction materials.

Obviously, the problem of estimating the course of pore pressure in a core with such a complex geometrical configuration would be still a challenge even for the best specialists in consolidation theory.

After more than thirty years of operation, the performance of the structure remains highly satisfactory.

The convenience of separating the two static and waterproofing functions must have been considered by ancient builders, as Jean Kerisel comments regarding the SADD-EL-KAFARA (cf. Fig. 15); as a matter of fact, in the cross section of that artifact one can observe an embryonal distinction between a central core of fine grained material and two rockfill shells [80].

Like monuments, which offer marvellous examples of restoration and superelevation (Fig. 44), dams can be raised for the purpose of compensating the progressive reduction of the reservoir's capacity due to sedimentation. The modern VERNAGO (Fig. 45) and DON STURZO (Fig. 46) dams, predisposed for that purpose, were raised in 1962 and 1970, respectively.

An importance, at least equal to that of the dam itself, do bear the outlet works (bottom outlets, spillways), through which the water is conveyed either in the open air or underground. The function of these works, which are some-

Fig. 56. Detail of a rammed-earth building (toulou) from Fujian ("happy architecture"), a province of the Southern part of the Republic of China [50]

Fig. 57. The GARCIA embankment dam on the Belice Destro River, Palermo

times not properly called "ancillary", is the protection of the dam body from the consequences of sudden floods which can be catastrophic for embankment dams. In Italy, the bottom outlets are designed for a discharge of the order of

Fig. 58. A view of the GARCIA reservoir at the foot of the Entella's Rock

Fig. 59. The ruins of the Calatrasi Castle in Sicily [129]

the tenth m^3/s for the purpose of rapid reservoir draw down. The spillways have a capacity up to the order of some thousand of m^3/s, according to the hydrology of the catchment area. For the mouths of the spillways, fascinating solutions have been invented, resembling cathedrals, as in the example of Fig. 47, representing the *vertedouro* (spillway) under construction for a Brazilian dam.

The so called "morning glory" or "bell mouth" spillway also deserve to be considered "monumental". Figure 48 shows the large mouths of two spillways under construction. In Fig. 79, one can see the open channel downstream of

Fig. 60. The Calatrasi Bridge near the Rocks of Maranfusa in Sicily (photo by C. Gambino, 2002)

the free overflow spillway located on the left abutment of the ZOCCOLO dam.

Pleasant architectural features can be safeguarded by improving details of the dam body and of the ancillary works. In Fig. 49, one can admire the beautiful ashlar work of the CIGNANA dam (1928) in Val d'Aosta, Italy. In Valtournanche, Fig. 50 shows the small weir BRUSSON (1928), surmounted by a covered passageway, where the gates drives are installed.

The elegant metallic footbridge at VILLAROSA near Enna, Sicily (design F. De Miranda) allows the access to the intake tower from the left abutment (Fig. 51). In 1700, in the German region of Oberharz, the same service was entrusted to a wooden trestle (Fig. 52).

The modern guard house at ROSAMARINA near Palermo, Sicily, houses the dam warden and the control centre of the data transmitted by the instruments installed in the dam body (Fig. 53). The niches for the access to instruments form, with the downstream slope of rockfill, a modern artistic composition (Fig. 54). The comparison between the rigid concrete artifacts and the softer embankment shows the constrains imposed by Soil and Rock Mechanics to the elevation. Concrete dams are clearly recognizable in the landscape, the features they engrave coming into sight almost overbearingly. Embankment dams, on the contrary, are less spectacular and conform almost humbly to the morphology of the site by mitigating the slope of steep hills, by filling narrow gullies or by smoothing abrupt asperities.

Similarly, some monuments must submit to the constrains of natural ground and conform to local situations with ingenious architectural solutions (Fig. 55).

Fig. 61. The fascinating rural path from the village Carlantino to OCCHITO's dam site near Foggia, Italy, as it appeared in 1953 (photo by R. Jappelli)

"We cannot command nature except by obeying her..." (Francis Bacon)

According to F.L. Wright "a house should not be supported by the ground but should rest on it; its design proceeds from the terrain". An interesting link between *natura naturata* and *natura artificiata* among different types of structures using earth as a building material can be found in the amazing rammed-earth constructions traditionally raised by the Chinese [50] and by other populations (Fig. 56).

6   Landscape

In Italy, the landscape is described in excellent publications accompanied by very beautiful images. In the most recent works [132,133], the features of the industrial landscape and the relationship with nature are also considered. Some years ago, ENEL dealt with this issue in a book full of original ideas and considerations, with special reference to hydroelectric power plants and dams [41]. On the other hand, the Italian landscape is also strongly influenced by agricultural features. A reference to the systematic history of the agricultural landscape can be found in the book by Emilio Sereni, who studied the "morphology that in the course and for the purpose of his agricultural

Fig. 62. OCCHITO's gorge from the right bank of the Fortore River in 1953 (photo by R. Jappelli)

Fig. 63. The OCCHITO embankment dam on the Fortore River today

productive activity, man systematically impresses with consciousness to the natural landscape" [125].

Fig. 64. The crest of the GOILLET gravity dam in Val d'Aosta, Italy (1947) from the right bank at the elevation of 2500 m above sea level, at the foot of the Cervino Mountain in western Alps [41]

The large and small artifacts which defend the territory from the effects of unregulated river courses are today part of this artificial landscape of antropic settlements.

An authoritative opinion on the subject can be found in a Leopardi's literary masterpiece entitled "Elogio degli uccelli" ("Laudation of the birds"): "...ora in queste cose, una grandissima parte di quello che noi chiamiamo naturale, non è; anzi è piuttosto artificiale: come a dire, i campi lavorati, gli alberi e le altre piante educate e disposte in ordine, i fiumi stretti infra certi termini e indirizzati a certo corso, e cose simili, non hanno quello stato né quella sembianza che avrebbero naturalmente. In modo che la vista di ogni paese abitato da qualunque generazione di uomini civili, eziandio non considerando le città, e gli altri luoghi dove gli uomini si riducono a stare insieme, è cosa artificiata, e diversa molto da quella che sarebbe in natura..." ("...now, among these things, a great part of what we call natural, actually it

Fig. 65. The MONGRANDO reservoir on the Ingagna River near Vercelli, Italy, from the crest of the concrete section of the dam (1998)

Fig. 66. The RAVASANELLA reservoir near Vercelli, Italy, from upstream (1998)

is not so; on the contrary, it is rather artificial; so to say, the cultivated land, the trees and the other plants that we have reared and arranged in order, the rivers constrained within certain terms and addressed to a defined course and similar matters, do not retain the state nor the resemblance they would have by nature. In a way that the sight of every village inhabited by any generation of civil people - likewise considering the cities and the other sites where men reduce themselves to live together - is of an artificial matter, and very different from what it would be in nature..." G. Leopardi [84])

Fig. 67. The elegant arch dam, PIEVE DI CADORE, at the edge of beautiful green woods [51]

The features of the GARCIA dam on the Belice Sinistro river (Fig. 57) almost mingle with the countryside at the feet of the rough Entella Rock, modelled in the *gessoso-solfifera* formation of Central Sicily (Fig. 58). A little farther, in the vicinity of the Maranfusa Rock, one can see the remains of the Castle (Fig. 59) [129] and the elegant single-span arch masonry Arabic-Norman footbridge of Calatrasi (Fig. 60). If original, the Greek inscription at the base of one of the banks dates its construction back to 1160.

In the fifties a muleback footpath (Fig. 61) was the only access from the Carlantino village (Daunia, Italy) to the OCCHITO site, which appeared as in Fig. 62. A few years later, the site was invaded by the great reservoir ($V = 300 \cdot 10^6 \ m^3$), which supplies water to the northern *Apulia*. The impact of the large embankment dam on the desolate landscape is not at all negative, as can be seen in a recent picture (Fig. 63). The site is close to the epicentre of the earthquake that struck the site in 2002, but the dam was not damaged.

The crest of the GOILLET dam built in 1947 in Val d'Aosta at an elevation of 2500 *m a.s.l.*, resembles a bridge to overpass the asperities of the valley bottom at the foot of Cervino Mountain (Fig. 64).

Flora and fauna contribute greatly to the pleasantness of the landscape around reservoirs (Fig. 65– 68). The ENEL 500 MW pumped storage hydro-electric power plant, built in the decade 1977-1987 conforms to the esplanade of the PONTE DIDDINO near Siracusa, Sicily (Fig. 69) forming a large basin beside the river bed, which respects the course of the ANAPO River,

Fig. 68. The downstream slope of the POMA embankment dam on the Iato River near Palermo, Sicily; a flock of sheep graze on the grassy dam facing [6]

the Greek water-work Galermi and Salvatore Quasimodo, who dedicated the following verses to the river:

"On your banks I hear the gurgling water,
my Anapo; on its mourning
a loud rustling
wails in the memory...
...
Peaceful animals,
The air drops
Drink in a dream"
(S. Quasimodo, L'Anapo, 1932-36)

"At landslide time,
Leaves and cranes burst out in the air..."
(S. Quasimodo, Cavalli di luna e di vulcani, 1936-42)

The heron can be seen circling on the basins and resting on the top of the intake towers (Fig. 70). Nearby, at Belvedere, one can admire the remains of the impregnable Eurialo Castle built by Dionisio between 401-397 b.C.

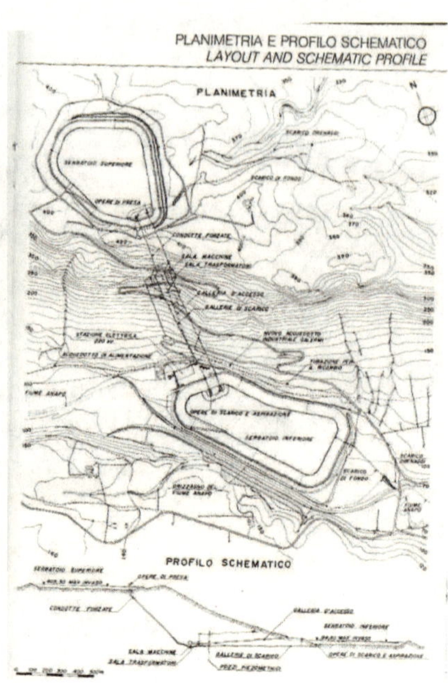

Fig. 69. The lower reservoir of PONTE DIDDINO near Syracuse, Sicily, of the Anapo pumped storage hydroelectric plant of 500 MW

with the purpose of defending the city of Syracuse (Fig. 71). A little farther, following the almost unknown valley bottom, one can ascend the river up to the amazing Necropolis of Pantalica (Fig. 72).

The reservoir created by the SANTA ROSALIA embankment dam, which was built with the resistant limestone rock locally used in traditional dry masonry structures (Fig. 73), penetrates into the narrow valley of the Irminio River, carved in the Iblean plateau; further downstream one finds the splendid old city of Ragusa Ibla.

In the environs of the BILANCINO reservoir (Fig. 74), which serves the primary purpose of mitigating the floods of the Arno River and protecting the city of Florence, the landscape of the Mugello in Tuscany is dotted with important historical sites. They include the Cafaggiolo Castle, first built by Michelozzo for the Medici, the Villa Le Maschere, one of the most important of the eighteenth century in the area, and the silent Monastery Bosco ai Frati, which guards a splendid wooden Christ by Donatello (Fig. 75).

At the conclusion of this rapid overview of landscapes and dams, I suggest that we look at the site of CASTELSANVINCENZO and outskirts in Abruzzi, as it appeared at the time (1954), when the construction of the dam had just begun (Fig. 76–78) and the ZOCCOLO dam in the green Val d'Ultimo (Bolzano) (Fig. 79).

Fig. 70. The upper reservoir of the ANAPO pumped storage hydroelectric plant with its intake towers

Fig. 71. Remains of the Eurialo Castle, built by Dionisio between 401 and 397 b.c. at Belvedere, Syracuse, Sicily [107]

Unfortunately, in Italy, the construction of some important dams has been suddenly interrupted due to unreasonable conflicts and left unfinished with irretrievable consequences on the landscape and the safety of the territory. As a matter of fact, it is at least surprising that man strongly opposes new

Fig. 72. Pantalica's necropolis

constructions, even when these definitely improve the landscape and the environment, demonstrating at the same time an irresistible, perhaps atavic, attachment to some small portions of the territory, irrespective of its actual significance and without regard for possible improvements.

7   Performance

Large dams are outstanding artifacts that man is inclined to safeguard, like monuments, for the benefit of future generations. However, the relationship of dams and appurtenances with the territory is far more complex than that of monuments. In fact, on the one hand, dams and monuments share the quality of "monumentality", which justifies safeguarding measures; on the other, the impact of dams on the territory requires a high and qualified surveillance for the primary purpose of safety.

Design and execution of dams and weirs are featured by a long but not unlimited duration, that man attempts to extend with various measures. This long arch of time is dotted by gradual or sudden events and phenomena, often with undesired effects, evolving and interacting in different ways, which do not always obey to well defined laws, and that must be kept under careful control. These events and phenomena offer the best training to civil engineers. On this

Fig. 73. The elementary dry masonry walls dividing the property in the Iblean plateau

basis, a story of dam performance, in which physical and mental workshops alternate and intersect, can be produced.

The *conception* of artifacts is supported by long-term research, studies, comparisons, choices, warnings and investigations. Their *execution* is marked by big contracts, variants, quality control, checks and tests. The *acceptance* is signed by the reservoir's filling and withdrawal, measurements, inspections, monitoring, concern and instructions. Its *operation* is accompanied by surveillance, guidelines, routine and exceptional maintenance; the eventual *abandonment* involves enduring decisions and binding designs. During this long-lasting cycle of *birth*, *life* and *death*, which greatly surpasses the course of man's active life, owners and contractors alternate, clients change, the destination of the plant transforms to comply with everchanging demand, solutions follow one after the other, archives - when not dispersed - do not acquire in order and clearness, and law and regulations grow in number and overlap. Men (owners, designers, engineers, managers, suppliers, consultants, experts, experimenters, inspectors, assistants, wardens, workers,...) alternate; each of these protagonists contributes only for a short period of such a long-lasting process.

Each stage of the process is marked by pauses of interruption, meditation, interpretation, hesitation, in series and/or parallel with the main flow of *conception, execution, acceptance, operation* and with alternate recourse to theories, experiments, experience, strategies and engineering judgment.

With such a premise, it is easy to conceive the difficulty that must be faced by those who are called to express a global opinion for safety assessment on

Fig. 74. The BILANCINO embankment dam and reservoir on the Sieve River near Florence in the imagination of the inhabitants of Mugello

the base of relevant fragments of knowledge. Such a task may be compared to the work of an archaeologist, who expects to reassemble tiny broken pieces, which gain significance only if put together with the patience and competence necessary to reproduce the original design (Fig. 80).

Table 1. List of the hollow buttressed gravity dams in operation in Italy

Name	PR	Compl. date	Height (m)	Reservoir vol. ($10^6\ m^3$)	Crest elev. (m a.s.l.)	Seismic class S
BAU MUGGERIS	NU	1949	58.7	63.00	802.0	-
POGLIA	BS	1950	49.4	0.59	632.4	-
LAGO INFERNO	SO	1944	37.0	4.33	2088.0	-
LAGO DI TRONA	SO	1942	53.0	5.48	1808.0	-
PANTANO D'AVIO	BS	1956	59.0	12.84	2379.0	-
SABBIONE	VB	1953	61.0	44.83	2461.6	-
ANCIPA	EN	1953	104.5	30.40	952.4	9
LISCIA	SS	1962	65.0	108.00	180.0	-
PONTE VITTORIO	BI	1956	36.0	0.53	710.0	-
MALGA BOAZZO	TN	1956	53.5	12.84	1226.5	-
MALGA BISSINA	TN	1957	81.0	62.37	1790.0	-
VENEROCOLO	BS	1959	26.0	2.64	2539.4	-

Fig. 75. Wooden Christ by Donatello housed at the "Bosco ai Frati" Monastery in Mugello, near Florence

The fifty year-old ANCIPA[4] ($H = 104.4\ m$) is a gravity dam with hollow buttresses founded on a bedrock of polygenic conglomerates and stratified sandstones (Fig. 81). This type of dam was proposed in Italy [91–94] after the second world war with the purpose of accomplishing the work with the minimum volume of concrete [66] (Table 1). However, after a number of interesting projects, this structural type has been abandoned due to technical reasons concerning the unsatisfactory performance of the dam body, and the costs that such a solution would require in terms of handwork and formworks, which today would be much higher than the presumed saving in concrete volume.

The ANCIPA dam is severely cracked. Due to the fractures, the mechanical response of the dam is presumably quite different from the performance originally evaluated under the hypothesis of the material's continuity. Different sources conflict on the origin of the phenomenon. Arredi et al. [4] maintain that cracking is syngenetic with the construction, which ended in 1953; while, Spagnoletti, who in 1962 was head of the Dam Monitoring Office of the Edison Society, gives us a somewhat different version [128]. The results

---

[4] The ANCIPA dam stands out for its height, close to the maximum of this type in the world. In Italy, it is the only dam of this type in a seismic area; VENEROCOLO is situated at the highest elevation; LAGO DI TRONA is the oldest; while the maximum reservoir volume is created by the LISCIA dam.

Fig. 76. The CASTELSANVINCENZO embankment dam near Isernia, Italy, under construction in 1956; the Mainarde chain of mountains rises in the background (photo by G. Chiolini)

of the rich set of data (displacements, rotations, deformations, temperature) collected during the first seven years by the instruments installed in the dam body, were correlated to the reservoir levels and to the temperatures (measured outside and inside the hollow cells). The measured values are compared to the values computed with the aid of the theory of elasticity, leading to an evaluation of the concrete elasticity modulus (back analysis).

The Author concluded that the prevailing stresses in the dam body were of thermal origin, but that the "performance of the artifact was absolutely normal, fully representing the theoretical model"; without so much of a hint to the cracks. The same Author expressed the opinion that, in hollow gravity dams, "the easy dissipation of the hydration heat prevents the concrete from reaching high temperatures", and concluded that "the main cause of the measured displacements is due to the variation of external temperature" and that "the phase difference between the measured displacements and the water levels in the reservoir is due essentially to thermal actions ...the displacements due to the reservoir oscillation should be attributed approximately for a half to the compressibility of the ground and for a half to the compressibility of the dam body ...; ... the stresses of thermal origin by far prevail on those originated by the hydrostatic thrust..."

The comparable features in the general framework of the cracks, which do not seem to depend on their elevation, have persuaded some authors to

Fig. 77. The cable-way used for the transport of the material from the quarry on the Iemmare River to the construction site of the CASTELSANVINCENZO embankment dam (photo by G.Chiolini)

Fig. 78. Remains of the San Vincenzo Abbey of the XIII century (?) near CAS-TELSANVINCENZO [131]

confirm that the phenomenon is due to both the shrinkage of concrete and the seasonal temperature variations, and to suppose that the ascertained crack distribution is a sort of "iatrogenous" consequence of an incorrect working process which was merely finalized to control shrinkage [3].

According to the same authors, the bad performance of the structure can be attributed to three causes: (a) the initial shrinkage of the concrete due to the lack of the modern corrective measures in placing concrete; (b) aggregate segregation in concreting; (c) thermal seasonal changes due to the particular exposure of the downstream facing. The ANCIPA case is considered extreme because of the coexistence of the three factors (a)+(b)+(c).

Fig. 79. The ZOCCOLO embankment dam in Val d'Ultimo, near Trento, Italy, from downstream; the spillway open channel in the foreground [63]

According to Arredi, as well, "the cracking was certainly originated from circumstances which go back to the execution ... the high cement content ... the difference in level between the foundation of the two buttresses ..." and ascribes "the evolution of the cracks with longitudinal extension to continuously variable states of tension consequent to temperature excursions..." [4]. Futhermore, the ascertained spread *alveolization* of concrete prevailing near the joints could be the consequence of an excessive water content in the concrete [3].

In anticipation of the difficult structural measures, non-structural provisions have been applied; among them, the simple control of the water level in the reservoir, with gradually increasing severity, in relation to the general evolution of the cracks in the dam body.

The phenomenon of the spreading of cracks in the concrete has been observed in other dams of the same type built in Italy soon after the Second World War. Solutions based on embankment dams, either earthfill or rockfill, with a central core or, lacking suitable materials, with a thin water barrier on the upstream face would be considered safer, also because of its higher resistance to ageing.

The NOCELLE zoned embankment dam ($H = 35\ m$) with central core was built with the current criteria on the Sila Plateau in 1927-31 to create

Fig. 80. Anfora K 8698 , of the Haltern 69 type, height 76 cm [62]

a reservoir ($V = 83 \cdot 10^6$ m^3) (Fig. 82) for hydroelectric production. The settlements of seven benchmarks placed on the crest and the discharge from the drains have been regularly measured since the end of construction.

In 1955, A. Croce and G. Baroncini drilled a boring through the core as far as the bedrock of weathered granite in order to ascertain the material's properties. The results of the research, that was rather unusual for the time, were published in the Proceedings of the IV International Congress of Soil Mechanics and Foundation Engineering, London, 1957, and in the journal "Geotecnica" [29].

The data collected during dam construction and twenty-five year operation were enlightened by the information obtained from the boring, with due allowance for the reservoir regimen. The conclusion was that settlements have increased very gradually with time and that the differences among the values at the various benchmarks are negligible (Fig. 82). In 1955, the average settlement was 14 cm; the phenomenon was still evolving, according to a curve close to the theoretical course of core consolidation. The slowness of the process was probably due to two different causes: (a) the lower portion of the core was built with a water content much higher than optimum; (b) the shells were formed with a non-fully draining material.

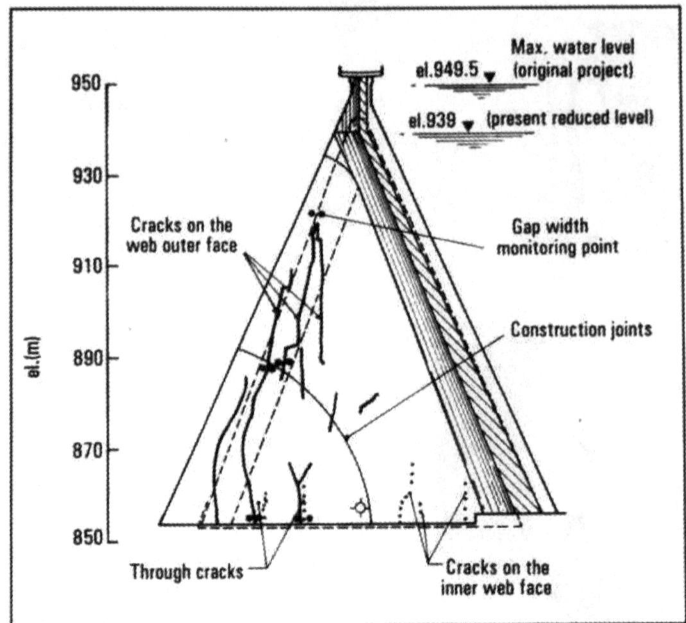

Fig. 81. Cracks in the buttresses of the ANCIPA hollow gravity dam near Enna, Italy [48]

The reason of the fully satisfactory performance of the dam has been attributed by the Authors to the central position of the core and to the favourable choice of the site. The same Authors express the view that the "method of drilling the dam body a certain time after construction with the purpose of extracting undisturbed samples to be submitted to laboratory tests is feasible" (with due caution) "and may furnish useful directions for scientific purposes and future improvement of design criteria."

The settlement measurement at the original benchmarks has been continued up to present. The results have been digested and represented in Fig. 82b as a function of time. The graph shows that after 70 years of operation the phenomenon is close to exhaustion [21]. Thus, the performance of the NOCELLE embankment dam, that is reported as one of the first and most significant examples of its type in Italy, is very satisfactory, notwithstanding some ascertained construction defects in the core.

The verification of an embankment dam by means of borings through the core and tests on samples has been adopted for the purpose of acceptance in other cases, as for example at the GARCIA dam, where the results were also very satisfactory. It is worth mentioning, however, that the operation is quite delicate and requires great caution.

Three important embankment dams still deserve to be mentioned in the present review. The ZOCCOLO dam ($H = 66.5\ m$, Figs. 79 and 83) is a ho-

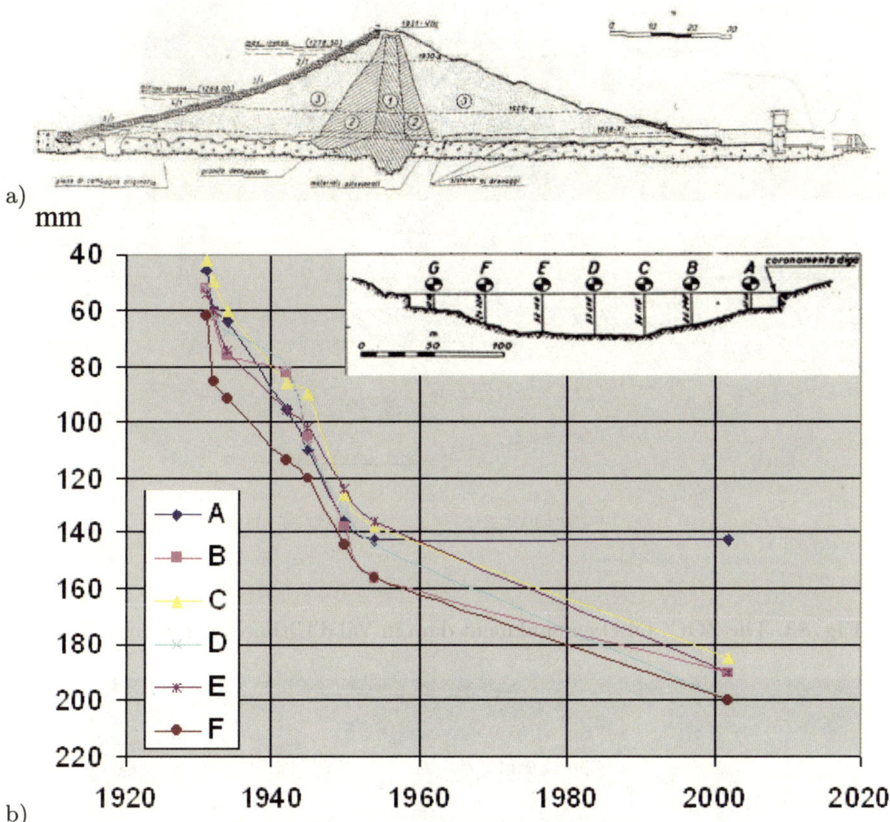

Fig. 82. The NOCELLE embankment dam after 70 years of service; a) maximum cross section: silt with sand; 2. silt and sand; 3. silty sand with gravel; b) measured settlement at seven benchmarks installed on the crest [21]

mogeneous embankment of coarse material with a bituminous conglomerate facing. The dam, built on the Valsura River (Bolzano, Italy) in 1961-65, is founded on a thick deposit ($k_{min} = 10^{-2} \div 10^{-5}\ cm/s$) of detritus and of soils of fluvioglacial origin that are interbedded by levels of lacustrine silt. The bedrock is located at a depth of about 150 m. The water barrier function in the foundation is entrusted to a diaphragm formed by compenetrated piles $\varnothing$ 600 mm driven to a depth of 50 m through the permeable cover soils. The embankment is supplied with an abundant system of drains, which are designed to convey 400 l/s.

Following the first reservoir filling, a gradual increment of the piezometer's levels in the dam body together with a global increase of leakage from 120 to 240 l/s was observed. These phenomena, albeit within the limits of design values, aroused some concern, as they were accompanied by a limited water

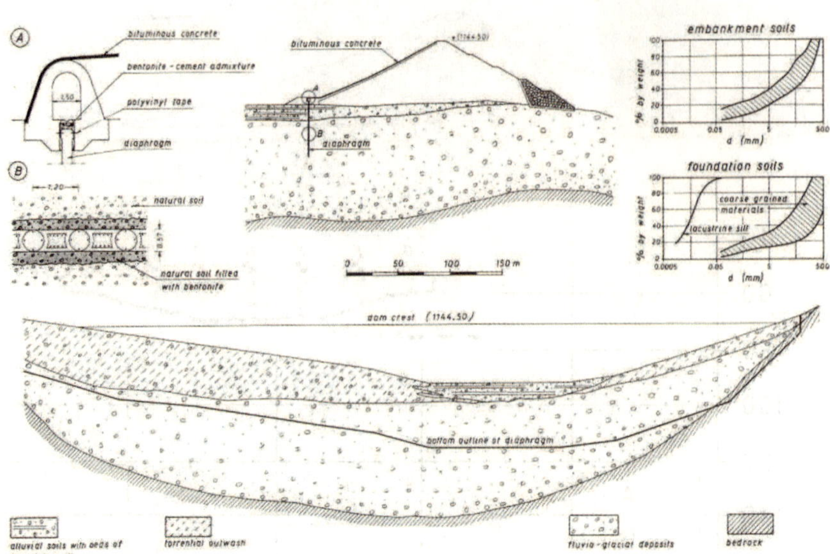

Fig. 83. The ZOCCOLO embankment dam in Val d'Ultimo, near Trento, Italy [28]

Fig. 84. The MONTE COTUGNO embankment dam on the Sinni River near Potenza, Italy [46]

appearance at midheight of the downstream slope. The doubt was that the deep diaphragm could be affected by syngenetic building deficiencies and/or acquired defects consequent to irregularly variable settlements $(20 \div 70 \; cm)$.

After great consideration, it was decided that, in the short term, the phenomenon would be of little concern. However, the reliability of the water barrier was improved by a curtain of cement and bentonite that was grouted

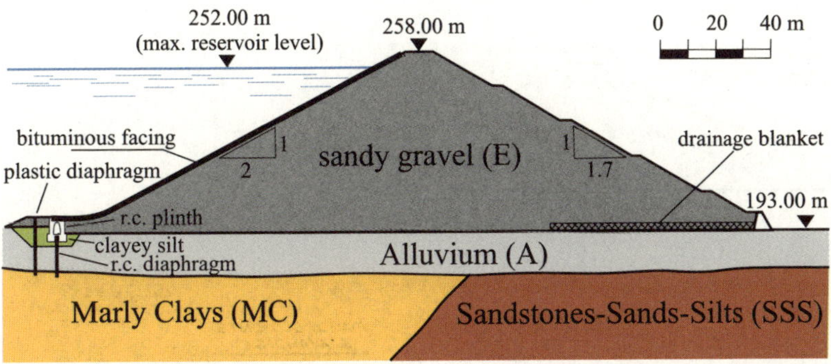

Fig. 85. Maximum cross section of the MONTE COTUGNO homogeneous embankment dam [17]

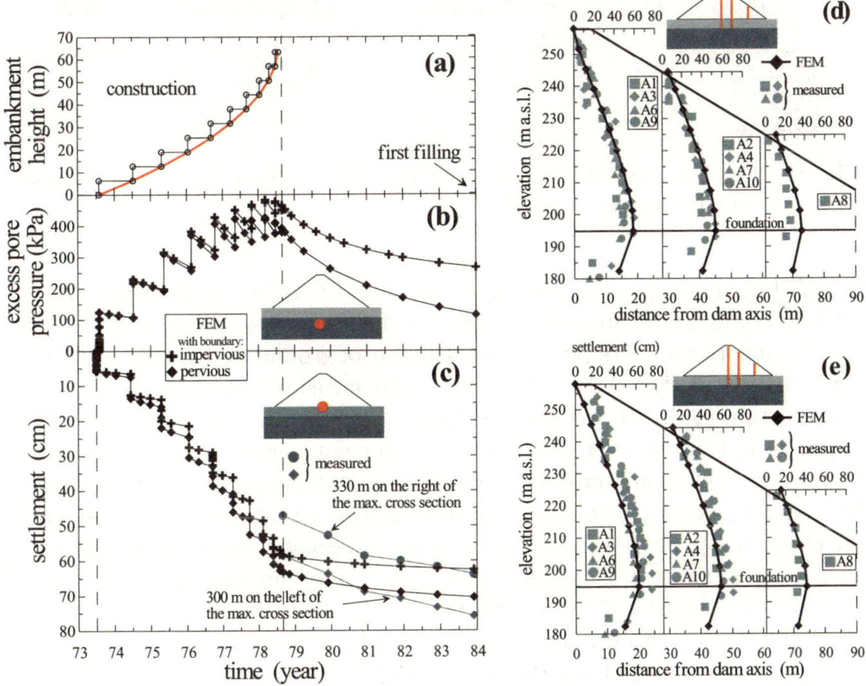

Fig. 86. Measured and calculated settlements of the MONTE COTUGNO embankment dam near Potenza, Italy: c) as a function of time; d) at the completion of the work; e) at the beginning of first reservoir filling [18]

upstream of the existing diaphragm. Successively, the performance of the dam was satisfactory. So, the design criteria, inspired by Arrigo Croce [30], were fully confirmed.

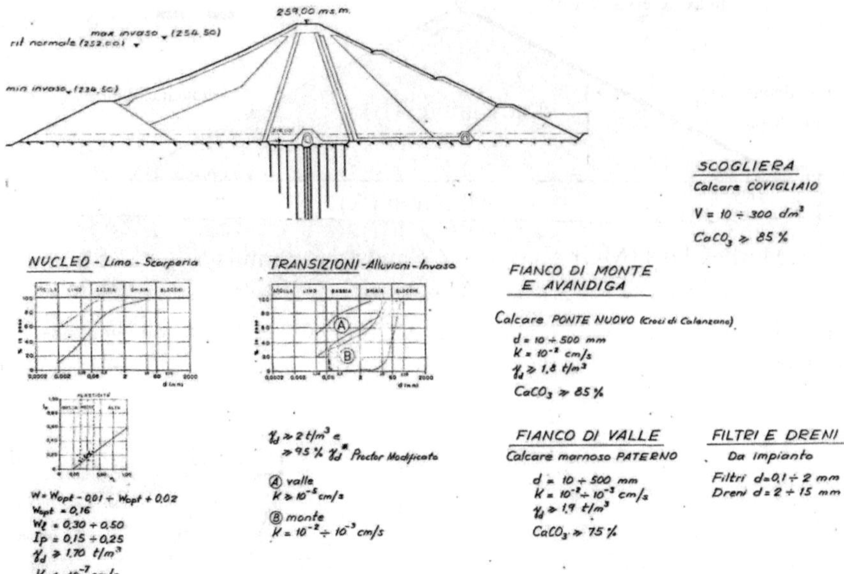

Fig. 87. The BILANCINO embankment dam on the Sieve River near Florence: design maximum cross section [77]

With its homogeneous embankment ($V = 12 \cdot 10^6 \ m^3$), the MONTE CO-TUGNO dam on the Sinni River supplies the largest earthfill in Europe (Figs. 84, 85). In a recent work [18], the settlements of ground (alluvial deposit, $12 \ m$ thick, on a stiff clay bed) and embankment measured from the beginning of construction (1973) to the start of the first reservoir filling (1983) have been digested and interpreted with the finite element method (Fig. 86), simulating the embankment's gradual construction and the ground's consolidation, by coupling the seepage with the deformation of the solid skeleton under an elasto-plastic hypothesis [19]. A good agreement between experimental and computational results has been obtained by an appropriate fitting of the mesh, of the boundary conditions and of the single steps of the analysis, which required a shrewd choice of the model parameters.

The performance of the BILANCINO dam in Tuscany (Fig. 87–89) during the experimental reservoir filling has been recently simulated under the assumption that the seepage process was initiated under the pore pressure values existing before the experiment [78]. The satisfactory efficiency of the grout curtain was confirmed by theory and experiments (Fig. 90).

It is quite difficult to propose synthetic remarks to reveal the performance tendencies of the vast set of large dams, heritage of future generations, that will have not shared the long painstaking course of efforts, connected with the birth of single artifacts. Therefore, it is not without hesitation that I have

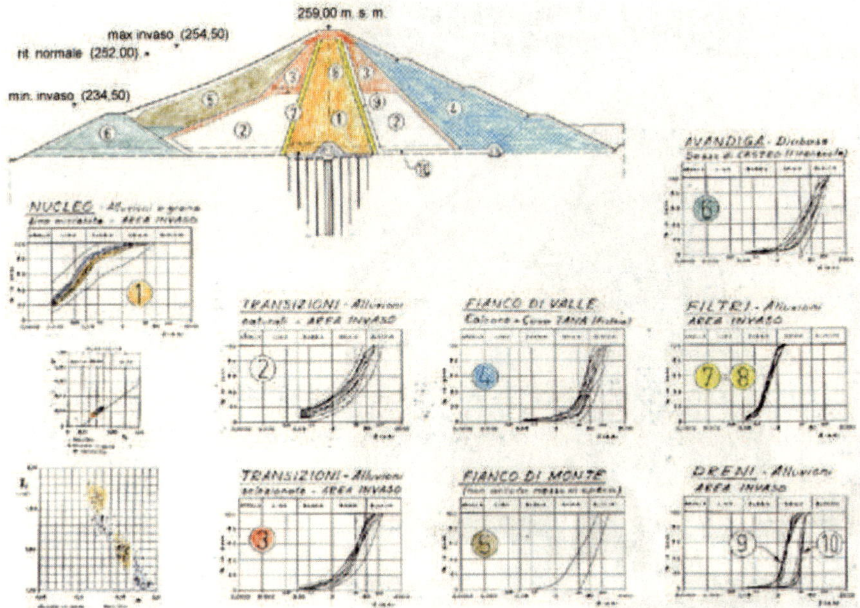

Fig. 88. The BILANCINO embankment dam on the Sieve River near Florence: as built maximum cross section, May, 1990 [77]

tried to express the following comments that I submit to your attention. These remarks originate from fifty years of experience, in addition to the considerations of preceding authoritative colleagues:

1. The dams in Italy have exhibited an outstanding performance, especially considering their advanced average age, approximately fifty-year old (cf. NOCELLE);
2. Embankment dams have revealed themselves generally safer than concrete dams and more conformable to the complex geotechnical situations prevailing throughout the territory, especially in the South (cf. POMA);
3. Gravity dams have shown a better performance than structures relying on thin elements, the later being exposed to earlier ageing and requiring significant restoration measures (cf. ANCIPA);
4. The deterioration of concrete in contact with water has often demanded protection measures of the upstream facing (cf. PIANA DEI GRECI);
5. Embankment dams with a bituminous concrete upstream facing have performed very well, apart from some cases of easily reparable *physiological* cracking localized approximately at the junction of the facing with the plinth on the occasion of the first filling of the reservoir (cf. CASTELLO);
6. With the progress of research on material properties and placing techniques it has been possible to accept the use of poor quality materials in large sections of embankment dams (cf. DISUERI);

Fig. 89. The BILANCINO dam on the Sieve River near Florence: work in progress

7. The methods in use to face landslide movements are considered satisfactory, especially around the abutments of embankment dams, provided
   they are adopted at the design stage with a long-term strategy (cf. CAM-
   POLATTARO);

8. In contrast to the displacements of the dam body, the ground settlements
   are often long-lasting, yet for decades, even in the case of coarse grained
   soil (cf. MONTE COTUGNO);

9. The pore pressure in fine-grained core materials dissipates in a relatively
   short time, often during the progress of the work (cf. BILANCINO);

10. Automatic monitoring improves rapidly and results are satisfactory; how-
    ever, individual situations still require manual measurement, especially
    for the surveillance of the slopes (cf. RAGOLETO);

11. Resistance to seismic action can be qualified satisfactory, on the whole, as
    demonstrated by the good performance of a number of dams on the occa-
    sion of some major earthquakes (Belice, Irpinia, Friuli, Umbria, Molise),
    followed by some non-essential measures on the appurtenances (cf. AC-
    CIANO);

12. Free overflow spillways are considered more reliable than gated spillways,
    the latter entrusted to complex electromechanical equipment requiring
    difficult maintenance (cf. ARANCIO).

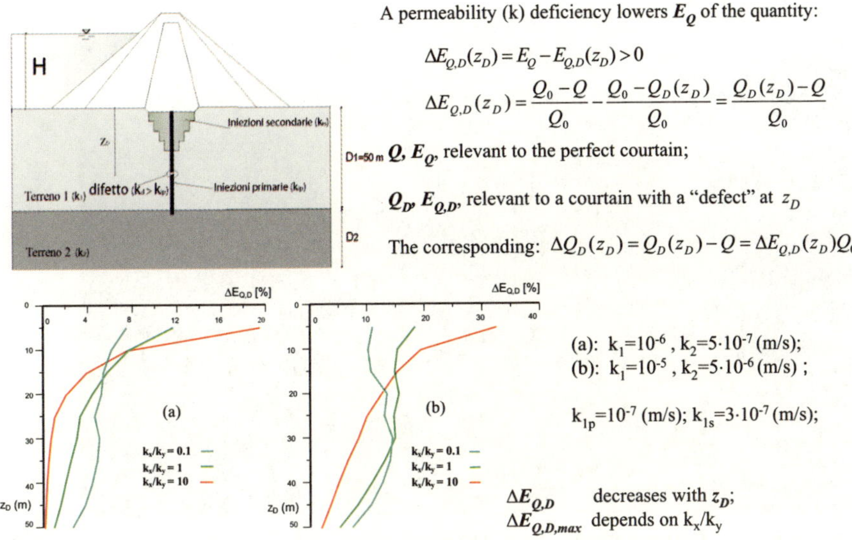

A permeability (k) deficiency lowers E_Q of the quantity:

$$\Delta E_{Q,D}(z_D) = E_Q - E_{Q,D}(z_D) > 0$$

$$\Delta E_{Q,D}(z_D) = \frac{Q_0 - Q}{Q_0} - \frac{Q_0 - Q_D(z_D)}{Q_0} = \frac{Q_D(z_D) - Q}{Q_0}$$

Q, E_Q, relevant to the perfect courtain;

Q_D, $E_{Q,D}$, relevant to a courtain with a "defect" at z_D

The corresponding: $\Delta Q_D(z_D) = Q_D(z_D) - Q = \Delta E_{Q,D}(z_D)Q_0$

(a): $k_1 = 10^{-6}$, $k_2 = 5 \cdot 10^{-7}$ (m/s);
(b): $k_1 = 10^{-5}$, $k_2 = 5 \cdot 10^{-6}$ (m/s);

$k_{1p} = 10^{-7}$ (m/s); $k_{1s} = 3 \cdot 10^{-7}$ (m/s);

$\Delta E_{Q,D}$ decreases with z_D;
$\Delta E_{Q,D,max}$ depends on k_x/k_y

Fig. 90. Efficiency E_Q of the BILANCINO dam grout curtain [78]

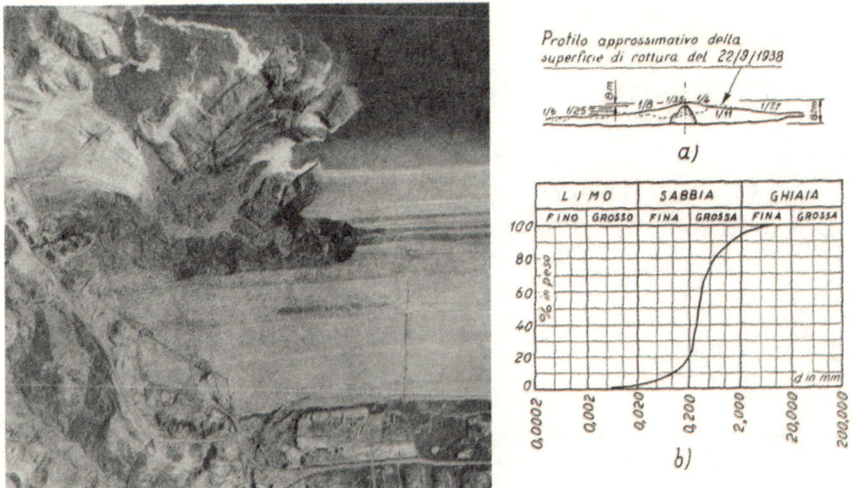

Fig. 91. The failure by liquefaction of the FORT PECK, USA, embankment dam in 1938 [97]

8   Concerns

My juvenile review [67] on embankment dam failures is now half century old; however, the general framework of *undesirable* phenomena has not substantially changed. Events that have marked dam history, like the FORT PECK failure in 1936 [97] (Fig. 91–93) in the U.S.A. (embankment $V = 100 \cdot 10^6 \ m^3$,

Fig. 92. The FORT PECK dam, USA, in service after the reconstruction of the right abutment [136]

Fig. 93. The spillway of the FORT PECK dam from upstream [136]

reservoir $V = 25 \cdot 10^9 \ m^3$), the MALPASSET (arch $H = 66 \ m$, 1959) or the VAJONT ($H = 262 \ m$, 1963 [124]) tragedies are in the sad memory of every specialist of this topic. Nowadays the situation is quite different, because (a) knowledge is increased by far; (b) surveillance has become deeper and more competent; (c) the profession has been gradually pervaded by the

Table 2. Classification of ageing causes for Italian dams [52]

		n. of cases
In foundation:	loss of strength under permanent or repeated actions	11
	ageing of grout curtains and drainage systems	3
	chemical reactions resulting in swelling	3
In dam body:	shrinkage, creep and reaction leading to contraction	17
	degradation due to chemical reactions of materials with environment	48
	loss of strength under permanent or repeated actions	53
	poor resistance to freezing and thawing	23
	ageing of upstream facings	23
	ageing of structural joints	1
	total:	182

sound concept that the anticipation of the consequences of incidents must be accompanied by the measures to apply in case these incidents actually occur. However, concerns do not decrease. The following is a short review of the main sources of concern.

1. *Ageing* is due *strictu sensu* to the complex phenomena accompanying or following the transformations which the materials and the ground undergo as time goes by; in a more general sense, the concept is extended to the state of the whole dam-slopes-reservoir system and appurtenances. Studies about ageing are among the most complex in Civil Engineering. The diagnosis attempts to differentiate the *causes* from the *effects* and the latter from the *consequences* that are often revealed only after many years. Among ageing causes, the *singenetic*, that is the original ones, depending on design and execution options, should be distinguished from *acquired* causes, which are induced by operation and/or maintenance.
   In 1994, ICOLD has conventionally, and perhaps too severely, assessed that a dam reaches "old age" five years after the first reservoir filling. According to the former definition, almost all Italian dams have a venerable age. Concrete dams are somewhat older than embankment dams. In spite of the small difference in age, embankments reveal a much higher resis-

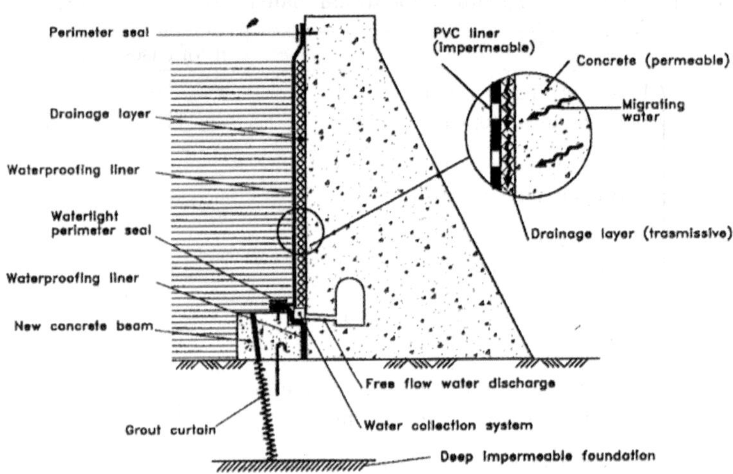

Fig. 94. Refurbishment of a gravity dam with the application of a geocomposite facing on the upstream slope [123]

tance to ageing than their concrete sisters. A proof of this can be found in ICOLD Bulletin n. 93 [52], where the defects of the major Italian concrete dams are compared to a single case (ZOCCOLO) of ageing due to internal erosion of the foundation soil. The ageing causes of concrete dams are classified in Table 2.

(a) In a general sense, the gradual decrease of compressive stress at the upstream heel can be considered a long-term consequence of ageing of concrete dams. In arch dams, a zero compression stress at the upstream border is considered admissible as a service limit state. However, a long-term consequence, the failure of the grout curtain, could follow this contact defect. Moreover, the related modifications of the stress distribution may contribute to the malfunctioning of the drainage system and to the increase of pore pressure on the foundation surface (PLACE MOULIN) (cf. [52]).

(b) Although more gradually than in concrete dams, embankment dams are also affected by some sort of ageing. The most serious phenomenon is internal erosion, which is nearly always concealed. Modern research shows that the phenomenon can reveal itself in different ways that are not yet fully explained (migration of fine particles, suffusion, hydraulic fracturing, piping,...). Elementary computation methods do not guarantee full safety. The danger is strictly related to the cracking susceptibility of the core material. This tendency depends on the type of materials, the state of stress and the placing technique. The analysis of numerous case histories in Europe shows that internal erosion can reveal itself after many years and sometimes dozens of years

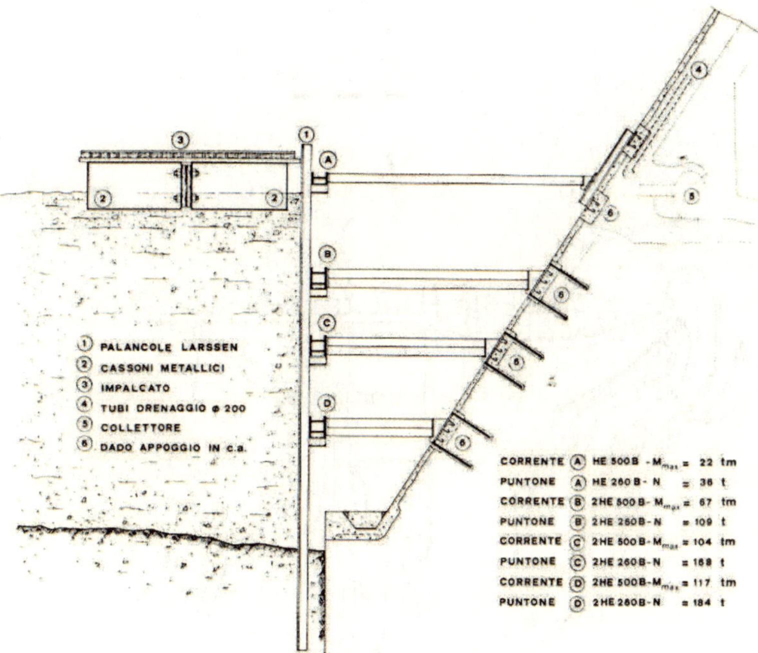

Fig. 95. The installation of a metallic facing on the upstream slope of the PIANA DEI GRECI dry masonry dam after the Belice earthquake in 1968. The work required special measures to overpass the thick deposit of soft mud (10 m) in the reservoir [138]

of satisfactory performance [43]. The ZOCCOLO dam is an emblematic case in Italy.

2. The gradual silting up of the reservoir is an obvious concern because of the relevant decreasing volume of the water resource. Moreover, the phenomenon hampers the correct operation of the bottom outlet gates, that are essential for safety. In cases where the solid transport is not yet controlled by suitable defences in the catchments area, the phenomenon has gained so great an importance to become a concern for the survival of the reservoir (ABATE ALONIA, COMUNELLI, DISUERI) or for a heavy restriction of its functionality (POZZILLO).

3. The technique followed in measuring the quantities controlling safety has certainly evolved with respect to the means prevailing at the time of instruments' installation. However, these are becoming old and in many cases unreliable; nor is their substitution always possible for reasons which are not difficult to imagine. The existence of the same problem elsewhere is not an encouraging circumstance: "... A large number of the world's dams contain no instruments and inspecting engineers may have to resort to various forms of in situ testing. In Britain, many dams built before

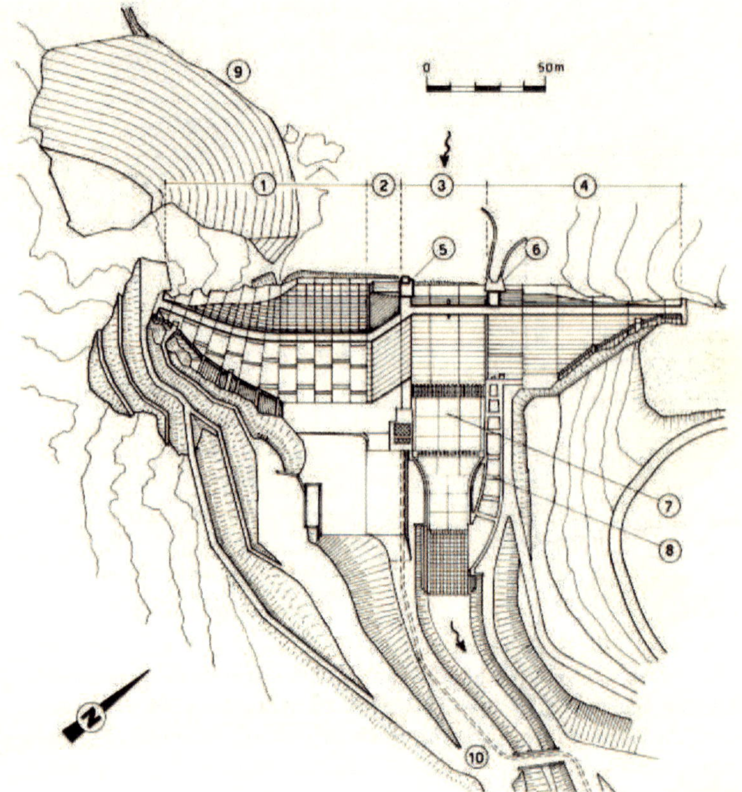

Fig. 96. Planimetry of the RAVASANELLA near Vercelli, Italy: 1. precast dam; 2. an element of transition; 3. gravity dam with spillway; 4. gravity dam without spillway; 5. intake tower; 6. outlet mouth; 7. sedimentation pond of the spillway; 8. sedimentation pond of the bottom outlet; 9. rockfill; 10. irrigation pipe [90]

1950 had cores of puddled clay and the BRS has begun studies of their conditions..." (A.D.M. Penman [112])

4. A special remark deserve some non-technical questions raising apprehension, which may be more serious than those concerning the mentioned technical topics.

(a) If it is true that for the expressed reasons embankment dams will have the upper hand on their concrete sisters, the further diffusion of the former is severely limited by the generalized and often unjustified opposition to the opening up of new quarries of construction materials for the dam body. Such an opposition, which in the instance was not supported by sound environmental and landscape reasons, was the source of very serious delays and economic distortions in the construction program of the BILANCINO dam in Tuscany (cf. Sect.

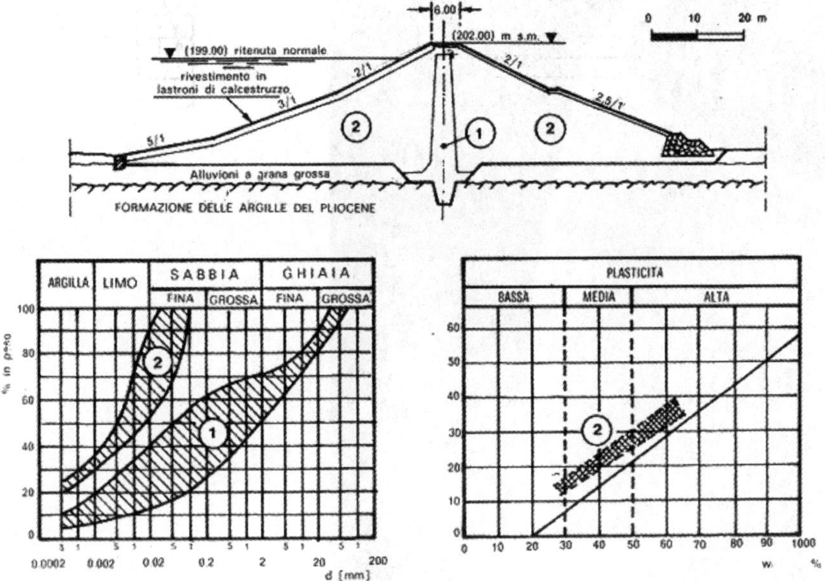

Fig. 97. Maximum cross section of the ABATE ALONIA embankment dam (1956) on the Rendina River near Foggia, Italy; 1.Mixture of coarse gravel (30%), coarse sand with fine gravel (50%) and silt with fine sand and clay (20%); 2. Silt, clay and fine sand [75]

7), with deep concern and personal distress of the technicians involved in the work. In the final solution, the cross section of the dam body resembles a Harlequin dress formed by different materials (Fig. 88), the reciprocal compatibility of which has been checked through patient experiments and computations under the different operation conditions [77].

(b) The sharp slack, or rather the interruption, lasting since the eighties, in the rate of dam construction has determined a gap of a full generation in the transmission of experience in this branch of Civil Engineering. The efforts of some universities to fill this gap are not sufficient. They still impart theoretical lectures about the computation of stresses and deformations in a dam body with the finite element method of analysis or venture, at most, as far as simulating the consequences of possible defects. Many of the colleague engineers, who participated in the front line in the studies and in the construction of some of these public works, retired or passed away without leaving any trace of their enlightened activity. Therefore, the maximum concern, which is great also in the restoration of monuments, is the damage that could be caused to the patrimony by so called ex-

Fig. 98. The BEAUREGARD arch dam in Val d'Aosta, Italy (1960), has been submitted in the last decades to a severe water level limitation due to cracks caused by the thrust from a large landslide against its left abutment [41]

perts, approaching dams without the necessary caution and wisdom suggested by long-term experience.

(c) Strictly associated to the preceding reason of concern is the scattering of archives related to single works. This occurence, that I have ascertained with great disappointment in different circumstances, has not been originated solely by the sudden interruptions of the public works, but also by other traumatic events in the country, e.g. the sudden cancellation of the Development Fund for the South, the transformations of the companies for the production of the electrical energy, the sudden interference of the magistracy, which after many years has not yet returned to the managers many valuable documents for the reconstruction of the complex events of important old and recent works.

(d) Last, but not least, the drop of public consensus should be mentioned. Erroneous information propagated by a not fully aware press has produced a real distortion in the so-called public opinion to the point that the tribute of 50 weekly victims distributed along our highways is being accepted almost as unavoidable, while dams are considered a great danger to public safety, instead of a benefit. Save that then one declares a scandal the delay in the full operation of the works

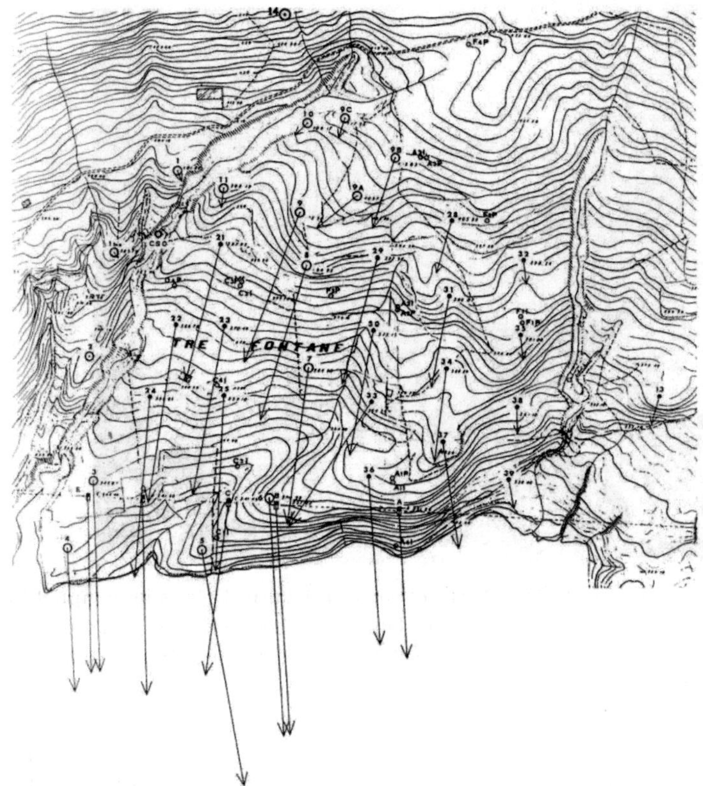

Fig. 99. Planimetry of the landslide in the right bank of the RAGOLETO reservoir and vectors of the average displacements between 1972 - 1994 [100]

that actually lag behind, just because of the consequences of the above mentioned complains. Enduring this situation, the difficulty of a general agreement for the completion of uncompleted works, also extends to the maintenance necessary for the safeguard of existing ones.

9   Measures

In Italy, a number of dams have reached a state of degradation requiring corrective measures. As in the realm of monuments, one can distinguish different levels and types of measures, that literature designates *rehabilitation*, concerning statical, water barrier, draining, filtering ... functions. The measures are spread or localized in the dam body, in the foundation, in the outlet works or in the natural slopes upstream, downstream or on the abutments; as for their amount, one can distinguish between protection, safeguard, restoration,

Fig. 100. The work at the ALTO ESARO near Cosenza, Italy, was interrupted in 1987 in consequence of a rupture with a mechanism in the rock formation during the excavation at the left abutment of the concrete gravity dam ($H = 120\ m$) [10]

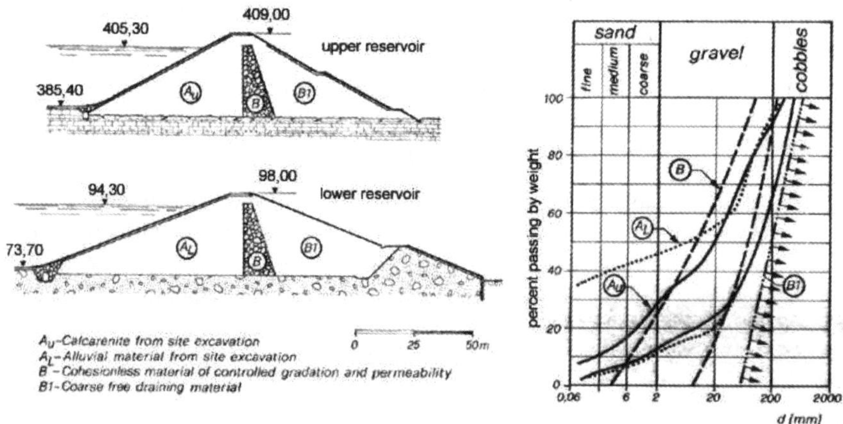

Fig. 101. Cross section and construction materials of the embankment of the upper and lower reservoir of ANAPO pumped storage hydroelectric power plant near Syracuse, Sicily [76]

integration, substitution, refurbishment. Measures are justified by hydraulic, structural or geotechnical reasons. The following are examples chosen among Italian cases.

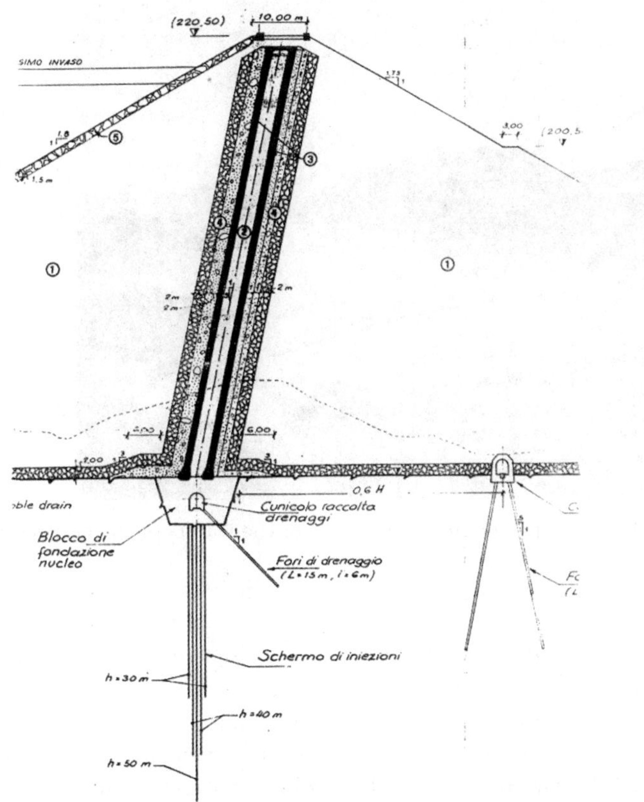

Fig. 102. Double central diaphragm with a sandwich drain for the planned GUALTIERI dam near Messina, Sicily [14]

At the site of ANCIPA dam, a difficult structural work has been recently initiated with the purpose of restoring the reservoir to its original efficiency level, which had been reduced for many years by a non-structural measure of serious limitation of the water level (cf. Sect. 7). In order to mitigate the severe thermal action prevailing at the site, a large protective structure described as a *thermal screen* is being installed on the downstream slope of the dam. The insulation panels are deemed to significantly reduce the thermal gradient which at the site reaches almost 30° and is considered responsible for the crack spreading in this hollow buttressed gravity dam. The screen will be supported by a heavy heel concrete block.

The original continuity of the very high structure will be restored by grouting the cracks in the buttress webs with cement and epoxy resins and by filling the honeycombings [48]. The work is very specialized and as far as possible it shall be executed without further lowering the reservoir level,

Fig. 103. The old MONCENISIO dam emerges out of the waters of the new reservoir [44]

which already severely penalises the seasons of heavy shortage of water. The completion will require from 3 to 5 years.

The life time of a dam mainly depends on the efficiency of the drainage system. The drainage system, in turn, depends on the efficiency of the water barrier system, which is composed of grout curtains, upstream blankets, cut-offs and diaphragms. Drainage is entrusted to boreholes, draining diaphragms or drifts in the dam body and in the abutments.

In concrete dams founded directly upon bed rock, where water seeps chiefly along the joints, the common solution is to use cement grouted curtains with admixtures (bentonite, ash, pozzolana) and sometime with resins. According to the situations, the gradually occurring curtain deterioration can be attributed to original (as would be the case of an unsuited grouting technique) or acquired defects (e.g. as a consequence of stresses inducing displacements in the joints or even rock fracturing; an erosion or the passing in solution of the mixture under the action of water movement; the chemical action of an aggressive water).

The degradation of the drains originates chiefly from inadequate dimensions, particular climatic circumstances or progressive obstruction. In both cases (curtains and/or drains degradation), the heaviest consequence of the defects is a pore pressure increment on the foundation surface.

The strict dependence of the water barrier-drainage performance on the state of stress in the foundation can be observed in the example of PLACE MOULIN [52]. The construction of this important arch gravity dam ($H = 155\ m$) in Val d'Aosta was terminated in 1964. The foundation cut-off was

Fig. 104. Views of the territory around the MONCENISIO reservoir on the occasion of a drawdown; the remains of a larch grove/wood [44]

created before concreting the dam body by grouting the crystalline schists of the bed-rock with cement mix to a depth of about 100 m; consolidation grouting to improve the bonding between the dam body and the bed-rock was then performed to a depth of about 20 m. At the reservoirs first filling (1967), the instruments monitored a slight detachment of the downstream dam foot from the bed-rock and an unusual increment of the drain discharge and of the pore pressure on the foundation surface. Grouting, drains and instruments were reactivated. The phenomenon reoccurred after 10 and again after 19 years on the occasion of events of rapid reservoir filling. Grouting was repeated, the first time with resins also, a material that probably is losing efficacy with time. Therefore, a new grouting campaign with fine cement has been proposed [52].

At the site of ZOCCOLO, after having monitored the dam's performance during the first series of experimental fillings, a new grouting curtain upstream from the existing pile diaphragm was prudentially created [30].

Time may reveal, in some occasions, the need for a measure concerning the permeability of the dam body. The problem may derive from ageing or from an accidental action, such as a seismic one.

Fig. 105. Views of the territory around the MONCENISIO reservoir on the occasion of a drawdown; devices for the lifting of old gates from the Savigliano Workshops [44]

Fig. 106. Views of the territory around the MONCENISIO reservoir on the occasion of a drawdown; a partly destroyed old ospice from the WWI days [44]

The water barrier function within the body of concrete dams is entrusted to the concrete itself, through which water seeps at a very low rate. However, it has been recently noticed that an originally good relation between concrete and water may deteriorate with time [20,22,123]. Here again, as when the

Fig. 107. Ruins from the old village Fabbrica of Careggine in the VAGLI reservoir in Garfagnana, Italy [40]

phenomenon happens in embankment dams, it is difficult to decide whether the prime cause of the small cracks fostering a more intense water circulation depends on the state of stress or the chemical attack associated to the original seepage through the sound concrete.

At the CIGNANA dam ($H = 58\ m$), terminated in 1928, the remedial measure in 1986 was the application of a geomembrane on the upstream face as shown in the schematic drawing of Fig. 94. At PIANA DEI GRECI, the original concrete facing was partially dislocated by the Belice earthquake in 1968. The solution was the installation of a new metallic facing over the entire upstream slope [138]. The execution of a work of this type requires a free draw-down of the reservoir and this measure is certainly not convenient on the occasion of a water emergency. At PIANA DEI GRECI, the problem was complicated by the existence of a thick layer of silt hindering access to the upstream dam heel. The difficulty was overcome with the installation of a strong sheet-piling supported by struts against the dam facing (Fig. 95).

Fig. 108. The SAN VALENTINO belltower arising from the reservoir created by the homonymous embankment dam in Val di Resia, Italy (photo by R. Jappelli, 1951)

At the RAVASANELLA dam, near Vercelli, originally conceived as a concrete gravity dam, it was necessary to change this choice during the execution works and to adopt a different solution for a section of the dam body (Fig. 96). At the site of this dam ($H = 50\ m$), the jointed and altered bed-rock of volcanic origin revealed itself less resistant and more compressible in comparison with the originally estimated values, particularly at the right abutment. After updated geotechnical assessments [90], the original solution for the main and for the left sections was confirmed; but the right wing of the dam body was changed into a more deformable section built with large concrete blocks with gravel filled joints. The solution was formerly adopted by Claudio Marcello in the fifties for a number of important dams (FANACO, Agrigento; POZZILLO, Enna; PIANPALU', Trento; EL FRAYLE, Perù). Alternate solutions for improving the ground with grouting measures were considered of uncertain results and more costly.

At RAVASANELLA, the shape and dimensions of the blocks and related contact problems have been studied by means of the finite element method

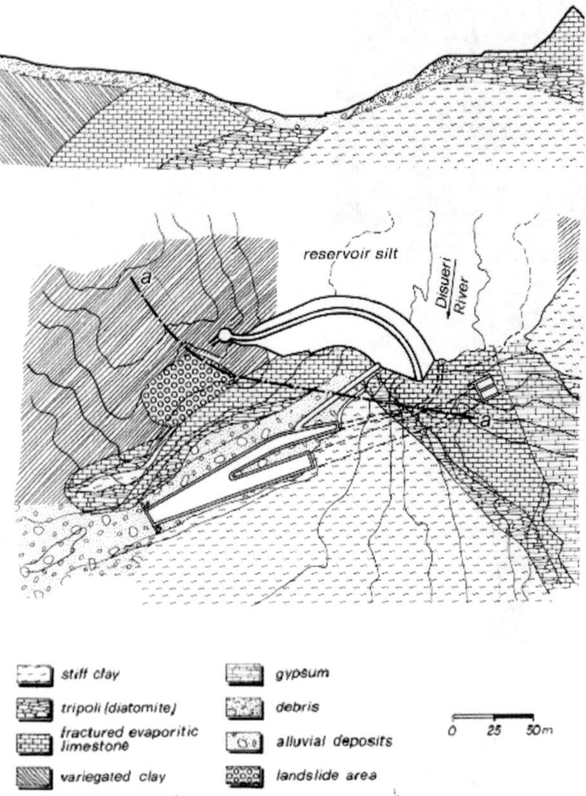

Fig. 109. The ground profile at the DISUERI gorge and the planimetry of the old dam [13]

of analysis, demonstrating the convenience of the solution also as far as the ground is concerned. The water barrier function has been entrusted to a geomembrane on the upstream facing instead of the metallic revetment adopted elsewhere. A convenient structural member was installed as a transition between the two different sections of the dams. As for INGAGNA [36], the peculiar architectural features of this dam, reveal the constraints imposed by the ground to the dam body. The water gathered in both the INGAGNA and the RAVASANELLA reservoirs are addressed to the irrigation of the well-known Vercellian rice-fields in Piemonte, Italy.

In different cases, the proposed interventions should be better calibrated in the light of modern knowledge in the field of Soil Mechanics. ABATE ALONIA was one of the first embankment dams built in Italy after the Second World War. At that time the dam body was conceived by the designer with a thin central core composed of a gravel and sand mix which was declared "impermeable" with the addition of bentonite; for the shoulders, silt with

Fig. 110. Longitudinal cracks at the crest of the old DISUERI dry masonry dam after 50 years of service [35]

sand and clay was used, which was available in situ in large quantities (Fig. 97).

The proposed remedy to control the embankment displacements that are considered too large is described as a "consolidation (of the core) by injecting cement grout and chemical mixes ... supporting fills raising the dam to conform with new standards" [27].

However, the deformations today attributed to bad execution of the work, may be originated by an unconventional design choice, perhaps unaware, of a permeable core, which behaves like a draining diaphragm with regard to the fine grained shells (shoulders) and the consolidation of which, after half a century, can hardly be considered exhausted.

The most serious defects, involving immediate risk and/or requiring long time for repair, are temporarily kept under control by the owners and/or by the National Safety Agency for Dams (RID) with *non-structural* measures. These remedies include the consequential risk analysis, the emergency plans,

Fig. 111. Agricultural produce in the irrigated area of DISUERI

Fig. 112. The RIDRACOLI double curvature arch dam ($H = 100\ m$) on the Bidente River in Romagna, under construction (L'Acquedotto della Romagna, 1991)

the methods for an anticipated warning of undesiderable events, an efficient ordinary maintenance, a wise regulation of the reservoir operation, a reinforced surveillance, the continuing education and mainly, if necessary, the restriction of reservoir's water elevation [58,73].

Fig. 113. The RIDRACOLI dam, in service since 1982 [54]

The large arch-gravity BEAUREGARD dam built in the fifties in Val d'Aosta (Fig. 98), is severely cracked since long ago; the phenomenon is caused by a thrust on the left abutment, originated by a large landslide extending upstream of the dam site for some kilometers along the left versant [12]. An adequate remedy to this difficult local situation, originated by an erroneous allowance of the dam's support on the left abutment, has not been found yet.

An interesting example of a convenient, though sophisticated, regulation with a non-structural method facing a landslide that moves slowly into the reservoir, is offered by the RAGOLETO [100–105]. A large portion of the right versant of this reservoir, on the Dirillo River (Ragusa), is affected by a slow movement about one mile upstream from the dam site (Fig. 99). The phenomenon affects a detritic deposit, the volume of which is about $6 \cdot 10^6 \ m^3$, resting on a stratified and jointed limestone formation. The displacements revealed themselves since the first reservoir filling (1963) and became more evident with time. In 1972, the slope was placed under systematic surveillance with topographic surveys of vertical and horizontal displacements.

According to the elevation and to the water level raising rate in the reservoir, the average rate of displacement of the landslide has attained values of

Fig. 114. The RIDRACOLI reservoir [54]

Fig. 115. Restoration of dry masonry farmhouses promoted by "Consorzio Acque per le Provincie di Forlì e Ravenna" (The Water Cooperate for the Provinces of Forlì and Ravenna) in the area of RIDRACOLI dam [54]

some centimeters with a maximum of about ten per month. Computations, considering different possible stabilization methods, have shown that the related increments of the safety margin of the slope are always too small to secure an adequate benefit in relation to the cost.

The problem was solved with recourse to a *non structural* method, accepting the small loss related to a temporary partial exploitation of the reservoir, with the aim of progressively improving the understanding of the phenomenon. The noticeable crop of data collected during thirty years of observation has shown a correlation between the rate of landslide displacement and the rate of water level in the reservoir. As a result, the reservoir's operation was adjusted by constraining the rate of water level. An estimation of the presumable starting time of the phenomenon and the approximate prediction of 2050 for its completion with a global displacement of the order of about 4 m was made [68,69,100,113]. It is worth to note that, unlike the case of BEAUREGARD, where the slide affects the dam's abutment, the slope at RAGOLETO does not concern the dam directly.

Execution plans are not rarely subjected to long interruptions as a consequence of sudden, unpredicted and/or hardly predictable slope instability during abutment excavations. The construction of the ALTO ESARO concrete gravity dam ($H = 120\ m$) at the site of Cameli (Cosenza, Italy) has been interrupted since 1987 after the rupture which occurred in the left abutment in form of dislocation of a portion of a jointed bed rock [9,10] (Fig. 100). The height of the dam and the reservoir's volume should perhaps undergo a significant reduction.

The strict rules in force in Italy and the difficulty of predicting with reasonable certainty and under every respect the dam's performance has often persuaded authorities and designers to follow a strategy based upon multiple lines of defence, with special attention paid to the effects of seepage through the dam body and its foundation.

The two reservoirs (Fig. 101) of the large ANAPO pumped storage hydroelectric plant (cf. Sect. 6) are fully faced by a bituminous conglomerate revetment. In consideration of the locally significant seismic activity, the Superior Council of Public Works recommended that the design should be improved by measures suited to discharge the water gradually with due safety through the embankments in the case of a sudden break [76]. This prescriptive measure was satisfied with the provision of a large central draining core made of a cohesive auto-protective material capable of maintaining its structural continuity even in the event of large deformations.

In the final design of the proposed GUALTIERI dam near Messina [14] ($H = 100\ m$), in a zone of high seismic activity, the central diaphragm of bituminous concrete was conceived as a sandwich similar to the usual double facings of the same material; the two impermeable structures include a drain with tailrace in a culvert, in which leakage would eventually indicate the failure of the water barrier (Fig. 102).

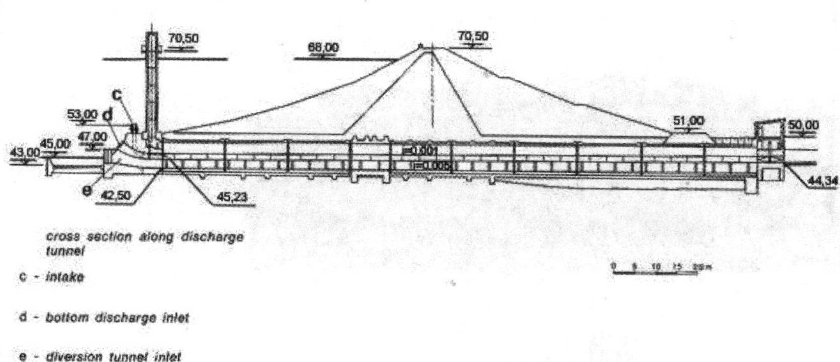

Fig. 116. The TRINITA' embankment dam under construction on the Delia River in Sicily in 1955. The outlet culvert crosses the dam foundation; one can note the reliefs at the extrados in order to improve contact of the concrete with the embankment core material [6] (photo by R. Jappelli, 1955)

10   Abandonment

As for every artifact, the time of dismission comes for dams as well: "Omnia tempus edax depascitur, omnia carpit..." ("The destructive time grasps and consumes everything..."), Seneca.

Fig. 117. The small Norman church, XII century, called "SS. TRINITA' di Delia" is sited near the homonymous dam (photo by R. Jappelli, 1955)

Abandonment is necessary when the original is replaced by a more important and functional modern structure or because the structure has reached the age of retirement. The abandonment of dams requires specific, accurate design work.

Sometimes, the remains of archaeological dams can be seen as has been shown elsewhere (Fig. 15). In other cases, the remnants are more recent. On the occasion of the decennial emptying, the still almost intact old dam emerges from the new MONCENISIO reservoir (Fig. 103). The empty reservoir offers romantic people the melancholic sensation that one experiences when past objects come back to the mind from time to time (Fig. 104–108).

"New ruins have not yet acquired the weathered patina of age ... It will not be for long. Very soon trees will be thrusting through the empty window sockets, the rose-bay and fennel blossoming within the broken walls..." (R. Macalulay [87]).

Fig. 118. The Selinunte Temples (photos by R. Jappelli, 1955)

This passage from the final paragraph "A note on new ruins" of the fascinating book "Pleasure of ruins" by Rose Macaulay, echoes Carl Sandburg, who in the poem "Grass" described with harsh efficacy the role of time in the representation of human calamities:

"Pile the bodies high at Austerlitz and Waterloo.
Shovel them under and let me work
I am the grass; I cover all.
And pile them high at Gettysburg
And pile them at Ypres and Verdun.
Shovel them under and let me work.
Two years, ten years, and passengers ask the conductor:
What place is this?
Where are we now?
I am the grass.
Let me work"
(Carl Sandburg, 1878-(?) [118])

Fig. 119. The site "Rocche di Cusa" where the builders of the Selinunte temples quarried the calcarenite (photo by R. Jappelli, 1955)

Fig. 120. A transportation vehicle at the time of the construction of the TRINITA' embankment dam (1956)

The old DISUERI dam is situated in a section of the Gela River, where different members of the *gessoso-solfifera* formation, variegated and tortonian clays intersect and alternate with each other in a disordered geological complex of very difficult identification and characterization (Fig. 109). Fifty years ago, the choice of the dam section was heavily conditioned by prejudice about the ground and by an *a priori* selection of the dam type. The designer

Fig. 121. The ITAIPU dam (earth: $17 \cdot 10^6 \ m^3$; concrete: $12 \cdot 10^6 \ m^3$) on the upper Paranà River at the Brazil-Paraguay border is the largest hydroelectric powerplant in the world ($12,600 \ MW$)

[25] rejected the traditional idea of a concrete gravity dam and preferred a dry masonry solution; the planimetry of the embankment was assigned an upstream convexity. The dam axis was slightly withdrawn upstream from the narrowest section of the valley in order to keep clear of the variegated clay, which was well known for its bad performance. The dam was in service until 1995. However, for many years a reduction measure to the reservoir volume had been in force, on the ground of cracks in the dam body (Fig. 110) presumably produced by a slide in the right abutment, where the dam axis was unfavourably oriented against its gradual movement [137]. This situation and the accumulation of a silt deposit of the thickness of about twenty meters in the reservoir, account for the dismission of the old artifact and the construction of a new earth dam, with higher water level elevation, in a section located at a short distance downstream of the original site.

The new DISUERI ($H = 74 \ m$; $V = 3 \cdot 10^6 \ m^3$) is a zoned embankment dam. The central core was built with a silty clay quarried out of an alluvial deposit; the material used in the shoulders was a soft evaporitic limestone quarried from the near M. Garrasia. The landslide stabilization was attained with a dam axis orientation opposite to the former by turning the concavity upstream. The right abutment was remodelled and supported by a wide backfill [35].

The volume comprised between the old and the new dam was also partially backfilled. The two new morning glory spillways with a capacity of $3,200 \ m^3/s$ are firmly founded on the limestone formation outcropping in the left side of the valley; they are located a little further downstream of the largely insufficient old spillways. Works were performed between 1988 and

1995. The constraint of carrying out the excavations without modifying the precarious equilibrium of the right bank and the condition of not discontinuing the limited operation of the old dam imposed preventive defence measures consisting in a row of large diameter shafts ($D = 12.5\ m$; $H = 18 \div 21\ m$) sinked to the undisturbed clay formation (Fig. 39) and unconventional procedures in the general excavation program. It was also necessary to adopt special safety measures at an unusual contact of the core with a large limestone bank outcropping across the dam section.

After fifty years of distinguished service, the old DISUERI dry masonry dam, which acted as cofferdam for the protection of the new construction site against floods, was dismissed with the regret of the ancient dam wardens. The dismission was accomplished by opening a wide breach in the dam body (Fig. 41). The reservoir has never ceased to supply water to the fertile Gela plane, where thanks to irrigation, the well known artichokes (Fig. 111) are reaped more than once a year for the benefit of gastronomes and for the advantage of those politicians, who from that humble product seek inspiration for their gradual strategies.

BEAUREGARD (Fig. 98) is now considered to be on the verge of dismission.

11   Final remarks

1. The artifacts described in this paper are not confined, as commonly believed, to dam bodies, *strictu sensu*; on the contrary, the dam is only a part of an articulated and complex system, in which the *natura artificiata* is strictly intermingled with the *natura naturata*. In this large system, a number of appurtenances are comprised: intake works; foundation soils; nearby slopes; reservoir and banks; direct and undirect catchments basin with related collecting works; outlets and release works (open channels, tunnels); access roads and ancillary works located in the territory upstream and downstream of the dam site. The whole of the water works and of the territory composes a *unicum*, differing from others in the solutions and the interconnection with nature. Likewise monuments, dams should be studied, built and operated within a context, in which the main artifact, foundation soils or rocks, reservoir, slopes and ancillary works are conceived as a system, from which the subsystems cannot be disjointed.

2. Dams are not short lived works, but unrepeatable testimony of the past, upon which the knowledge necessary to accomplish future building activity should be founded. They have a story, that should be reconstructed with the story of the system in which they are included. Some dams deserve the attribute of monumentality, which derives from their stories, from the characters of the natural site and architectonic features, from the peculiar relationship with water, soils and/or rocks and the nearby

territory. These systems of buildings and nature, which far from disfiguring, enrich the environment and landscape in otherwise deserted and desolated areas exert a peculiar charm over the visitor, deriving from their close relationship with man, just as happens with monuments.

3. The system should be studied, designed, built and operated, with the greatest care, not only the single parts, but for the correct operation of the whole. Some past failures should be attributed to the lack of a larger vision, which in the light of experience has revealed itself essential, for the operation of an organism to which the holistic theory ascribes more completeness and perfection than the sum of the subsystems, as in Biology and Medicine.

4. Experience shows that the great capability of survival of dams, as for monuments, is bound to a careful, continuous and competent surveillance during operation. The surveillance is entrusted with the very difficult task of controlling the performance of the complex system to reveal possible defects and to predict the system's short and long-term evolution should those defects come into existence. The safeguard of a dam requires continuous routines and exceptional maintenance with competent *structural* and *non-structural* measures. Competence presupposes as minimum requirements the ability to observe, remember and recount. As for monuments, the operation of water reservoirs should be carefully practiced in respect of the superior public interest and not for the economic benefit of a single private owner. In some cases, the distressing problem of the abandonment of a plant that cannot be operated safely anymore, due to original and/or acquired defects, must be faced.

5. Dams in Italy are often found in a natural context, which the dam and the reservoir contribute to improve with beneficial transformations. The RIDRACOLI dam (Fig. 112, 113), terminated in 1982, is a symmetric double curvature arch dam, controlled by an important computerised monitoring system of different quantities (internal stresses, displacements, pore pressure, temperature, dynamic properties). The dam is part of a large complex of artifacts built and operated with a unitarian criterion aiming at coupling hydroelectric production with hydropotable supply in Romagna[5] (Italy) through extensive measures directed to the improvement of the environment and the recovery of the territory (Fig. 114). Results demonstrate the actual possibility of conceiving and operating the dam and ancillary works with the precise aim of stopping the subsidence and the ingress of marine water consequent to the extraction of water from the subsoil; and, at the same time, enriching the territory with potabilization plants and aqueducts [54]. So, in the case of RIDRA-

---

[5] There is a saying in Romagna about the chronic shortage of water; it was told to travellers along the road to the sea: "should you find out where you are during your journey, just ask for a drink. If you are given a glass of water, you are in Emilia; should you get a nice glass of wine, you have arrived in Romagna [54]".

COLI, located in "sweet sunny Romagna", the enduring exchange of ideas among the local authorities has stimulated development with the result of a well-documented advancement of the quality of the territory under every environmental, social and economic aspect (Fig. 115).

6. The Italian territory is rich in monuments. Some of them are magnificent and well known; others, almost unknown, gain allure by their proximity to artificial reservoirs. In the vicinity of the TRINITA' dam on the Delia River (Sicily, Fig. 116) one can admire the small church of SS. Trinità di Delia in a green oasis of palm-trees and eucalypts (Fig. 117). If one moves to the south, along the route SS 115 among typical rural buildings across the Trapani territory, one can observe the magnificent remains of the temples of Selinunte above the horizon of the blue Mediterranean Sea (Fig. 118). A little farther towards west, the Rocche di Cusa site provides the watchful traveller with the amazing sight of the calcarenite quarries for the drums of the nearby temples, as if the workmanship had been suddenly interrupted upon arrival of the Carthaginians (Fig. 119). The same calcarenite was used fifty years ago to build (Fig. 120) the shells of TRINITA' dam (Fig. 116).

Acknowledgements

The Author is indebted to Paola Provenzano and Carlo Callari for their hard editing work.

References

1. Agnello G. (2001) Il vivaio di San Cusumano, il Piverium di Lentini, L'Architettura Sveva in Sicilia, Ed. Clio, Gruppo Editoriale Brancato.
2. Anidel, Diga di Ancipa a gravità alleggerita, ad elementi cavi, in calcestruzzo.
3. Appendino M. et al. (1991) Specific and general trends of the ageing of buttress dams as revealed by investigations carried out on Ancipa dam, XVII ICOLD, Q.65, R.22, Vienna, 373–404.
4. Arredi F. (1982) Costruzioni idrauliche, 1–4, UTET, Torino.
5. Aslan G., Akkaya C. (2002) Combining the necessary large dam construction in Turkey with preservation of the archeological heritage. International Conference Hydro 2002, November.
6. Bigalli F., Dolcimascolo F., Jappelli R., Pezzini R., Toti M., Valore C. (1980). Dighe in Sicilia. Volume Issued on the occasion of 48th ICOLD Executive Meeting. Rome, 6–9 October 1980
7. Baldovin G.,Barro F., Coen L., Lavorato A., Pezzini R. (1991) Rosamarina dam design and construction. Idrotecnica, 2.
8. Beausire P. (1971) Instants: Segeste et Eraclea Minoa, "Sicilia", Ed. Flaccovio.
9. Belloni L., Sordi P. (1993) Slope failure triggered by reshaping of the valley side at the Esaro Dam. In: Proc. International Symposium on Assessment and Prevention of Failure Phenomena in Rock Engineering, Istanbul.

10. Belloni L., Costantini P., Sordi P. (1996) Prove di taglio dirette in sito del tipo multistage per la caratterizzazione della roccia di fondazione della diga sull'Alto Esaro. Rivista Italiana di Geotecnica, 4.

11. Bianchi Bandinelli R. (1969) L'arte Romana nel centro del potere, Feltrinelli, Milano.

12. Bianchini A., Fornari F. (1997) Beauregard dam: left bank settlements and monitoring evolution. In: XIX ICOLD, C17, Florence.

13. Bigalli F., Jappelli R., Valore C., (1988) Geotechnical design of embankment dams on clay formations in Southern Italy. Idrotecnica, March-April.

14. Bigalli F., Jappelli R. et al. (1990) Progetto della diga Gualtieri sul T. Sicaminò.

15. Braida S. (1988) Il castello di Fawara. Studi di restauro, Memorie del Centro di Cultura di Cefalù V/2/1988–1992, Incontri e iniziative.

16. Brezina P., Satrapa L., Skokan T. (1998) The reconstruction of the collapsed bituminous concrete facing, the Morávka Dam, Czech Rep. In: Symposium Rehabilitation of Dams, New Delhi, INCOLD, November.

17. Callari C., Jappelli R. (2004) Comportamento a breve e a lungo termine della diga di Monte Cotugno sul fiume Sinni. In: XXII Convegno Nazionale di Geotecnica "Valutazione delle condizioni di sicurezza e adeguamento delle opere esistenti", Palermo, 22–24, September 2004, Pàtron Editore, Bologna

18. Callari C., Jappelli R., Raggi F. (2004) Coupled finite element analysis of the construction of a large italian embankment dam founded on stiff clay, Numerical Models in Geomechanics. In: NUMOG IX, G.N. Pande & S. Pietruszczak, eds., Balkema, Rotterdam,615–622.

19. Callari C., Fois N., Cicivelli R. (2004) The role of hydro-mechanical coupling in the behaviour of dam-foundation system. Computational Mechanics, WCCM VI, Z.H. Yao et al., eds.,Sept. 5-10, 2004, Beijing, China, 2004, Tsinghua University Press & Springer-Verlag, **I**, 411 (abstract), CD 1.

20. Cancelli A., Cazzuffi D. (1994) Environmental aspects of geosynthetic application in landfills and dams, Key note lecture. In: V International Conference Geotextiles, Geomembranes and Related Products, Singapore, September.

21. Catalano A., Federico F., Jappelli R. (2004) Caratteristiche e comportamento della diga di NOCELLE dopo settant'anni di esercizio. In: XXII AGI, Palermo.

22. Cazzuffi D. (1987) The use of geomembranes in Italian dams. Water Power and Dam Construction, March.

23. Chinese Nat. Com. Large Dams(1987) Large dams in China. History, Achievements, Prospects.

24. Consorzio Acque Forlì e Ravenna (1991) L'Acquedotto della Romegna.

25. Contessini F. (1951) La diga di Gela in muratura di pietrame a secco. L'Energia Elettrica, 2, February.

26. Cooke J.B. (1984) Progress in rockfill dams. Journal of Geotechnal Engineering, ASCE, vol.110, 10, October, ASCE, paper 19206.

27. Cotecchia V., Monterisi L., Simeone V. (1998) The deterioration of earth dams. An example: the Rendina dam. In: Symposium. Rehabilitation of Dams, New Delhi, INCOLD, November, 271–278.

28. Croce A. (1963) Recent developments in the design and construction of earth and rockfill dams, CNR, Rome.

29. Croce A., Baroncini G. (1957) Caratteristiche dei terreni e comportamento della diga dell'Arvo durante venticinque anni di esercizio. Geotecnica, 3.

30. Croce A., Linari C., Motta A. (1978) Schermo di iniezioni per il ripristino della tenuta di fondazione della diga di Zoccolo. In: XIII Convegno Nazionale di Geotecnica, Merano.

31. Croce A., Martinelli D., (1978) Il sovralzo della diga di Vernago e le opere in fondazione durante 20 anni di esercizio. In: XIII Convegno Nazionale di Geotecnica, Merano, 1978, 1, 137.

32. Cuccia S. (1965) I bagni arabi di Cefalà Diana. I.T.E.S., Catania.

33. Dello Vicario E., Petaccia A., Savanella V. (1997) Caratteri descrittivi delle grandi dighe italiane, Servizio Nazionale Dighe, Ufficio Idraulica.

34. De Loyola Brandao I. (2001) Cronicas da vida lindeira of people and dams, Ed. Dorea Books and Art, Sao Paulo, Brasil.

35. Di Berardino P., Jappelli R., Percopo E. (1997) Replacing a dry masonry dam without service interruption, XIX ICOLD, Q75, R43, 579–611.

36. Di Maio S., Callari C. (2004) Comportamento delle dighe di calcestruzzo e di rockfill costituenti lo sbarramento dell'Ingagna. In: XXII Convegno Nazionale di Geotecnica "Valutazione delle condizioni di sicurezza e adeguamento delle opere esistenti". Palermo, 22-24 September 2004, Pàtron Editore, Bologna, 497–505.

37. DTK (2001) Historische Talsperren und Wasserkraftanlagen, Dresden.

38. Eco U. (2002) Bustina di Minerva, L'Espresso, 6 June.

39. EDF (1996) Aménagement du Mont-Cenis.

40. Emiliani V. (2000) Il paesaggio tra passato e presente, in T.C.I.: Il paesaggio italiano, idee, contributi, immagini, 191–202.

41. ENEL (1998) Paesaggi elettrici: territori, architetture, cultura, Marsilio Ed.

42. ENEL (1984–87) Le dighe di ritenuta degli impianti idroelettrici italiani, Rome.

43. European Club Icold (2000) Internal erosion in embankment dams.

44. Evangelista G. (1998) Mont-Cenis, les couleurs du temps, Ed. Il Punto, Torino, giugno.

45. Garbrecht G.(2001) The Sadd-el-Kafara, the World's Oldest Dam, Historical Dams, Ed.H.Fahlbusch, Dek Top Publ., ICID Central Office.

46. Gatto(1999) La città dei Sassi T.C.I.: Attraverso l'Italia del Novecento, immagini e pagine d'autore, 258–259.

47. Gentili G.V. (1971) La Villa Imperiale di Piazza Armerina, Min.P.I., Ist.Pol. Stato.

48. Giuseppetti G. ed altri (1997) Design for the rehabilitation of Ancipa dam. In: International Journal Hydropower and Dams, Vol. IV, 2.

49. Glaeser E., Juengel E. (2001) Innovative methods for the renovation of masonry dams classified as historic monuments illustrated by the case of the Carlsfeld barrage. In: Proc. DTK, Workshop Modern Techniques for Dams, etc., Vol. I ICOLD 69th Exec Meeting, Dresden, September, 141.

50. Greco C. (2003) Le case di terra del Fujian, Collana BABELE coordinated by R.Pavia e M.Ricci, Meltemi Ed., Rome.

51. Hydropower and Dams (1997) cover: PIEVE DI CADORE dam.

52. ICOLD (1994) Ageing of dams and appurtenant works, review and recommandations. Bulletin 93.

53. ICOLD (1994) Technical Dictionary on Dams.

54. ICOLD (1995) Barrages et environnement, Ridracoli: une réalization exemplaire. Bulletin 100.

55. ICOLD (1999) Embankment dams with bituminous concrete facing, Bulletin 114.

56. ICOLD (2000) Le barrage-poids, un barrage d'avenir, Bulletin 117.

57. ICOLD (2000) Rehabilitation of dams and appurtenant works, Bulletin 119.

58. ICOLD (2001) Non structural risk reduction measures - Benefits and costs of dams, August.

59. IMPREGILO (1956–1981) Venticinque anni di lavoro nel mondo.

60. INCOLD (1998) Rehabilitation of dams, Symposium, November.

61. IRCOLD (1995) Ancient dams in Iran.

62. Isler H.P. (1986) Monte Iato, la sedicesima campagna di scavo. Sicilia Archeologica, Dir. V. Tusa, XIX, 62.

63. ITALSTRADE (1963–71) Illustrazione dei principali lavori eseguiti nel periodo 1963–71, Milano.

64. ITCOLD (1997) Dams in Italy, Firenze.

65. ITCOLD (1999) Processi di invecchiamento di dighe e loro fondazioni, Quaderno 7, March.

66. Jappelli R. (1948) Il criterio economico nel calcolo delle dighe a contrafforti alleggerite, Laurea thesys.

67. Jappelli R. (1954) Fenomeni di rottura nelle dighe di terra. Geotecnica 4.

68. Jappelli R., Musso A., Umiltà G. (1977) An observational approach to a creep problem in debris due to reservoir filling. In: International Symposium on Geotechnics of Structurally Complex Formations, Vol. II, Capri.

69. Jappelli R., Musso A. (1981) Slope response to reservoir level fluctuations. In: X ICSMFE, Stockholm.

70. Jappelli R. (1994) Commenti al Seminario Internazionale Geoarcheologia nella Sibaritide, Risultati e prospettive, Sibari, 10-11 December.

71. Jappelli R. (1998) Lavori Interrotti: motivi ed iniziative per il completamento. Giornata ITCOLD, Rome, 22th May.

72. Jappelli R. (2002) Le système sol-monument. Propositions pour une approche integrale, Colloquium Lagrangianum. Ecole National Ponts et Chausses, Paris, février.

73. Jappelli R. (2003) Il problema della vigilanza, Giornata ITCOLD.

74. Jappelli R., Dolcimascolo F. (1964) Sul comportamento della diga della Trinità durante sei anni di esercizio. Geotecnica, 5.

75. Jappelli R., Pellegrino A. ed altri (1981) Materiali per la costruzione di dighe di terra e di pietrame in Italia. Quaderno 2, Sott. Mat. del C.I.G.D.

76. Jappelli R. ed altri (1988) Impervious facing and large central drain for the embankment dams of a pumped storage plant. In: XVI ICOLD, San Francisco.

77. Jappelli R., Di Maio S. (1990) Ricerca progettuale in corso d'opera sul tema materiali da costruzione del Bilancino, Visita del C.I.G.D., 4 May.

78. Jappelli R., Federico F., Marchetti A. (2002) Effectiveness of grout courtains for seepage control. In: V European Conference on Numerical Methods in Geotechnical Engineering, Paris 4 - 6 September.

79. Jappelli R., Indelicato S., Pace E., Ferla M. (2002) Verifiche di sicurezza della diga di Piana dei Greci.

80. Kerisel J. (1987) Down to earth, Foundations past and present: the invisible art of the builder, Balkema.

81. Kerisel J. (1997) Geotechnical problems in the Egypt of Pharaohs. In: Proc. International Symposium on Geotechnical Engineering for the Preservation of Monuments and Historic Sites, Naples, October. 1996, Ed. C. Viggiani, Balkema.

82. Kroenig W. (1987) Vedute di luoghi classici della Sicilia, Il viaggio di P. Hackert del 1777, Sellerio, Palermo, Bibl. Siciliana di Storia e Letteratura.

83. Lazzarini P., Bister D.(1998) EDF's recent experience in upgrading spillways. The cases of les Mesces and Motz Dams. In: Symposium Rehabilitation of Dams, New Delhi, INCOLD, November.

84. Leopardi G., Elogio degli uccelli, 1824

85. Lemperiere F., Bister D. (1998) Risks and hazards of dam failures: low cost remedies. In: Symposium Rehabilitation of Dams, New Delhi, INCOLD, November.

86. Lotti C. (2002) Due parole ancora sul Vajont. L'Acqua, 3.

87. Macaulay R. (1953) Pleasure of ruins. Weidenfeld a. Nicolson, London.

88. Mangiagalli L. (1921) Dighe in muratura a secco e sbarramento sul T. Hone. L'Elettrotecnica, 33,705–711.

89. Mangiagalli L. (1925) L'impianto idroelettrico dell'Alto e Medio Belice della Società Generale Elettrica della Sicilia, L'Energia Elettrica, 5, 417–471.

90. Marcello A., Eusepi G., Olivero S., Di Bacco R. (1991) Ravasanella dam on difficult foundation. In: XVII ICOLD, Q66, R21, Vol. III, 363–380.

91. Marcello C. (1950) Moderner Talsperrenbau in Italien, Scheweizerische Bauzeitung, Vol. 68, n. 33, 34, 35.

92. Marcello C. (1950) Barrages modernes en Italie. Bulletin Technique de la Suisse Romende, 4/11, 297 and 18/11, 313.

93. Marcello C. (1952) La diga di Ancipa sul T. Troina, Tecnica e Ricostruzione, Catania, n. 2 . 3, 45.

94. Marcello C. (1957) The cellular gravity concrete dam and some applications of this new design. Journal of Japanese National Commission Large Dams, n. 3.

95. Marcello C. (1957) Un nuovo tipo di diga per terreni fortemente compressibili. Geotecnica, 4, 164–176.

96. Mason P., Scott C., Molyneux J.D. (2003) Ageing of dam components and dam safety regimes. In: XXI ICOLD, Q.82, R.35, Vol.II, 599, Montreal.

97. Middlebrooks T.A. (1940) Fort Peck Slide. In: Proc. ASCE, Vol. 66, n.10, December.

98. MIN.LL.PP./C.Sup./Serv.Dighe (1961) Note tecniche in occasione VII ICOLD, Rome.

99. Moscati S. (1997) Italia: la storia e il destino. In: Conferenza Università di Rome Tor Vergata.

100. Musso A. (1997) Limitazione nell'uso dei pendii con ridotto margine di sicurezza. Interventi di Stabilizzazione dei Pendii, CISM, Udine.

101. Musso A., Provenzano P., Selvadurai A.P.S. (2003) Stability of motion of detrital reservoir banks. In: ISEC-02 Second International Structural Engineering and Construction Conference, Rome, 24-27 September, 2003

102. Musso A., Provenzano P. (2003) Identificazione dello stato di attività delle frane mediante Reti Neurali. In: International Workshop-Living with landslides, Anacapri, October, 27–28, 2003

103. Musso A., Provenzano P., Selvadurai A.P.S. (2003). Assessment of a Landslide Activity: the Ragoleto case history. In: Colloquium Lagrangianum, Montpellier November, 20-23, 2003

104. Musso A., Provenzano P. (2003). Forecasting of landslide movements by Artificial Neural Networks. In: Convegno annuale della Italian Society for Computer Simulation, Cefalù, 27–28 November, 2003

105. Musso A., Provenzano P. (2004). A sliding block model to analyse the stability of motion of detrital reservoir banks. In: IX International Symposium on Landslides, Rio de Janeiro, Brazil,June 2, 2004

106. Nanda V. K. (1998) Monitoring of health of Bhakra dam. In: Symposium Rehabilitation of Dams, New Delhi, INCOLD, 201–209, November.

107. Nebbia U. (1955) Castelli d'Italia, Istituto Geografico De Agostini, Novara.

108. Niccolai F. (1941) Concetti autarchici nella progettazione di dighe italiane, L'Elettrotecnica 8, p. 181–191, fig.29.

109. Noetzli F. A. (1932) Pontalto and Madruzza arch dams, Western Construction News, August.

110. Omran M.E., Abbas H. (1998) Strengthening, heightening and leakage control measures for an ancient Fariman Dam in IRAN. In: Symposium on Rehabilitation of Dams, New Delhi, INCOLD, November.

111. Panvini R. (19?) Gli insediamenti nel territorio dalla preistoria all'età tardoantica, Museo Archeologico Regionale, Gela (CL).

112. Penman A.D.M. (1986) On the embankment dam. Geotechnique 36, 3, 303–348.

113. Provenzano P., Musso A., Romagnoli F. (2004) Influence of failure propagation on slope movements. In: Journées de Géologie, La géologie appliquée aux grands ouvrages, Ax-les-Thermes, 2004

114. Quasimodo S. (1942) Ed è subito sera, Erato e Apollion, Lo Specchio, A. Mondadori.

115. Rajpal A.K., Corso R.A. (1998) Dam safety programs and privatization. In: Symposium Rehabiltation of Dams, New Delhi, INCOLD, November.

116. Ricci A. (1996) I mali dell'abbondanza, Considerazioni impolitiche sui beni culturali. Ed. Lithos-Cester.

117. Rissler P. (2001) Sanierung alter Talsperren, Rehabilitation of old gravity dams.

118. Sandburg C. (1878–?) The grass - in Guanda (1949), Poesia americana contemporanea e poesia negra, edited by C. Izzo, 150.

119. Schewe L.D. (1998) Rehabilitation of an old German Masonry Dam, Symposium Rehabilitation of Dams, New Delhi, INCOLD, November.

120. Schmidt M. (2001) The mining ponds of the Upper Harz, ICID, Historical Dams, Ed. Fahlbusch .

121. Schnitter N. (1994) A history of dams, the useful pyramids, Balkema, Rotterdam.

122. Sciascia L. (1999) I fiumi della Trinacria, in T.C.I. Attraverso l'Italia del Novecento, immagini e pagine d'autore, p.290.

123. Scuero A.M., Vaschetti G.L. (1998) Remedial works for seepage and alkali aggregate reactions control by using watertight geomembranes. In: Symposium Rehabilitation of Dams, New Delhi, INCOLD, November.

124. Semenza E. (2001) La storia del Vajont raccontata dal geologo che ha scoperto la frana, Tecamproject, Ferrara.

125. Sereni E. (1961) Storia del paesaggio agrario italiano, Laterza (sesta ediz. 1993).

126. Shenouda W.K. (1998) Investigations and remedial measures to safeguard the old Asswan Dam. In: Symposium Rehabilitation of Dams, New Delhi, INCOLD, November.

127. Shenouda W.K. (1999) Benefits and concerns about Aswan High Dam. In: Workshop on Benefits and Concerns about Dams: Case Studies, Antalya, Ed. Turfan M.

128. Spagnoletti S. (1962) Sul comportamento di un tipo di diga a gravità alleggerita a elementi cavi. Il comportamento rilevato nella diga di Ancipa. L'Energia Elettrica 3 e 4.

129. Spatafora F., Calascibetta A.M.G. (1986) Monte Maranfusa, un insediamento nella media Valle del Belice, Sicilia Archeologica, Dir. V.Tusa, XIX, 62.

130. Sundaraiya E. (1998) Rehabilitation of dams under DSARP. In: Symposium Rehabilitation of Dams, New Delhi, INCOLD, November.

131. Touring Club Italiano (1948) Abruzzi and Molise.

132. Touring Club Italiano (1989) Sicilia, Guida "Rossa" (per geografia, storia, monumenti, bibliografia).

133. Touring Club Italiano (2000) Il paesaggio italiano, idee, contributi, immagini.

134. Tronconi L. (1951) Impianto per la preparazione del calcestruzzo con dosaggio a peso adottato per le dighe di Ancipa (E.S.E) e di Cecita (SME), Giornale del Genio Civile,2, p.142, February.

135. Tusa V. (1986) Che cosa è l'Archeologia?, Sicilia Archeologica XIX, 62.

136. USCOLD (1988) Development of Dam Engineering in USA, Pergamon Press.

137. Vecellio P. (1960) La frana presso la diga di Gela, Geotecnica 3-4, 1960.

138. Vecellio T., De Pellegrin P. (1969) Intervento eccezionale sul paramento di una diga. Industria delle Costruzioni.

139. Zanchi G.B. (1952) L'impianto idroelettrico dell'Ancipa. Costruzioni, Tecnica ed Organizzazione dei Cantieri, 4, p.191, September.

Approach of Mechanical Behaviour and Rupture of Cohesive Granular Media. Validation on a Model Medium

Jean-Yves Delenne, Moulay Saïd El Youssoufi, Jean-Claude Bénet

Laboratoire de Mécanique et Génie Civil,
cc 048, Université Montpellier II,
34095 Montpellier cedex 5, France

Abstract. This paper deals with an experimental and numerical investigation of mechanical behaviour and rupture of cohesive granular media. Cohesion is characterised experimentally, from a reference medium made of aluminium cylinders glued between them with an epoxy resin. For each type of loading (tension/compression, shearing and couple), experiments give a force-displacement relation, as well as a failure criterion. This local mechanical behaviour is then introduced in a numerical code based on a discrete element method. Finally, from a comparison between numerical and experimental compression test on macroscopic granular samples, we present a validation of the mechanical approach. The approach presented here enables to analyse the localisation of the deformation as well as the initiation and propagation of fractures in granular media.

1 Introduction

The objective of this work is to investigate the mechanical behaviour and rupture of granular media using a cohesive granular media approach. The granular medium is represented by a set of solid particles. Interactions between particles take into account contact repulsion, sliding friction and cohesion laws. Thus, the macroscopic mechanical behaviour of these granular media is at the limit between solid and fluid behaviours. Depending on their cohesion state, they can have the rigidity of a solid or run out as a fluid. This generic problem can have multiple industrial applications in many different areas, such as civil engineering when dealing with soil mechanics [1,2], or process engineering in order to optimise the compaction of salt tablets [3] and pharmaceutical pills [4]. This description has been also used to represent the fractionation process of wheat [5,6].

The mechanical behaviour and rupture of granular media are generally modelled at the macroscopic scale using the concept of Representative Elementary Volume (REV). The weak point with these approaches is their incapacity to take into account the mechanisms that take place at the particle scale (strains localization, initiation and propagation of fractures...).

Numerical methods such as "Contact Dynamics", "Event Driven", "Discrete Element Method", based on a discrete description of the granular media [7,8] have led to a better understanding of the influence of local scale contact and friction on the macroscopic non-cohesive material behaviour [12,10]. Taking into account cohesive phenomena [11] have been proved to be necessary in many cases when dealing with different natural and industrial materials (soils, compacted powders, fractionation processes...). In these materials, the forces are transmitted not only by contact and friction between grains, but also by "cohesive bonds" which ensure the cohesion of the macroscopic sample without confinement. In the literature, several models and techniques have been proposed in order to handle these cohesive bonds. An exhaustive bibliography of the cohesive interactions remains a difficult task due to the rapid evolution of this domain. Classical models which only take into account a normal solicitation for the cohesive bond [12] can be extend to obtain a shearing strength [13–15]. Discrete models taking into account cohesion at local scale were used to compute capillaries [16,17] or cementation effects [18–22]. In this study, an experimental validation of a cohesive granular media approach is presented. It is illustrated for a two-dimensional reference medium made with aluminium cylinders.

A brief description of the mechanical modelling developed is given in the next section. The numerical description used in this work, is an extension of the classical "discrete element method" [23]. Particular attention is given to cohesion between grains, which is the focus of this work. At the local scale, cohesion is characterised experimentally on a pair of cylinders glued together with an epoxy resin. This pair of cylinders is subjected to different kinds of loading (tension, shearing...) using a tensile machine and specific experimental devices [24,25]. The measured cohesion law is then implemented in the numerical code. This code is used with different macroscopic granular media made of a set of cohesive particles in regular arrangement. At the same time, experiments are carried out on samples which have similar geometrical configurations, allowing some comparisons with numerical predictions. This comparison is made on the basis of the macroscopic mechanical behaviour and the kind of rupture observed.

2 Mechanical Modelling

The mechanical approach used here is based on a Lagrangian method, where kinematics and deformations are defined with respect to a reference state. Using a discrete element method [23], an extension to the cohesive case is presented.

The mechanical model is limited to a two dimensional problem, thus, grains are represented by discs, for which the kinematics is described in a global system of axes $(O, \overrightarrow{x}, \overrightarrow{y})$. A disc i is characterised by the position of its center C_i, a rotation angle θ_i, and a reference position that generally refers

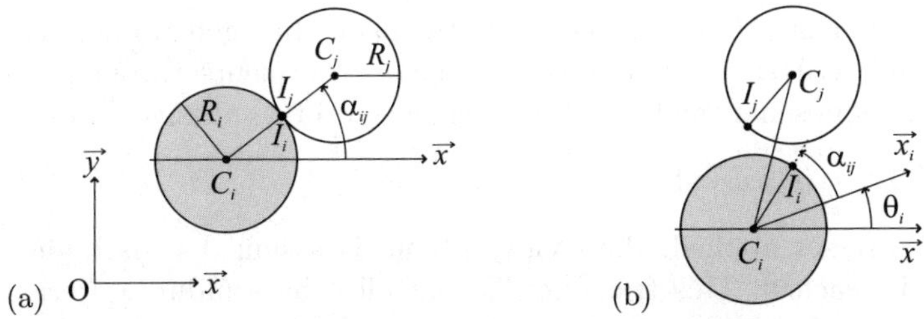

Fig. 1. Initial (a) and deformed (b) state characterisation

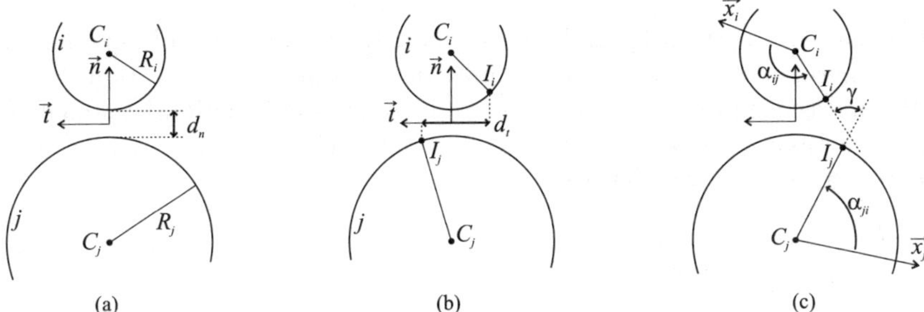

Fig. 2. Degrees of freedom at the local scale: (a) normal displacement; (b) tangential displacement; (c) rotation

to the initial position. In this reference configuration, all rotation angles are taken equal to zero ($\theta_i|_{t=0} = 0$) (Fig.1a). A cohesive bond is defined between two geometrical points I_i and I_j, belonging respectively to disc i and disc j. The position of I_i is referenced by an angle α_{ij} defined in a local axis system $(C_i, \vec{x_i}, \vec{y_i})$ attached to the disc i.

In the reference state, which corresponds to the initial configuration, points I_i and I_j overlap (Fig.1a). When the granular medium is submitted to a macroscopic deformation, the positions of these two points I_i and I_j are known at every time (Fig.1b). This enables to define the local displacement of a cohesive bond in a local frame $(\vec{n}, \vec{t})$ associated with this bond, where the unit vector $\vec{n}$ (Fig.2) is defined by $\vec{n} = \overrightarrow{C_j C_i} / \left\| \overrightarrow{C_j C_i} \right\|$ and the unit vector $\vec{t}$ is orthogonal to $\vec{n}$. The local displacement is characterised by a normal component d_n, a tangential component d_t and a rotation angle γ as described in Fig.2 and equation (1). To derive these expressions, the displacements are assumed to be small.

$$d_n = \left\| \overrightarrow{C_i C_j} \right\| - (R_i + R_j) \qquad d_t = \overrightarrow{I_i I_j} \cdot \vec{t} \qquad \gamma = \theta_i - \theta_j \qquad (1)$$

All mechanical actions between two grains can be represented by a force and a moment. In the two-dimensional case, it means taking into account a normal force F_n along the $\vec{n}$ direction, a tangential force F_t along the $\vec{t}$

direction and a moment M_γ perpendicular to the plane. In a general manner, these actions are written as functions of local displacements through the operator ψ that represents the force-displacement law of a cohesive bond

$$(F_n, F_t, M_\gamma) = \psi\left(d_n, d_t, \gamma\right) \tag{2}$$

Depending on stresses applied, the cohesive bond is assumed to fail when the yield load is reached. This transition is controlled by a failure criterion written as a loading function

$$\kappa\left(F_n, F_t, M_\gamma\right) = 0 \tag{3}$$

These general functions (ψ and κ) are characteristics of the cohesive bond. Their experimental determination is fully detailed in the next section.

Beyond the rupture of a cohesive bond, contact and friction laws control the interactions between two grains. A repulsion law is used to take into account the contact interaction

$$\begin{cases} F_n = -k_n d_n & \text{if} \quad d_n \leq 0 \\ \quad F_n = 0 & \text{otherwise} \end{cases} \tag{4}$$

where k_n is the normal stiffness of contact.

Dry friction interactions between grains are described through the regularised Coulomb's law. The friction force acting between two grains in contact takes the form

$$F_t = -\min\left(|k_t v_t|, \mu\, F_n\right) sgn\left(v_t\right) \tag{5}$$

where the tangent shear stiffness k_t, is used to avoid discontinuity [23,8], v_t is the sliding velocity of the bodies in contact and μ is the friction coefficient.

Based on the local mechanical behaviour of interactions between grains presented above, the general algorithm dealing with a collection of particles considers the following stages:

1. from the geometrical configuration at a time step, local relative displacements (d_n, d_t, γ) of each bond are calculated,
2. these displacements are used, through the cohesion law (2), to compute the local interactions between each pair of grains,
3. from the resulting forces acting on a particle, the equations of motion are used to determine the acceleration quantities of each grain,
4. a "velocity verlet" scheme [26] is used to integrate the equations of dynamics over a time step in order to calculate the displacement of each particle and thus to obtain the new configuration at the next time step.

In the next section, an experimental investigation of the mechanical behaviour of a cohesive bond is presented. The objective is to characterise the cohesion law (2) and the failure criterion (3) from experimental results at the particle scale.

3   Experimental characterisation of cohesion

The objective of this section is to characterise the bond between grains in terms of mechanical behaviour (elasticity, viscosity...) and rupture point (yield load) [24,25]. The cohesion law will be implemented in the numerical code presented in the previous section. In this study a simplified mechanical modelling of the cohesive bond is desired rather than an elaborated and detailed description. The use of simple mechanical modelling and geometry at the local scale can produce complex non linear behaviour at the macroscopic scale.

Cohesion characteristics are determined experimentally on a "doublet". This doublet is made of two aluminium cylinders glued together with an epoxy resin (Epolam 2010). The length of cylinders (60 mm) is sufficiently large with regard to the diameter (8 mm) in order to correspond to a two-dimensional case. Different kinds of systems of loads are applied on this doublet as illustrated in Fig.3. These systems of loads are referred as tension, compression, shearing and moment. For each kind of stress, specific experimental devices have been designed, and used with a tensile machine that gives the force as a function of displacement (Fig.3). The experimental procedure involves a constant loading velocity ($V = 5.10^{-3}\ mm\,s^{-1}$) until failure is observed. A series of ten trials is performed for each kind of loading. Since low variability is observed inside a series, only the averaged results for each kind of system of loads are presented in Fig.4.

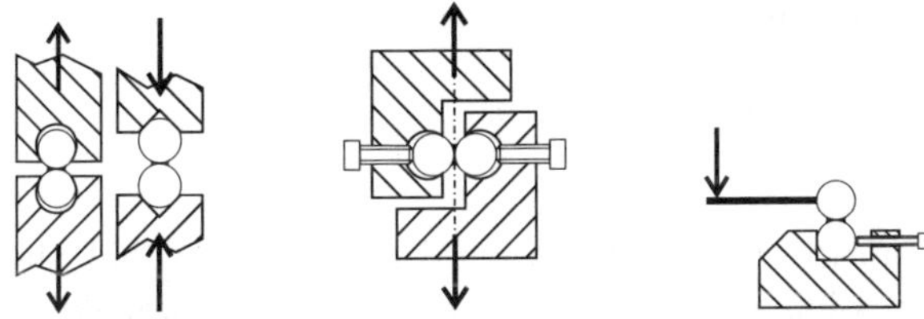

Fig. 3. Experimental mechanical devices designed to characterise a cohesive bond

One can note that the first part of all plots is relatively smooth and linear. Based on this observation, a first approximation by an ideal elastic behaviour is fairly sufficient. In the shearing case, one can see in Fig.4 that the cohesive bond has a non-linear behaviour. This ductile aspect beyond failure will play a significant role on the macroscopic behaviour. However, a more elaborated model will bear algorithmic difficulties, and will not be taken into account at this point.

The ideal elastic model is affected by two parameters: the slope of the curve (a stiffness coefficient) and a rupture point (the yield load). Experimental stiffness and yield load values obtained from these results are summarised

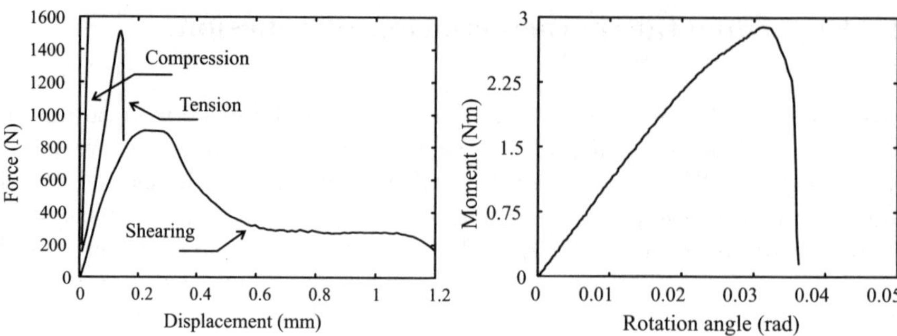

Fig. 4. Characterisation of local cohesion: tension, compression, shearing and moment

in table 1. The four values of the stiffness obtained for each system of loads govern the local cohesion law (2) written as

$$
\begin{bmatrix} F_n \\ F_t \\ M_\gamma \end{bmatrix} = \begin{bmatrix} \psi_n & 0 & 0 \\ 0 & \psi_t & 0 \\ 0 & 0 & \psi_\gamma \end{bmatrix} \begin{bmatrix} d_n \\ d_t \\ \gamma \end{bmatrix}
\tag{6}
$$

Table 1. Experimental stiffness and yield load values

Loading systems	stiffness	yield load
tension	$\psi_n^- = 10.8 \times 10^6 \ N.m^{-1}$	$F_n^{rupt} = 1500 \ N$
compression	$\psi_n^+ = 200 \times 10^6 \ N.m^{-1}$	no rupture
shearing	$\psi_t = 4.2 \times 10^6 \ N.m^{-1}$	$F_t^{rupt} = 900 \ N$
moment	$\psi_\gamma = 101 \ N.m.rad^{-1}$	$M_\gamma^{rupt} = 2.9 \ N.m$

With regards to the rupture, these experimental results show that the fracture strength in compression is much larger than the one obtained in tension. Thus, failure in compression will not be considered here and an infinite value is assumed.

In this experimental characterisation of cohesion, the doublet is subjected to only one loading at a time. Obviously, in a granular medium, the doublet is subjected simultaneously to several of the previously mentioned loadings, such as tension/shearing or compression/shearing. Indeed, it requires determining a more elaborated failure criterion in order to extrapolate the above results to combined loading. The three kinds of loading that occur, i.e., tension-compression, shearing and moment, can be considered as the three axis of a three-dimensional plot. In this loading space, a failure criterion is represented by a surface defining a region where cohesion is maintained (inside the surface) and a region where rupture is observed (on the surface) (Fig.5). This surface is constrained to pass through the 5 points correspond-

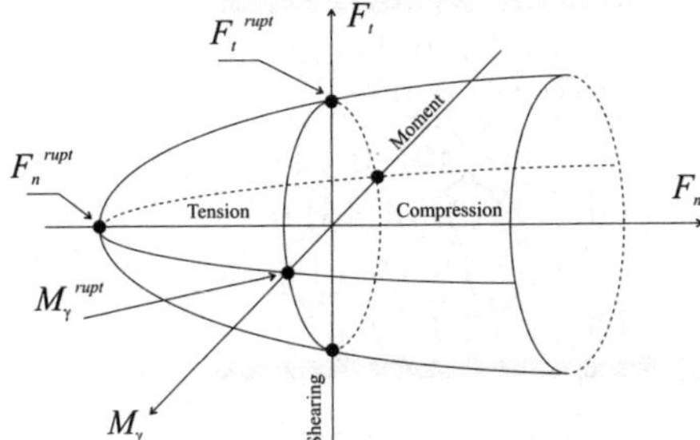

Fig. 5. Failure criterion

ing to the rupture in the case of simple loading (tension, shearing, moment).
Different shapes have been tested in order to choose a failure criterion. The
parabolic surface represented in Fig.5 has been retained. This criterion is
defined using the function ζ given by

$$\zeta = \left(\frac{F_n}{F_n^{rupt}}\right) + \left(\frac{F_t}{F_t^{rupt}}\right)^2 + \left(\frac{M_\gamma}{M_\gamma^{rupt}}\right)^2 - 1 \tag{7}$$

The cohesion between the two grains is maintained as long as $\zeta < 0$ and if
the value of ζ reaches 0, the cohesive bond fails.

The cohesion law (6) and the rupture criterion (7) are implemented in a
computer code which allows to estimate the macroscopic behaviour of a granular assembly. A validation, at the macroscopic scale, based on the comparison between experimental and numerical results is proposed in the following
section.

4 Comparison between numerical and experimental behaviour of a cohesive granular medium

Our objective at this point is to compare the macroscopic mechanical behaviour obtained from the numerical procedure presented in section 2, with
experimental results, so as to validate our approach without adjustable parameters. The comparison will be done in terms of the force-displacement
curve, the yield load and the failure pattern.

The reference media are made of aluminium cylinders glued between them
using the same technique and the same epoxy resin as for the doublets (see
section 3). Two series of macroscopic samples are used. The first series is
characterised by a regular arrangement of 33 cylinders (Fig.6a), while the
second one shows some defects created by withdrawing one of the cylinders 12,
13, 18 or 19 (Fig.6b). The experimental procedure consists in a compression
test until failure at constant loading velocity between two parallel plates.
Nine trials were carried out on samples of the first series. Figure 7 gives the

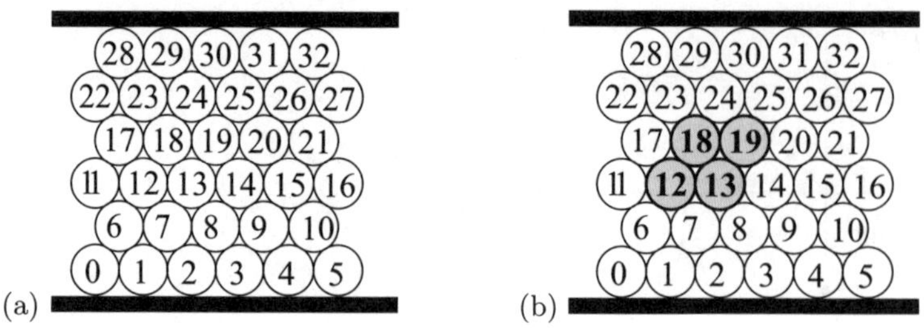

Fig. 6. Macroscopic samples: (a) regular; (b) positions of defects

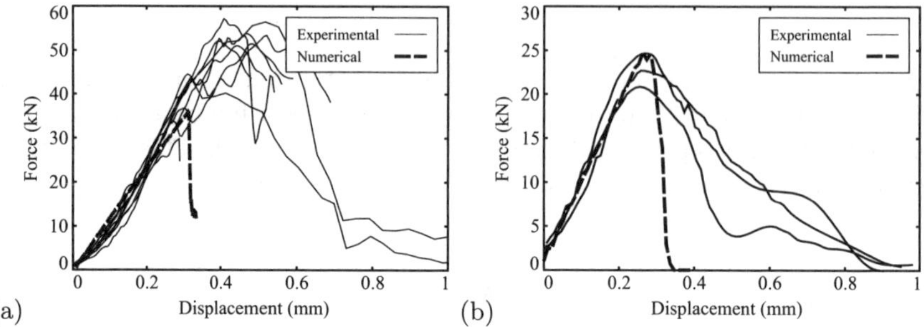

Fig. 7. Experimental and numerical compression tests carried out on macroscopic granular samples; (a) without defect, (b) with a defect in position 18

experimental force-displacement curves obtained, a reasonable repeatability can be observed between each trial.

Regarding to the numerical simulations, equations (6) to (7) are used to compute cohesion interactions with parameters given in table 1. The dry friction coefficient between grains is taken equal to $\mu = 0.3$. Concerning the grains in contact with the lower/upper plate, friction characteristics have been determined experimentally, the coefficients obtained are $\mu_u = 0.5$ for the upper plate and $\mu_l = 0.25$ for the lower plate. The macroscopic behaviour obtained from numerical simulation is also reported on Fig.7a.

This figure shows a good agreement between numerical computation and experimental results. The experimental and numerical macroscopic stiffness coefficients are very close (about 118×10^6 $N.m^{-1}$). Nevertheless, the numerical simulation seems to underestimate the macroscopic yield load. This should be due to the simplification assumed in the characterisation of local cohesion laws, which neglects the ductile aspect beyond failure, as mentioned in section 3 (Fig.4). This point is of great importance when local shearing is more strongly involved, as the main mechanism responsible for local failure. Similar numerical and experimental fracture patterns are obtained [24,25]. In both cases, two shear bands are identified and a cohesive triangular core is observed after failure. Similar experimental and numerical compression tests have been carried out over samples including defects in the form of missing cylinders. Results are given in the case of a missing grain in position 18 (Fig.6b). A very good agreement between numerical and experimental macroscopic mechanical behaviours is obtained (Fig.7b). In this case, both macroscopic characteristics, i.e., stiffness (about 95×10^6 $N.m^{-1}$) and yield load, are well estimated with the cohesive granular media approach. One can also notice that similar fracture patterns are obtained [24,25]. Other good results are obtained with the last three series, where the defect is in position 12, 13 or 19. Simulations on regular samples and on samples with defects allow a comparison of stiffness coefficients and yield loads. One ends naturally in weaker values in the case with defect.

5   Conclusion

An experimental and numerical study of the mechanical behaviour and the rupture of cohesive granular media has been presented. From particle-scale experiments, local cohesion laws have been formulated, and then introduced in a numerical tool that simulates the behaviour of a collection of grains. A validation has been proposed based on a comparison of numerical prediction with results from experimental compression tests. One positive point of the approach developed in this paper is that it takes into account combined loading (tension, compression, shearing, moment) and allows to focus on the mechanical behaviour and the fracture of the cohesive bonds. It is well adapted to introduce some more complicated behaviours such as the plasticity or a damage at the local scale. Therefore, since the numerical tool allows one to localise the failure initiation and to follow its propagation, it seems to be a way to revisit the mechanics of failure from a different point of view.

References

1. Masson S., Martinez J. (2001) Micromechanical analysis of the shear behaviour or a granular material. Journal of Engineering Mechanics **127**(10), 1007–1016.
2. Nouguier C. (1999) Simulation des interactions outil-sol. PhD thesis, Université Montpellier II, France.
3. Saix C., El Youssoufi M.S. (1996) Thermo-hygro-mécanique de milieux granulaires compactés. Application aux sels sous forme de pastilles. Rapport de synthèse. Convention de recherche CSME/UMII/CNRS, France.
4. Masteau J.C., Thomas G. (1999) Modelling to understand porosity and specific surface area changes during tabletting. Powder Technology **101**, 240–248.
5. Haddad Y., Bénet J.C., Delenne J.Y., Mermet A., Abecassis J. (2001) Rheological behaviour of wheat endosperm - Proposal for classification based on the rheological characteristics of endosperm test samples. Journal of Cereal Science **34**, 105–113.
6. Mabille F., Haddad Y., Delenne J.Y., Bénet J.C. (1999) Experimental study of the rheology and the cracking of granular media with cementation. In: Kishino (ed) Powder and grains, Swets and Zeitlinger, The Netherland, 63-66.

7. Oda M., Iwashita K. (eds). (1999) Mechanics of granular materials - An introduction, Balkema, Rotterdam.

8. Cambou B., Jean M. (eds). (2001) Micromécanique des matériaux granulaires, Hermes Sciences Publications, Paris.

9. Moreau J.J. (1994) Some numerical methods in multibody dynamics: application to granular materials. Eur. J. Mech. A/Solids, **3**, 93-114.

10. Radjai F. (1999) Multicontact dynamics of granular systems. Computer Physics Communications, **121/122**, 294-298.

11. Vermeer P.A., Diebels S., Ehlers W., Herrmann H.J., Luding S., Ramm E. (Eds.) (2001) Continuous and discontinuous modelling of cohesive-frictional materials, Springer Verlag, Berlin.

12. Bortzmeyer D. (1997) Mechanical properties and attrition resistance of porous agglomerates. In: Behringer and Jenkins (Eds.) Powders and grains, Balkema, Rotterdam, The Netherlands, 121-124.

13. Greening D.R., Mustoe G.G.W., DePoorter G.L. (1997) Discrete element modeling of fabrication flaw precursors in the compaction of agglomerated ceramic powders. In: Behringer and Jenkins (Eds.) Powders and grains, Balkema, Rotterdam, The Netherlands, 113-116.

14. Radjai F., Preechawutipong I., Peyroux R. (2001) Cohesive granular texture. In: Vermeer et al. (Eds.) Continuous and discontinuous modelling of cohesive-frictional materials, Springer Verlag, Berlin, 149-162.

15. Preechawuttipong I. (2002) Modélisation du comportement mécanique de matériaux granulaires cohésifs. PhD thesis, Université Montpellier II, France.

16. Lian G., Thornton C., Adams M.J. (1998) Discrete particle simulation of agglomerate impact coalescence. Chemical Engineering Science **53**(19), 3381-3391.

17. Mikami T., Kamiya H., Horio M. (1998) Numerical simulation of cohesive powder behavior in fluidized bed. Chemical Engineering Science **53**(10), 1927-1940.

18. Potapov A.V., Campbell C.S. (1997) The two mechanisms of particle impact breakage and the velocity effect. Powder Technology **93**, 13-21.

19. Kun F., Herrmann H.J. (1999) Transition from damage to fragmentation in collision of solids. Physical Review E **59**(3), 2623-2632.

20. Magnier S.A., Donzé F.V. (1998) Numerical simulations of impacts using a discrete element method. Mechanics of Cohesive-Frictional Materials **3**, 257-276.

21. Brara A., Camborde F., Klepaczko J.R., Mariotti C. (2001) Experimental and numerical study of concrete at high strain rates in tension. Mechanics of Materials **33**, 33-45.

22. Pisarenko D., Gland N. (2001) Modeling of scale effects of damage in cemented granular rocks. Phys. Chem. Earth (A) **26**(1/2), 83-88.

23. Cundall P.A., Strack O.D.L. (1979) A discrete numerical model for granular assemblies. Geotechnique **29**, 47-65.

24. Delenne J.Y., El Youssoufi M.S., Bénet J.C. (2002) Comportement mécanique et rupture de milieux granulaires cohésifs. C. R. Mecanique **330**, 475-482.

25. Delenne J.Y. (2002) Milieux granulaires comportement solide - Modélisation, analyse expérimentale de la cohésion, validation et applications. PhD thesis, Université Montpellier II, France.

26. Allen M.P., Tildesley D.J. (1986) Computer simulation of liquids. Oxford university press, Oxford.

Phase Change of Volatile Organic Compounds in Soil Remediation Processes

Ali Chammari, Bruno Cousin, Jean-Claude Bénet

Laboratoire de Mécanique et Génie Civil,
cc 048, Université Montpellier II,
F-34095 Montpellier Cedex 5, France

Abstract. Liquid-vapor phase change is a central phenomenon in remediation of soils contaminated by volatile organic compounds. An original experimental study leads to validate a law of phase change of heptane in a soil. The dependence of the phenomenological coefficient to liquid content and temperature is established. A simulation of self-removal of heptane in a soil, involving phase change of pollutant and diffusion of the vapor, is compared with experimental results. Finally, treatment by venting of a column of contaminated soil is studied experimentally in order to improve the understanding of the phenomena involved. The results obtained are likely to direct future modelisation of the processes.

1 Introduction

Volatile Organic Compounds (VOCs) are commonly involved in soil pollution cases [1]. Their elimination may occur by self-removal process or by vapor extraction technique. In the first case, the phenomena involved are phase change of contaminant from liquid to gas and diffusion of the vapor in gaseous phase from the subsurface to the atmosphere. In vapor extraction (or venting) method, pumping creates a global movement of the gas phase. Phase change of the liquid is therefore the limiting phenomena. The study of those two processes belongs to the field of heat and mass transfer in heterogeneous media. Their modelisation requires mass balance equations and a set of closure relations such as diffusion and filtration laws as well as the expression of the liquid-vapor relationship. The later can be assumed to be an equilibrium condition. The vapor pressure is directly deduced from the isotherm of desorption. Such hypothesis should be consistent with experiments if the velocity of the vapor of contaminant is extremely small i.e. if the phase change is not the limiting phenomenon. Nevertheless, experiment shows that this hypothesis is rapidly invalidated. Moreover, for non hygroscopic medium and for isothermal conditions, it follows from equilibrium assumption that the vapor pressure is uniform. Consequently, the transfer of vapor cannot be described by Fick's law that use gradient of concentration. Obviously, it appears that phase change must be taken into account within the framework of non equilibrium thermodynamics. The purpose of this paper is, at first, to present an experimental validation of a law of non equilibrium phase change of a liquid in a soil and the determination of the corresponding phenomenological coefficient. Next, experiments of self-removal process are presented with the results of numerical simulations. Thereafter, we expose the first experimental results obtained with VOCs removal by venting.

2   Phase Change of a Volatile Organic Compound in Soil

2.1   Law of Phase Change of a Pure Liquid in a Porous Medium

The law of VOCs phase change in a soil is founded on the following description of the problem : the system consists of an inert solid phase, a pure liquid phase and a gaseous phase composed by air and vapor. Mass balance equations for the liquid and the vapor include a term of phase change from liquid to gas. According to linear Thermodynamics of Irreversible Processes, the corresponding phenomenological relation is [1]

$$J = -L \frac{(\mu_v - \mu_l)}{T} \qquad (1)$$

$J \ [kg.m^{-3}s^{-1}]$ is the mass of contaminant which transfers from liquid to gaseous phase per unit of medium volume and time. $L \ [kg.K.s.m^{-5}]$ is the phenomenological coefficient of phase change. For a given medium, it depends on the liquid content and the temperature. $T \ [K]$ is the absolute temperature, μ_v and μ_l are the mass chemical potential of liquid and vapor respectively.

Assuming the perfect gas hypothesis to express the chemical potentials and by means of an original approach based on the introduction of a virtual transformation, Bénet [3] wrote expression (1) for non isothermal case as follows

$$J = \frac{\partial \rho_{veq}}{\partial t} + \nabla \Phi_{veq} - L \frac{R}{M_l} \ln \frac{P_v^*}{P_{veq}} \qquad (2)$$

$\rho_{veq} \ [kg.m^{-3}]$ is the virtual mass density of the vapor at equilibrium with the liquid, $\Phi_{veq} \ [kg.m^{-2}s^{-1}]$ is the virtual mass flux of vapor induced by the liquid transformation, $R \ [J.mol^{-1}K^{-1}]$ is the perfect gas constant, $M_l \ [kg.mol^{-1}]$ is the molar mass of the contaminant, $P_v^* \ [Pa]$ is the partial vapor pressure and $P_{veq} \ [Pa]$ the equilibrium vapor pressure.

2.2   Experimental Study of Phase Change of Heptane in a Soil

The purpose of the experimental study is to verify the previous law and to determine the coefficient L. To this end, experiments are lead in static conditions so that the macroscopic fluxes can be ignored. The experiment consisted in measuring the temperature, T, and the total pressure of the gaseous phase , P_g^*, of a small volume of soil which contains pure liquid, its vapor and air. From a non-equilibrium initial state for which $P_v^* \neq P_{veq}^*$, the kinetic of pressure measures show a return to an equilibrium state. In static conditions, and with respect to perfect gas law, the mass balance of vapor in integrated form allows the phenomenological coefficient to be determined [1]

$$\left[\frac{P_v^* - P_{veq}}{T}\right]_t = L\frac{R^2}{\eta_g M_l^2}\int_t^{t_{eq}}\ln\frac{P_v^*}{P_{veq}}dt \tag{3}$$

η_g is the volume of the gaseous phase per unit of volume of soil. t_{eq} is the time when equilibrium is reached, i.e. when $P_v^* = P_{veq}$.

The characteristics of the silty sand soil used are given in table 1. So as for the used VOC, which was heptane as one of the components of gasoline.

Table 1. Soil and VOC characteristics

SOIL	
mineralogy	calcite 50% quartz 40% clay 10%
Atterberg limits	$W_l = 25\%$ $W_p = 14,5\%$
real mass density $\rho_s^*[kg.m^{-3}]$	2650
dry apparent specific gravity $\rho_s[kg.m^{-3}]$	1500
porosity	0,43
HEPTANE as volatilorganic compound	
molar mass $M\ [kg.mol^{-1}]$	0.1
saturation vapor pressure $P_{vs}(T)\ [kPa]$	$10^{(6.02167-\frac{1264.9}{T+216.544})}$
diffusion coefficient in air $D_{va}\ [m^2.s^{-1}]$	$0.824\ 10^{-5}$
apparent mass density $\rho_l\ [kg.m^{-3}]$	675.19
latent heat of evaporation (à $20°C)[kJ.kg^{-1}]$	365.5

The sample of contaminated soil occupied the entire volume of a 12 cm^3 cylindrical cell. The cell and its accessories were immersed in a thermostatic bath. The initial thermodynamic non-equilibrium was obtained by replacing the gaseous phase of the soil, first extracted by vacuum, by a vapor-free one. The temperature and total pressure of the gaseous phase of the sample were recorded. After closure, vapor partial pressure of heptane rised to its equilibrium value and could be evaluated at each moment as follows $P_v^*(t) = P_g^*(t) - P_a^*$ where the air partial pressure P_a^* is equal to $P_a^* = P_g^*(t_{eq}) - P_{veq}^*$. The isotherm of desorption for heptane has not been established experimentally. We assume in this paper that it can be deduced from the isotherm of desorption of water in the same soil by means of an ordinate-parallel affinity with a ratio equal to $w_c^{water}/w_c^{heptane}$. w_c^{water} and $w_c^{heptane}$ are the liquid content corresponding to the entrance in the hygroscopic domain respectively for water and heptane.

At the end of an experiment the data are collected by a computer and exploited with a spreadsheet. That allows us to represent the first member of equation (3) versus the integral of the second member. All experimental curves obtained showed two distinct parts. A linear part and a non-linear one. The non linear part corresponds to states the farest from equilibrium.

Those curves allow us to conclude to the validity of the phenomenological relation (1) in the linear part. The slope of the linear part enables the phase change coefficient L to be calculate according to relation (3).

Figure 1 shows the values of the experimental phase change coefficient of heptane versus the liquid content expressed relatively to the dry mass of solid. The three curves correspond to temperatures of 30, 50 and 80°C. Note that similar liquid content dependence has been found for water in soil [3]. The liquid content corresponding to the maximum value of the coefficient is close to $w_c^{heptane}$ previously defined. The decrease of the coefficient for lower values of liquid content is due to the increase of the energy of bonds between liquid and solid when liquid content decreases in hygroscopic domain. For higher liquid content, the decrease of the coefficient can be explained by the decrease of the connectivity of the gaseous phase when liquid content increases in non-hygroscopic domain. The coefficient reaches zero value when the gaseous phase is entirely occluded.

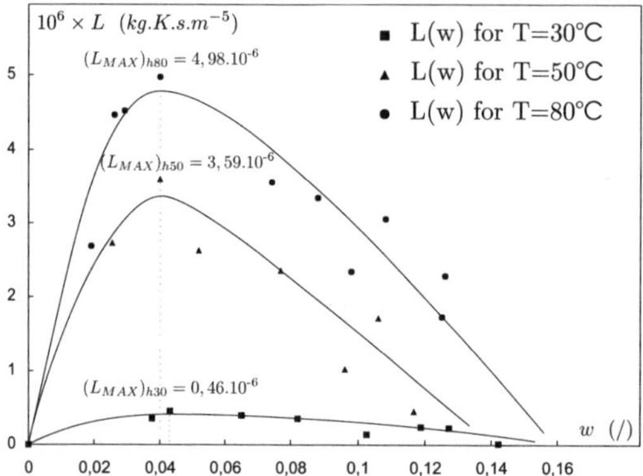

Fig. 1. Experimental phase change coefficient L

Figure 1also shows the evolution of the maximum value of the coefficient vs temperature. The increase of the coefficient is higher when the temperature rises from 30 to $50°C$ than from 50 to 80°C. This kind of observations is likely to direct choices when thermo-activated venting treatment of a contaminated site is considered.

3   Self-removal of a contaminant in a soil

Natural diffusion of hydrocarbon vapor from the contaminated zone to the surface can significantly contribute to soil remediation. This so called "self removal process" is the purpose of this section. A numerical simulation is proposed using the phase change coefficient previously determined. Then, we present some experimental results obtained with heptane. Finally, a comparison between simulations and experiments is performed.

3.1   Simulation

Based on the phenomenological model presented below, whose main equations are recalled in table 2, a one directional simulation of heptane self-removal from a soil is performed.

Table 2. model equations

Liquid mass balance	$\frac{\partial}{\partial t} w = -\frac{J}{\rho_s}$
Vapor mass balance	$\frac{\partial}{\partial t} P_v^* = \frac{RT}{\eta_g M_l}\left(-\frac{\partial}{\partial z} J_v + J\right)$
Phenomenological relation of phase change	$J = -L\frac{R}{M_l}\ln\frac{P_v^*}{P_{veq}}$
Phenomenological relation of diffusion	$J_v = -D\frac{\partial}{\partial z} P_v^*$

The diffusion coefficient of the vapor in the gaseous phase of the soil D can be related with the diffusion coefficient in air D_a. Among different relations, we chose the expression proposed by Penman [9] that is $\frac{D}{D_a} = 0.66\eta_g$. In addition the pressure of vapor at equilibrium P_{veq} is expressed by $P_{veq} = a(w)\,P_{vs}(T)$, where the saturating vapor pressure of heptane is given in table 2 and $a(w)$ is the activity of the liquid given by the isotherm of desorption and from which it is deduced a polynomial approximation

$$a(w) = 10^4\left[55.46\,w^4 - 6.18\,w^3 + 0.19\,w^2 + 5.47\,10^{-4}w\right] \tag{4}$$

The phenomenological coefficient of phase change at 30°C can be approximated from the curve in Fig.1 by a polynomial expression

$$L = 10^{-4}[6.34w^3 - 2.06w^2 + 0.17w] \tag{5}$$

The above expression of the coefficient L is available close to the equilibrium i.e. when the ratio P_v^*/P_{veq} is close to 1. Nevertheless, in practical applications, phase change occurs in conditions that are outside of this linear domain. In these situations, the coefficient L increases significantly as the partial pressure of VOC vapor P_v^* decreases. Thus, a first interpretation of experimental results far from equilibrium is used in the case of water [10]

$$\text{if } P_v^*/P_{veq}^* \leq 0.93 \text{ then } L = -18.4\,10^{-8}\left(\frac{P_v^*}{P_{veq}^*}\right) + (L_{\text{équilibre}} + 17.1\,10^{-8}) \tag{6}$$

where L_{eq} is the coefficient given by equation (5). In this study we fitted phase change coefficient far from equilibrium for heptane from relation (6). Finally experiments are setted at $T = 30°C$.

As showed in Fig.2a, simulation is performed in one directional geometry. The space axis is divided in two sections

- soil: from $z = 0$ to $z = h$
- air: from $z = h$ to $z = H$

The initial and boundary conditions are the following

- at $t = 0$
    - $w = w_0$ for $0 \leq z \leq h$
    - $P_v^* = 0$ for $h \leq z \leq H$
- at $z = 0$, $J_v = 0$ $\forall t$
- at $z = H$, $P_v^* = 0$ $\forall t$

Integration is performed using Newton method with a time implicit scheme.

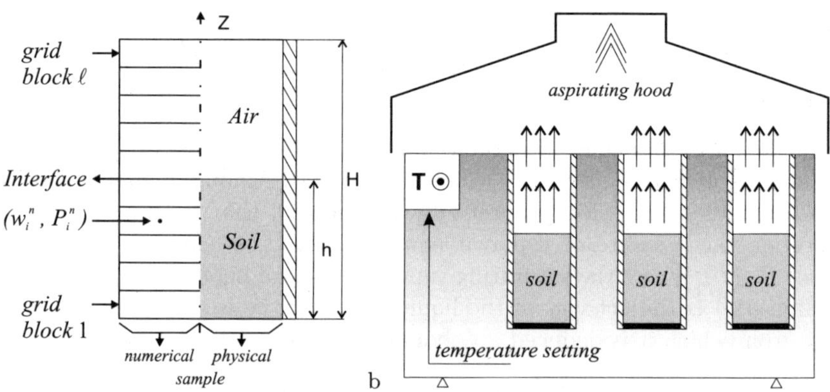

Fig. 2. Heptane self-removal principle a) simulation and b) experiments

3.2   Experimental study

Soil samples contaminated with heptane were placed in cylindrical tubes as shown in Fig.2. Samples were 81,4 mm in diameter and 100 mm height. Tubes were placed under an aspirating hood which allows to assume the heptane vapor pressure being equal to zero at the top of the tubes. Temperature of the samples was regulated at 30°C by means of a thermostatic bath. The initial homogeneous liquid content was about 10%. Vapor of heptane diffused in the air surrounding the sample to the top of the tube. Samples were regularly weighted to determine the global pollutant removal kinetics. Furthermore, some samples are cut in sections for the purpose of establishing the liquid content profile at a given time. Figure 3a shows experimental kinetics of heptane removal. Final liquid content can be explained at least in part by heptane equilibrium content since no heptane vapor is present in the atmosphere. Liquid content profiles at different times are given on Fig.3b. This figure shows that the removal of heptane does not operate as a front propagation process. The VOC content decreases from the first hours at all level of the sample.

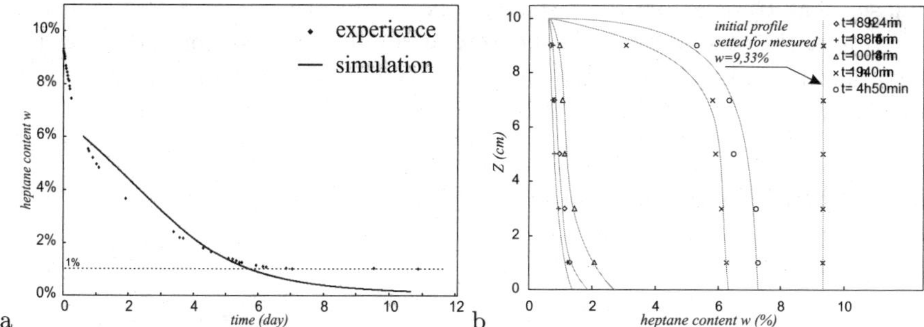

Fig. 3. VOC self-removal test results
a) experimental and simulated kinetics, b) experimental profiles

Figure 3a also illustrates the comparison between experimental and simulated heptane removal kinetics. The general decay of heptane content versus time is well depicted by the model. However, it appears some notable differences. Wheras the simulated kinetic converges to zero liquid content due to the boundary condition at $z = H$. In the same time experiemental data tend to a value about 1% of heptane at equilibrium. As said formerly in sec.2.2, the isotherm of desorption for heptane in this soil has not been established experimentally and we don't have yet informations about heptane content at equilibrium in the used soil. The model, indeed, does not take into account the water which cannot be distinguished from heptane in the experimental determination of liquid content. Likewise, the slope of the calculated kinetic at the beginning of the process is not in agreement with the experiment. Possible transport of liquid heptane in the soil, not taken into account by the model, can explain this difference. Moreover, the assumption of strictly diffusive transport of vapor in the atmosphere surrounding the soil can be abusive. Light convective movements of the air can significantly increase the velocity of heptane vapor from soil surface to the top of the tube and consequently increase the heptane removal. Experimental VOC content profiles on Fig.3b shows that liquid content at the surface of the sample ($x = h$) seems to reach the equilibrium rapidly. This can also be attributed to an activated vapor transport by convection in the air of the tube. Finally, the differences between experiments and theoretical simulation can be also attributed to the coefficient of phase change far from equilibrium that has been inspired by a study of phase change of water in soil.

4   VOC Removal by venting

This section is devoted to presenting the first results of an experimental study of venting. In comparison with the self-removal process, venting or vapor extraction introduces supplementary phenomena: filtration of gaseous phase, heat transfer between gaseous phase and other phases and heat conduction within the phases. No theoretical simulation is performed for lack of data about, in particular, the phenomenological coefficients involved in the description of the different phenomena.

4.1   Thin layers experiments

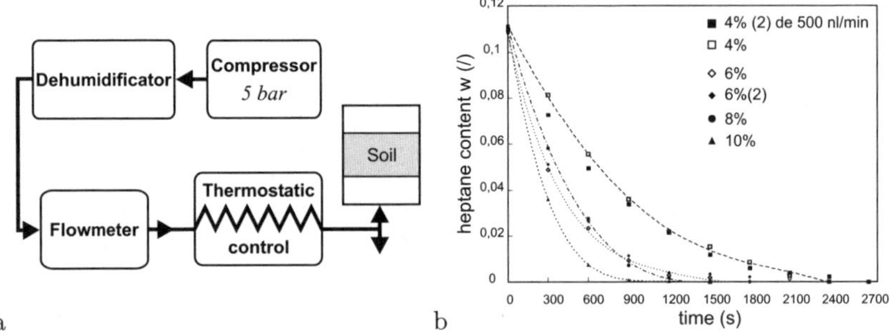

Fig. 4. soil vapor extraction. a) schematic test cell, b) thin layer kinetics

The experimental plant is schematicly represented in Fig.4a. Contaminated soil was contained in a cylindrical tube. The diameter was 81,4 mm and the height was 60 mm for experiments with thin layers. The air provided by a compressor was first dried and then heated to the process temperature. Air pressure was controlled by a flowmeter. The samples were weighted regularly to establishe the global kinetics of depollution. Holes drilled along the tube permitted to uptake soil at different levels in order to determinate profiles of liquid content. The charecteristics of the soil were the similar to those used in the previous experiments.

Thin layer experiments were intended to determine the influence of the treatment conditions on the removal rate. The influencing parameters are the air flux, its temperature and its saturation relating to VOC vapor that is quantified by the ratio P_v^*/P_{vs}. In this study, we only explore the influence of the air flux. Temperature is $30°C$ and the partial pressure of heptane vapor at the inlet of the sample is fixed to zero. The small thickness allows the homogeneity of the sample to be assumed. Figure 4b shows experimental kinetics for 60 mm height samples. The different curves are relating to different air flux: 0.077, 0.116, 0.154 and 0.193 $kg.m^{-2}s^{-1}$. This figure shows that an increase of the air flux logically induces an increase of the removal efficiency.

Exponential approximation of the experimental kinetics permits to calculate derivative functions of the form $dw/dt = k.w$. That makes possible to highlight the dependance of the coefficient k to the air fluxes F_a. The curve obtained suggests a linear type dependance of the form $k = A \cdot F_a$. The influence of temperature and saturation of air relating to vapor of heptane can be taken into account using an expression of the form

$$\frac{dw}{dt} = f_1(T) \cdot f_2(\frac{P_v^*}{P_{vs}}) \cdot A \cdot F_a \cdot w \tag{7}$$

Experiments have to be completed by the study of the influence of temperature and air saturation in order to propose an expression of $f_1(t)$ and $f_2(P_v^*/P_{vs})$.

4.2   Thick layers experiments

The purpose of experiments on thick layers was to observe the evolution of the profiles of liquid content in the course of time. The samples consisted of five layers of 60 mm thickness. We compacted each layer to obtain the chosen apparent specific gravity ($\rho_s = 1500\ kg.m^{-3}$). The inlet air flux was fixed at $0,116\ kg.m^{-2}s^{-1}$. Figure 5a shows some global heptane removal kinetics. The general aspect of the kinetics is similar to the one observed in classical drying processes [7]: a constant rate period, characterized by a linear decrease of the liquid content along the time, followed by a decreasing rate period. In our experiments, the initial rate of heptane removal varies from $1.5\ 10^{-5}$ to $2.5\ 10^{-5}s^{-1}$. Influencing factors can be identified such as the homogeneity of heptane content, the compaction regularity, the control of air flux, and variations of external temperature. The liquid content profiles represented in Fig.5b was measured after 30, 45, 60, 90, and 95 minutes of treatment. These profiles was established from separated experiments that affects the coherence of the curves. One can yet observe following general tendencies: first, the liquid content at the air inlet extremity of the sample, i.e. in z=0, reaches the equilibrium value ever since the first minutes of the treatment. Then the removal of heptane progresses by front propagation and by general decrease of the liquid content along the height of the sample. And finally the profiles show higher liquid content near the top of the sample than in the other parts.

The last point is partially explained by the increase of the vapor saturation of the air after having crossed through a part of the sample. But, in addition, we observed that the liquid accumulated in the upper part of the column of soil. This accumulation can have two physical origins: either an upward

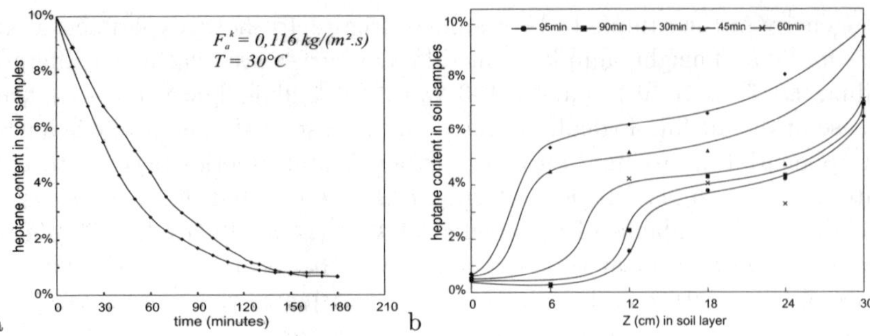

Fig. 5. Thick layers venting test a) kinetics, b) profiles

movement of the liquid mechanically induced by the air flux or a condensation phenomenon due to a decrease of temperature near the surface of the soil or both. Further experiments will have to precise these hypothesis. Theoretical models would have to take into account these phenomena.

5   Conclusion

Phase change appears to be a central phenomenon in remediation of soils contaminated by Volatile Organic Compounds. An original experimental device has been designed to probe the phase change of a liquid in a soil. We demonstrated that the law founded on Linear Thermodynamics of Irreversible Processes was verified near equilibrium and we have determined the phenomenological coefficient of phase change. The results show the dependence of the coefficient to liquid content and temperature. A theoretical model was established to simulate the self-removal of a VOC in a soil by phase change and diffusion of vapor. The model is able to correctly simulate the process. The comparison with experimental results incite to study the phase change phenomenon far from equilibrium and to verify the good agreement of the boundary conditions. Experimental study of venting was instrumental in showing the phenomena involved in the process such as liquid transport induced by gaseous flux and condensation of vapor. The latter phenomenon leads to the study of the reversibility of phase change in porous media and should be experimentally investigated.

References

1. Lecomte P. (1995) Les sites pollués, traitements des sols et des eaux souterraines, Lavoisier, Paris.
2. Bénet J.-C., Jouanna P. (1982) Phenomenological relation of phase change of water in a porous media: experimental verification and measurement of the phenomenological coefficient, Int. J. Heat and Mass Transfer, **25**(11), 1747-1754.

3. Bénet J.-C., Jouanna P. (1983) Non équilibre thermodynamique dans les milieux poreux non saturés avec changement de phase, Int. J. Heat and Mass Transfer, **26**(11), 1585-1595.

4. Ruiz T., Bénet J.-C. (2001) Phase change in a heterogeneous medium : comparison between the vaporisation of water and heptane in a unsaturated soil at two temperatures, Transport in Porous Media, **44**, 337-353.

5. Chammari A. (2002) Transfert gazeux dans les sols avec changement de phase - Application à quelques aspects de géotechnique environnementale, Thèse de doctorat, Université Montpellier II.

6. Mignard E. (1984) Contribution à l'étude des transferts en milieux poreux par la thermodynamique du non équilibre - cas de la dissolution des gaz dans les sols, Thèse de doctorat, INSA Toulouse.

7. Cousin B., Bénet J.-C., Auria R. (1993) Experimental study of the drying of a thick layer of natural crumb rubber, Drying Technology, **11**, n°6, 1401-1413.

Thermo-mechanical Behaviour of a Soil. Yield Surface Evolution

Moulay Saïd El Youssoufi, Christian Saix, Frédéric Jamin

Laboratoire de Mécanique et Génie Civil,
cc 048, Université Montpellier II,
34095 Montpellier Cedex 5, France

Abstract. This paper presents a study of temperature influence on the yield surface for saturated and unsaturated soils. It lies within the framework of characterization and modelling of soil behaviour on thermo-hydro-mechanical loading. In this study, clayey silty sand samples are subjected to mechanical consolidation tests at different temperatures and fixed suctions. These experimental tests make it possible to determine the yield surface evolution according to temperature ranging between $20°C$ and $60°C$. For the studied soil, the experimental study shows a thermo-extensible nature of the yield surface. An explanation of this result is proposed while gone by microscopic considerations in touch with capillary meniscuses evolution according to temperature and suction.

1 Introduction

This paper presents a contribution to the experimental study and the modelling of temperature influence on the yield surface (border surface between the elastic and plastic domains) of soils. Since the sixties, many studies have been interested in the temperature influence on the mechanical behaviour of saturated soils. In the clayey media case, these researches are generally based on problems involved by storages of radioactive waste in deep geological formation, but applications can concern all porous media subjected to various thermal gradients (geothermic structures). In the case of soils with lower depth, we can quote buried high-voltage cables, different waste storage and notably domestic. For these depths, the hypothesis of a completely saturated soil isn't always acceptable and it's necessary to take into account the effect of suction. In the isothermal unsaturated case, several authors have studied the suction influence on the mechanical behaviour of soils. For example, Alonso et al. [1] proposed a general constitutive model that takes into account the effect of suction in the mechanical behaviour of soils based on the Modified Cam-Clay model.

In the non-isothermal saturated case, several studies put clearly in evidence the temperature influence on the yield surface. Some authors used the Modified Cam-Clay model that appears to be a good estimate of the behaviour to take into account the temperature effect on the yield surface [4,5,2,6,3].

In the non-isothermal unsaturated case, Saix et al. [7] have studied experimentally the effect of a thermal hardening in elastic limits in vertical total stress for a given suction. Jamin et al. [8] were interested more particularly in the determination, for various temperature, of a yield surface $f(p, q, s) = 0$, where p is the net mean stress, q the deviatoric stress and s the suction. The first results of the experimental study show that this surface can be approximate by the Modified Cam-Clay model in the case of clayey silty sand and presents a thermo-extensible nature according to temperature for no null suction. This study has been pursued by adopting news loading paths in the aim of specifying the Alonso et al. [1] model with temperature. This evolution is studied through the temperature influence on the preconsolidation stress, the slope of the critical state line and the stiffness parameters.

In a first part, a bibliographical synthesis leads to emphasize the most essential results of the literature relating to the temperature influence on the model parameters in the saturated and unsaturated cases. Then, we present the studied soil, the experimental device and experimental process. Lastly, the experimental results are presented and are followed by a physical interpretation.

2 Yield surface model

2.1 Hydro-mechanical modelling

In the isothermal unsaturated case, Alonso et al. [1] proposed a constitutive model to describe the unsaturated soils behaviour. This model takes into account the Modified Cam-Clay yield surface developed in Cambridge University by Roscoe and Burland [9]. This yield surface depends on the net mean stress p, the deviatoric stress q and the suction s defined in axisymetric triaxial conditions by:

$$\begin{cases} p = \frac{1}{3}\left(\sigma_z + 2\sigma_x\right) - p_g^* \\ q = \sigma_z - \sigma_x \\ s = p_g^* - p_e^* \end{cases} \tag{1}$$

where σ_z = principal vertical stress, σ_x = principal radial stress, p_g^* = pressure of the gas phase and p_e^* = pressure of the liquid phase.

The equation of this yield surface in the stress space (p, q, s) is given by:

$$q^2 - M^2\left(p + p_s(s)\right)\left(p_0(s) - p\right) = 0 \tag{2}$$

where M = slope of the critical state lines not dependant on suction, $p_s(s)$ = parameter introducing the evolution of apparent cohesion with suction and

$p_0(s)$ = isotropic preconsolidation stress function of suction. In the case of a mechanical loading with constant suction, the relation between the isotropic preconsolidation stresses $p_0(s)$ for unsaturated conditions at a suction s and $p_0(0)$ for saturated conditions ($s = 0$) is written [1]:

$$\frac{p_0(s)}{p^c} = \left(\frac{p_0(0)}{p^c}\right)^{\frac{\lambda(0)-\kappa}{\lambda(s)-\kappa}} \tag{3}$$

where p^c = reference stress, κ = elastic stiffness parameter for changes in net mean stress, $\lambda(s)$ and $\lambda(0)$ = stiffness parameters for changes in net mean stress for virgin states on the soil with constant suction s for the former and in saturated condition ($s = 0$) for the latter, with:

$$\lambda(s) = \lambda(0)\left[(1-r)\,e^{-\beta s} + r\right] \tag{4}$$

where r = constant equal to the ratio $\lambda(s \to \infty)/\lambda(0)$ and β = parameter controlling the rate of increase of soil stiffness with suction. Equation (3), named Loading Collapse, plays an important role in the model. It shows an increase of the preconsolidation stress according to the suction, and also the phenomenon of collapse observed with the decrease of the suction (wetting path). In the case of an increase in suction with constant stress, Alonso et al. [1] give as assumption that the limit of the elastic domain is defined by:

$$s = s_0 \tag{5}$$

with s_0 = hardening parameter of the suction increase yield curve. This equation is called Suction Increase.

2.2   Thermo-mechanical modelling

In the non-isothermal saturated case, several authors have studied the temperature influence on the mechanical behaviour of clayey soils. In the oedometric tests case, we generally note a decrease on the preconsolidation stress when the temperature increases [10,11]. In the axisymetric triaxial tests case, Hueckel and Baldi [2], Tanaka et al. [3], Cui et al. [4] and Graham et al. [5] have shown that the ellipse of the Modified Cam-Clay model contracts with an increase of temperature with the result that the slope of the critical state line M remains constant so independent of temperature. Besides, other authors note a decrease of the critical state line M with an increase of temperature.

2.3   Thermo-hydro-mechanical modelling

Romero [12] has studied coupled effect between the suction and the temperature on the mechanical behaviour of Boom clay in drained condition. The results show, on the one hand, an increase of the yield surface with an increase of suction at constant temperature and, on the other hand, a decrease of this surface with an increase of temperature at constant suction.

However, Jamin et al. [8] obtained different results on clayey silty sand. Indeed, they noted that, in unsaturated conditions, the yield surface shows a thermo-extensible nature with an increase of temperature. Consequently, in the non-isothermal unsaturated case, the temperature is introduced into parameters in relation (2) of the Modified Cam-Clay yield surface which is defined within a general framework by:

$$q^2 - [M(T)]^2 \left(p + p_s(s;T)\right) \left(p_0(s;T) - p\right) = 0 \tag{6}$$

where $M(T)$ = slope of the critical state line depending on the temperature, $p_0(s;T)$ = preconsolidation stress depending on suction and temperature and $p_s(s;T)$ = parameter introducing the evolution of apparent cohesion with suction and temperature. This parameter defined by Gens [13] is given by:

$$p_s(s;T) = k\left(1 - f(T)\right)s \tag{7}$$

with $f(T)$ = function taking into account the variation of apparent cohesion due to the temperature and k = constant.

2.4   Temperature influence on stiffness parameters

In a general way, the tests carried on various soils do not show any temperature influence on the values of stiffness parameters. In the saturated case, Campanella and Mitchell [14] and Eriksson [10] undertook mechanical consolidation tests at various temperatures. Tidförs and Sällfors [11] carried out mechanical consolidation tests while varying the temperature during the loading on Bäckebol clay. In the unsaturated case, Saix et al. [7] studied the temperature influence on stiffness parameters evolution. All results show an independence of these stiffness parameters with respect to the temperature. Despax [15] also supposes an independence of the stiffness parameters with respect to the temperature in spite of the slight incidence observed. More recently, Graham et al. [5] take a slight increasing of elastic stiffness with an increase of temperature.

All the results presented in the literature do not allow proposing general conclusions. The variety of the studied soils, the experimental conditions, preparation and loading make necessary to lay out complementary tests to confirm or cancel the evoked tendencies.

3   Studied soil, experimental device and procedure

3.1   Studied soil

The soil used in this study is clayey silty sand, which particle size does not exceed 2 mm. A mineralogical analysis has shown that this soil is mainly constituted of calcite (50%) and quartz (40%). The clayey fraction represents only 10% and is essentially constituted of smectite, chlorite and phyllite [16]. This material, with a liquid limit $w_L = 25\%$ and a plasticity index $I_P = 10.5\%$, is similar to a clay of low plasticity according to the LCPC classification (Laboratoire Central des Ponts et Chaussées, France) and USCS (Unified Soil Classification System, U.S.A.). The reconstituted soil sample have, after the preparation, a dry density $\rho_d = 1500$ kg/m^3 (void ratio $e = 0.77$) and a water content $w = 18\%$ (degree of saturation $S_r = 63.5\%$). The interest of this soil type is to make it possible to obtain an unsaturated state for low values of suction.

3.2   Experimental device

The experimental characterization and the modelling of the soil behaviour under thermo-hydro-mechanical loading use essentially a thermal triaxial apparatus (Fig. 1).

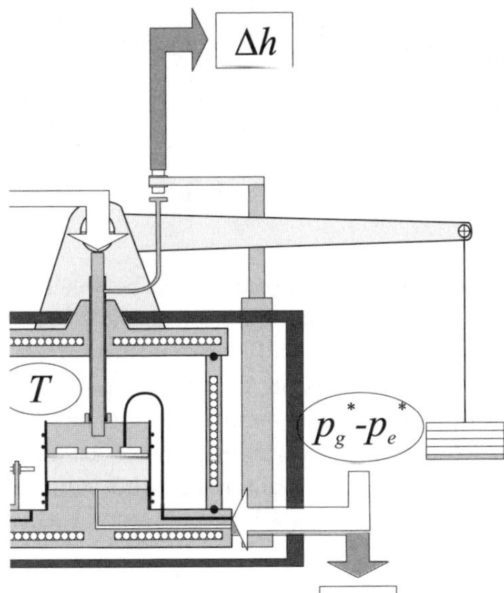

Fig. 1. Thermal triaxial apparatus

This experimental device allows reaching the volume soil sample changes of very slight slenderness (70 mm in diameter and 15 mm in height) during different thermo-hydro-mechanical loading paths. The fixed experimental variables are σ_x, $(\sigma_z - \sigma_x)$, p_g^*, p_e^* and T. The measured experimental variables are the height variation Δh, the radius variation Δr and the water volume variation ΔV_e going in or out of the sample. These measures, allow reaching the void ratio e and the water content variation Δw.

3.3 Experimental procedure

Experimental procedure (Fig. 2) consists in realizing thermo-mechanical consolidation tests with a constant suction s at different temperatures T. These tests follow loading paths [17] defined by a constant ratio $\eta = q/p$ between deviatoric stress q and net mean stress p.

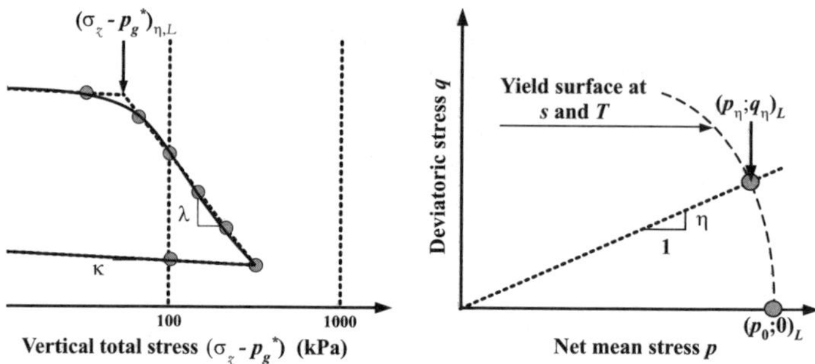

Fig. 2. Consolidation curve example. Yield surface in (p, q) plan

The values of η used in this study are included between 0 (isotropic loading) and $3/2$, the temperatures of different trials are 20, 30, 40, 50 and 60°C, and the constant suctions are equal to 0 kPa (saturated conditions) and 5 kPa (unsaturated conditions). Whatever the loading path η and the temperature T, the mechanical loading consists in imposing increments of vertical total stress $(\sigma_z - p_g^*)$ which takes the values between 10 and 300 kPa. From the consolidation curve obtained, we determine the value of preconsolidation vertical total stress $(\sigma_z - p_g^*)_{\eta,L}$ for the different ratios η and various temperatures T.

We deduce the preconsolidation stress couples $(p_\eta; q_\eta)_L$ using the following relation:

$$(p_\eta; q_\eta)_L = \left(\frac{3}{3 + 2\eta} \left(\sigma_z - p_g^* \right)_{\eta,L} ; \frac{3\eta}{3 + 2\eta} \left(\sigma_z - p_g^* \right)_{\eta,L} \right) \tag{8}$$

4 Results and discussions

Following the bibliographical synthesis presented in the section 1, tests were carried out on clayey silty sand, in saturated and unsaturated conditions. The latters allow to specify the temperature influence on the stiffness parameters at constant suction. Then we present the preconsolidation stress evolution according to temperature and propose an estimation of the yield surface in accord with the Alonso et al. model [1]. Lastly, we propose a physical interpretation of the experimental results.

4.1   Stiffness parameters evolution

The evolution of the elastic stiffness κ and plastic stiffness λ, in the saturated case (Fig. 3), shows an increase with the temperature for the plastic stiffness. In the unsaturated case, all measured stiffness in the tests carried out, show a slight incidence of temperature.

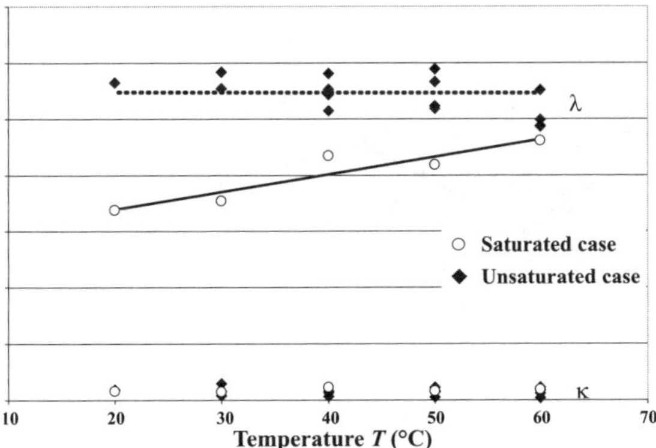

Fig. 3. Temperature influence on the stiffness parameters

4.2   Temperature influence on the preconsolidation stress

In the saturated case, the results show the independance of the preconsolidation stress $p_0(0;T)$ with an increase of temperature. In the unsaturated case, the results (Fig. 4) show, at constant suction $s = 5$ kPa, an increase in the preconsolidation stress $p_0(5;T)$ with the temperature T.

4.3   Yield surface evolution

From the obtained tests values and using the relation (8), we deduce the preconsolidation stress couples $(p_\eta; q_\eta)_L$. In the stress space (p, q), the thermo-extensible nature of the yield surface is clearly in evidence (Fig. 5).

From the relation (6), the values taken into account in the modelling of the yield surface are given in table 1. On the one hand, we note an increase in the preconsolidation net stress leads to the temperature influence on the

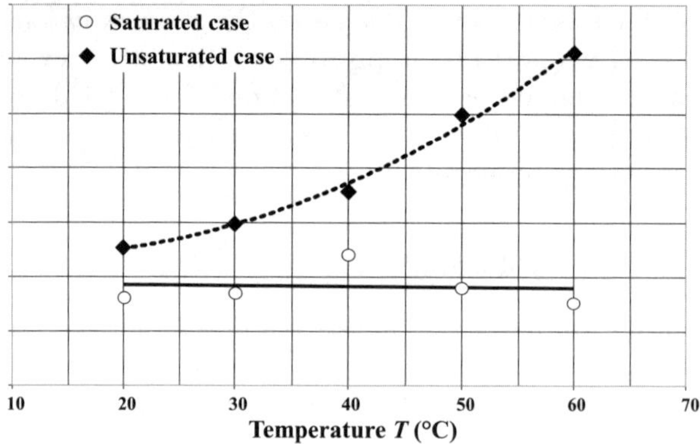

Fig. 4. Temperature influence on the preconsolidation stress $p_0(s;T)$

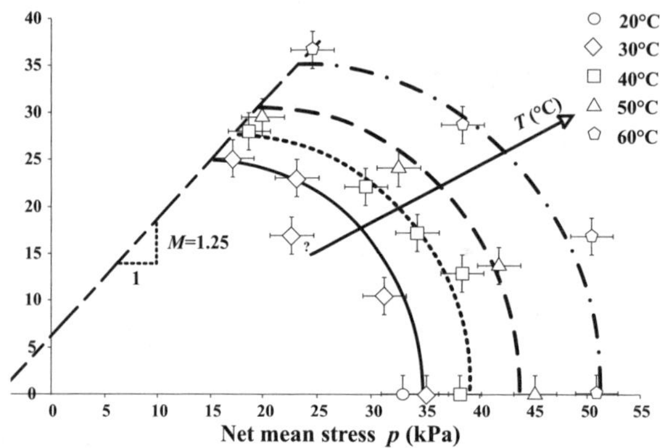

Fig. 5. Representation of the yield surface for different temperatures

preconsolidation vertical total stress (Fig. 4). On the other hand, it's turned out that the slope of the critical state line does not undergo notable variation whatever the test temperature. These results don't ensure to estimate the temperature influence on the apparent cohesion which has been supposed here constant whatever the temperature value.

Table 1. Parameters of Modified Cam-Clay model at constant suction $s = 5$ kPa

Temperature	30°C	40°C	50°C	60°C
M	1.25	1.25	1.25	1.25
$p_0(5;T)$	34.5	39	43.5	51

4.4   Results interpretation

In the unsaturated case, at ambient temperature, the undertaken experimental studies [1,18] show an increase of the preconsolidation net mean stress with an increase in suction. In the unsaturated clayey silty sand case for constant suction, our results also show an increase of the preconsolidation net mean stress with temperature [17]. According to the Jurin's law:

$$s = \frac{2\Gamma}{R} \; with \; \Gamma = \Gamma(T) \; and \; \frac{d\Gamma}{dT} < 0 \tag{9}$$

where Γ = surface air-water tension and R = meniscus radius, theses tendencies can have a physical interpretation. Indeed, at constant temperature, an increase in suction causes a decrease of the meniscus radius. In addition, at constant suction, an increase in temperature also causes a decrease in the meniscus radius; the surface air-water tension decreases according to temperature. Consequently, an increase in suction at constant temperature or an increase in temperature at constant suction leads to a decrease in the meniscus radius. This reduction of meniscus radius results in an increase of the cohesive forces between grains by capillarity, which causes to limit their rearrangements. Indeed, the arrangements are the main factors of soils behaviour irreversibility. This limitation corresponds to an increase of the preconsolidation net mean stress, and so to an extension of the yield surface.

5   Conclusion

Temperature influence (between $20°C$ and $60°C$) on the yield surface of clayey silty sand was studied. In the saturated case, we have noted the independance of the preconsolidation stress with an increase of temperature. In the unsaturated case, at constant suction ($s = 5$ kPa), the experimental results show the thermo-extensible nature of the yield surface for the studied soil. These results appear to be in opposition with those proposed by Romero [12] who observed, for Boom clay, a thermo-shrinkage nature of the yield surface whatever the suction value. This opposition lets to think that the clayey fraction of the studied soil is a determining factor in the yield surface evolution. Indeed, Romero [12] studied a clay (Boom clay) whereas the clayey fraction of clayey silty sand, studied here, is only 10%. Similar tests on a material, whose clayey fraction would be intermediate, should bear new informations on a possible transition between the thermo-shrinkage and thermo-extensible natures of the yield surface.

References

1. Alonso E.E., Gens A., Josa A. (1990) A constitutive model for partially saturated soils. Géotechnique **40**(3), 405-430.

2. Hueckel T., Baldi G. (1990) Thermoplasticity of saturated clays: Experimental constitutive study. J. Geotech. Engrg. **116**(12), 1778-1796.
3. Tanaka N., Graham J., Crilly T. (1997) Stress-strain behaviour of reconstituted illitic clay at different temperatures. Engrg. Geol. **47**, 339-350.
4. Cui Y.J., Sultan N., Delage P. (2000) A thermomechanical model for saturated clay. Can. Geotech. J. **37**(3), 607-620.
5. Graham J., Tanaka N., Crilly T., Alfaro M. (2001) Modified Cam-Clay modeling of temperature effects in clays. Can. Geotech. J. **38**(3), 608-621.
6. Laloui L., Cekeverac C., Vulliet L. (2002) Thermo-plasticity of clays: A simple constitutive approach. In: Vulliet et al. (Eds.) Environmental Geomechanics, EPFL Press, Lausanne, Suisse, 45-58.
7. Saix C., Devillers P., El Youssoufi M.S. (2000) Eléments de couplage thermomécanique dans la consolidation de sols non saturés. Can. Geotech. J. **37**, 308-317.
8. Jamin F., El Youssoufi M.S., Saix C. (2002) Temperature influence on the yield surface for unsaturated soils. In: Auriault et al. (Eds) Poromechanics II. 2nd International Conference on Poromechanics, Grenoble, France. Balkema, Lisse, The Netherlands, 461-466.
9. Roscoe K.H., Burland J.B. (1968) On the generalized stress-strain behaviour of wet clay. In: Heyman J. and Leckie F.A. (Eds.) Engineering Plasticity, Cambridge University Press, Cambridge, England, 535-609.
10. Eriksson L.G. (1989) Temperature effects on consolidation properties of sulphide clays. 12th International Conference on Soil Mechanics and Foundation Engineering, Rio de Janeiro, 2087-2090.
11. Tidförs M., Sällfors G. (1989) Temperature effect on preconsolidation pressure. Geotech. Testing J. **12**(1), 93-97.
12. Romero E. (1999) Characterisation and thermo-hydro-mechanical behavior of unsaturated Boom clay: An experimental study. PhD Thesis, Universitat Politecnica de Catalunya, Barcelona, Spain.
13. Gens A. (1995) Constitutive laws. In: Gens et al. (Eds) Modern Issue in Non-Saturated Soils, Springer-Verlag, Wien New York, 129-158.
14. Campanella R.G., Mitchell J.K. (1968) Influence of temperature variations on soil behavior. ASCE J. Soil Mech. and Found. Engrg. Div. **94**(SM3), 709-734.
15. Despax D. (1976) Influence de la température sur les propriétés mécaniques des argiles saturées. PhD Thesis, Université de Grenoble, France.
16. Devillers P., El Youssoufi M.S., Saix C. (1998) Transfert d'eau dans les sols faibles teneurs en eau. 16me Congrs Mondial de Science du Sol, Montpellier, France.
17. Jamin F. (2003) Contribution l'étude du transport de matire et de la rhéologie dans les sols non saturés différentes températures. PhD Thesis, Université Montpellier II, France.
18. Cui Y.J., Delage P. (1996) Yielding and plastic behaviour of an unsaturated compacted silt. Géotechnique **46**(2), 291-311.

Water Transport in Soil with Phase Change

Ali Chammari[1], Bétaboalé Naon[2], Fabien Cherblanc[1], Jean-Claude Bénet[1]

[1] Laboratoire de Mécanique et Génie Civil,
cc 048, Université Montpellier II,
F-34095 Montpellier Cedex 5, France
[2] IUT-Université Polytechnique de Bobo-Dioulasso
01 BP 1091 Bobo-Dioulasso - Bukina Faso

Abstract. The description of water transport in soil at low water content proposed is based on the idea that liquid/gas phase change occurs with vapor diffusion in the gas phase. The phase change modelisation introduces a phenomenological coefficient characterized through an experimental investigation done on smallscale soil samples. Then, numerical simulations are compared to experimental drying kinetics carried out on macroscopic soil samples. Good agreement is obtained, which strengthens the hypothesis of transport in the gas phase associated with a nonequilibrium liquid/gas phase change.

1 Introduction

At high water content, the liquid flow in soils is generally well described by the Darcy's Law. In non saturated cases, capillary potential, associated to existing curved meniscus, plays a key role in this Law. As long as the water phase is continuous or partially continuous, the capillary pressure is defined and can be used. However, as water content decreases, liquid/gas meniscus disappear, and water is adsorbed on the solid interface. So the concept of capillary potential becomes meaningless and must be substituted by the chemical potential. In this domain of low water content, governing mechanisms of transport phenomena are not well known. Nevertheless several experiments [1–3] have shown that the water transport in this range of moisture can be explained by the diffusion of vapor in the gaseous phase. In this case the vapor is provided by the phase change that should be the determining and limiting factor of water transport in soil. Advances in this field must improve knowledge in some geomechanical and geoenvironnemental applications as studies on thermo-hydro-mechanical coupling in soils or water ressources management in dry land or remediation processes in contaminated soils.

The first part recalls the mathematical model used to describe transport phenomena in soil taking into account the phase change. In order to show all coupled phenomena the model is based on the Linear Thermodynamics of Irreversible Processes ideas (linear TIP). To use the proposed phase change law , it must be validated and the coefficient introduced clearly identified. That's the aims of the second part. Finally, the third part is dedicated to drying experiments on a wet soil. The kinetic carried out on macroscopic clayey silty sand samples is compared with numerical simulation based on the theorical model. The comparison able to consider the relevance of the model and the taken assumptions. It also points out details that must be cleared in futur investigations.

2   Water Transport Model in a Non Saturated Soil

Theorical model is based on a phenomenological description of all transport mechanism involved in the liquid transport in a soil. The soil is represented by a porous medium in which are distingushed, a solid phase, a liquid phase and a gaseous one composed of two constituants, air and vapor. Elementary phenomena taken into account are air and vapor diffusion and liquid/gas phase change. Others phenomena as filtration and heat exchanges are already included into the theorical model but not considered to validate the phase change approach in this study. The state variables chosen are the water content w defined as the ratio between apparent mass densities of liquid and solid, the partial pressure of water vapor in the gas phase, P_v^*. Thus the mass balance for liquid phase and the the vapor partial pressure balance are given by [1,3]

$$\frac{\partial w}{\partial t} = \frac{LR_g}{\rho_s M_l} \ln \frac{P_v^*}{P_{veq}^*} \tag{1}$$

$$\frac{\partial P_v^*}{\partial t} = -\frac{LR_g^2 T}{\eta_g M_l^2} \ln \frac{P_v^*}{P_{veq}^*} - \frac{D_v}{\eta_g} P_{v,kk}^* \tag{2}$$

where, in one hand, $L \left[kg.K.s.m^{-5}\right]$ is the phenomenological coefficient for liquid/gas phase change , $R_g = 8.314 \left[J.mol^{-1}.K^{-1}\right]$ is the perfect gas constant, $\rho_s = 1500 \left[kg.m^{-3}\right]$ is the apparent mass density of the solid phase and $M_l = 0.018 \left[kg.mol^{-1}\right]$ is the molar mass of water. In the other hand, η_g is the volumetric fraction of the gaseous phase and D_v is the diffusion coefficient of vapor into the porous medium. At last the vapor pressure at equilibrium, $P_{véq}^*$ is given by

$$P_{veq}^* = a(w)P_{vs}(T) \tag{3}$$

where $a(w)$ is water activity. At low water content it can be distinguished two ranges of w. Pendular and hygroscopic domains [4] whom boundaries are closely tied to intrinsic soil characteristics and liquid nature. The upper limit of the pendular domain is about 13% [5] and it is founded carrying out the cappilary retention curve in water in this soil. The hygroscopic domain extends from 0% to 6% [6]. This is deduced from the desorption isotherm curve of water of the soil. And from that the activity is calculated at the considered temperature as

$$
\begin{cases}
w < 6\% \; a(w) = 831847.520 \; w^4 - 92668.702 \; w^3 + \\
\qquad\qquad\qquad\qquad +2795.628 \; w^2 + 8.207 \; w \\
w \geq 6\% \; a(w) = 1
\end{cases}
\tag{4}
$$

$P_{vs}(T)$ is the saturation vapor pressure of water given by Riddick [7] as

$$
P_{vs}(T) = 98.1 \times 10\text{^}(5.978 - \frac{2225}{T + 273.15})
\tag{5}
$$

Finally the diffusion coefficient of water in the air, D_{va}, has been determined by DeVries [8] and its value is $D_{va} = 2.57 \times 10^{-5} m^2.s^{-1}$. To express the diffusion coefficient of water in the soil taking into account the tortuosity of the medium, D_{vs}, Penman [9] has proposed the following relation is used

$$
D_{vs} = 0.66 \; \eta_g \; D_{va}
\tag{6}
$$

3   Experimental Study of the Phase Change

The study focuses on transport of water in soil at low water content. Therefore it is limited to the water content range where the liquid phase is discontinuous. It corresponds to pendular and hygroscopic domains. As sad in the used clayey silty sand, the pendular domain extends from 6% to 13%. Phase change appears in equations 1 and 2 through the coefficient L. It depends on water content, temperature, air pressure and the soil characteristics. This coefficient is introduced by the phase change phenomenological relation included in the theorical model. The scope of the experimental study is to verify the validity of this relation and to determine the coefficient L [6,10]. Investigations are done on centimetric-scale samples of clayey silty sand, at different temperature. The principle of the experiments is to impose a thermodynamic imbalance between liquid and vapor phases in the interstices of the porous skeleton. A great depression in samples genenerates the extraction of almost the totality of the gaseous phase. This phase composed of air and vapor is replaced by dry air. From this point temperature T, and especially the total pressure of the gaseous phase P_g^* are measured. The obtained kinetics show that the system return to an equilibrium sate. This progression of variables is considered to be dued to phase change. For the set of experiments at $T = 30°C$, the results are presented in the Fig.1a for linear domain, i.e. near equilibrium. It shows the values of the experimental phase change coefficient of water in the used soil versus liquid content. The coefficient L tends to 0 for low ($\approx 0\%$) and high ($\approx 13\%$) water content, and shows a maximum for intermediate water content, interpreted as the transition between hygroscopic and pendular domains. Nevertheless, in practical applications, phase change occurs in conditions that are outside of the linear domain. In these situations,

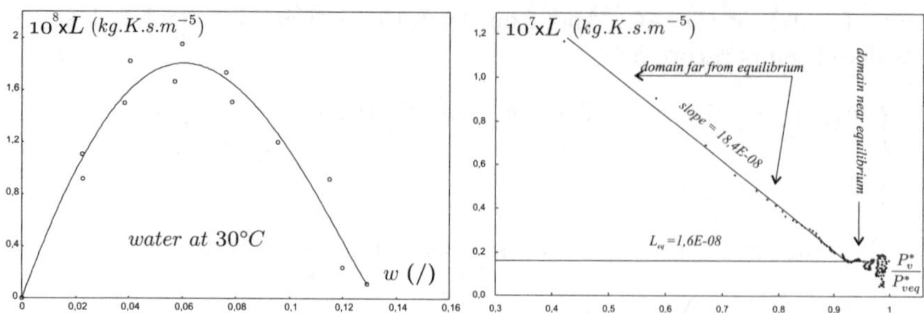

Fig. 1. Phenomenological phase change coefficient for water ($T = 30°C$). a) L near equilibrium, b) L far from equilibrium

the coefficient L increases significantly as the partial pressure of water vapor P_v^* decreases. Thus, a first interpretation of experimental results far from equilibrium [11] is proposed in terms of a linear dependence of coefficient L on the partial pressure of water vapor P_v^* as shown in Fig.1b. The analysis of this curve allow to distinguish two regions. In the first, where $P_v^*/P_{veq}^* \geq 0.93$, the phase change coefficient named L_{eq} is calculated using the results data showed in Fig.1a. From that curve the coefficient L_{eq} is approximated by the polynomial expression

$$L(w, 30°C) = 10^{-4} \left[1.407 \; w^4 - 0.249 \; w^3 - 0.035 \; w^2 + 0.006 \; w \right] \tag{7}$$

In the second part, that corresponds to the domain far from equilibrium, where $P_v^*/P_{veq}^* \leq 0.93$, the phase change coefficient increases in linear way as the ratio P_v^*/P_{veq}^* decreases. The adopted expression in this range is the following

$$\text{if } P_v^*/P_{veq}^* \leq 0.93 \text{ then } L = -18.4 \; 10^{-8} \left(P_v^*/P_{veq}^* \right) + \left(L_{eq} + 17.1 \; 10^{-8} \right) \tag{8}$$

Those results complete the theorical model that can be exploited numerically to simulate applied problems where water transport with phase change is orbserved. It's the scope of the next section. Drying experiments are done on macroscopic samples of wet soil. Then, numerical simulation based on the model are compared to experimental results.

4   Self Drying of a Soil at Low Water Content

4.1   Experimental Investigations

Cylindrical samples are obtained by compacting wet soil. The soil is the same one as in the study of phase change (sec.3), i.e. a clayey silty sand whom real mass density is $\rho_s^* = 2650 kg \cdot m^{-3}$. The mixture is compacted in a PVC pipe in order to reach an apparent mass density equal to $\rho_s = 1500 \; kg \; m^{-3}$. The height of the samples is $L = 10cm$ and their diameter is $D = 8.14cm$.

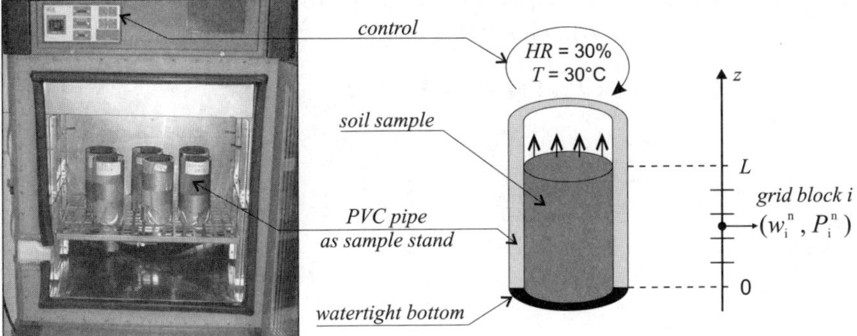

Fig. 2. Drying of a wet soil sample ($w_0 = 8\% - T = 30°C - HR = 30\%$).

Samples are placed in a cooled incubator at contolled temperature and humidity as illustrated in Fig.2a. They were regularly weighted outside the incubator. Mass measures permitted to calculate the averaged water content in the samples and to plot drying kinetics. For this study nine kinetics were achevied and a good reproducibility is obtained.

4.2   Numerical Simulation

Former experiments were simulated using a one-dimensional discretized mesh. The unknowns (w et P_v^*) were located to the center of grid blocks. An implicit scheme is used with constant time step dt in order to insure stability. For grid block i at time step n, the unknowns are named w_i^n and P_i^n (Fig.2b). So equations (1) and (2) becomes

$$\frac{w_i^{n+1} - w_i^n}{dt} - \frac{J\left(w_i^{n+1}, P_i^{n+1}\right)}{\rho_s} = 0 \tag{9}$$

$$\frac{P_i^{n+1} - P_i^n}{dt} + J\left(w_i^{n+1}, P_i^{n+1}\right)\frac{R_g T}{\eta_g M_l} + D_{vs}\frac{P_{i+1}^{n+1} - 2P_i^{n+1} + P_{i-1}^{n+1}}{(dz)^2} = 0 \tag{10}$$

where η_g is the porosity and J is the term of phase change and it is equal to

$$J = -L\frac{R_g}{M_l}\ln\frac{P_v^*}{P_{véq}^*} \tag{11}$$

both depend on unknowns and ware estimated at step $n+1$. A Newton method is chosen to solve the system because of the non-linearaty of the equations. According to the experiments, the initial conditions are at $t = 0$, $w_0 = 8\%$ and partial vapor pressure is deduced from relation 3 and equal to

$$P_{v_0}^* = P_{vs}\left(30°C\right) = 4.267\ kPa \text{ because } a\left(w_0 = 8\%\right) = 1 \tag{12}$$

Furthermore, boundary conditions fixed by the experiments are a non-flow condition at lower end of samples ($z = 0$) and a relative humidity $HR = 30\%$ on the upper face ($z = L$). This latter condition corresponds to:

$$P_v^* (x = L) = HR \times P_{vs} (30°C) = 1.280 kPa \tag{13}$$

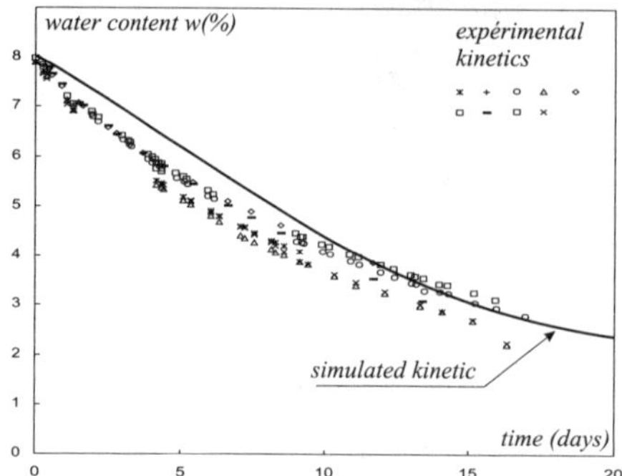

Fig. 3. Experimental and numerical drying kinetics.

4.3   Comparaison Between Experiments and Simulation

The numerical prediction based on the phenomenological model is compared to experimental drying kinetics on Fig.3. A good agreement is obtained. At the end of drying all curves converge to a water content close to $w = 2\%$ corresponding to the equilibrium value given by the desorption isotherm. This also confirm a good description of water transport in the hygroscopic domain. Moreover experimental drying rates seem to be higher than numerical one when $6\% < w < 8\%$. This can be due to a non homogeneous water distribution in soil. So, at the begining of the experiments some parts of the samples may contain water that still move by filtration while the liquid phase is assumed to be immobile in the theorical description. This lead to discrepancies observed for water content $w > 6\%$, i.e., outside of the hygroscopic domain. In this case the differences observed should come from the model that doesn't already take account of filtration in that range of water content.

5   Conclusion

The comparison between theoretical prediction and experimental results shows the relevance of this approach to describe water transport at very low water content. It emphasizes the significance of focusing on phase change far from equilibrium, even if further investigations are required. The study of the phase change phenomena have to be extended to a wider range of temperature and other soils with different characteristics. It's also necessary to modify some experimental setting in order to improve the control of the measures when the system is moving to equilibrium thus to better describe the phenomena in the non linear domain. Moreover the numerical model as used in this study could already treats applied problems involving phase change. First we put the interest on thermo-hydro-mechanical coupling in soils and water management in dry land but the theorical work presented could also be extrapolated to describe the natural attenuation of volatile contaminants such as light hydrocarbons trapped in the porous skeleton after conventional pumping. The phase change is thermally activated and increased by dry air injection (venting technique) bringing up some interesting perspectives to this work.

References

1. Bénet J.-C., Jouanna P., Phenomenological relation of phase change of water in a porous media: experimental verification and measurement of the phenomenological coefficient, International Journal of Heat and Mass Transfer, 25-11 (1982) 1747-1754.
2. Fras G., Macroscopisation des transferts en milieux dispersés polyphasiques - Application à l'étude de l'interface entre un milieu poreux et une atmosphère séchante, Thèse de doctorat d'Etat, Université Montpellier 2 (1989).
3. Ruiz T., Bénet J.-C., Phase change in a heterogeneous medium: comparison between the vaporization of water and heptane in an unsaturated soil at two temperatures, Transport in Porous Media, 44 (2001) 337-353.
4. Bear J., Buchlin J.M., Modelling and applications of transport phenomena in porous media, Kluwer Academic Publishers, (1989).
5. Saix C., Devillers P., El Youssoufi M.S., Eléments de couplage thermomécanique dans la consolidation de sols non saturés, Canadian Geotechnical Journal, 37 (2000) 308-317.
6. Ruiz T., Eléments de modélisation et contribution expérimentale à l'étude du transport réactif dans un sol non saturé, Thèse de doctorat, Université Montpellier 2 (1998).
7. Riddick J.A., Bunger W.B., Sakano T.K., Organic solvents, physical properties and methods of solidification, 4th Edition, Vol.II, John Willey and Sons (1986).
8. DeVries D.A., Kruger A.J., On the value of the diffusion coefficient of water vapor in air, Proceddings of The International CNRS Conference, Transports phenomena with phase change in porous media or colloids, CNRS eds., Paris, (1966) 18-20.
9. Mignard E., Contribution à l'étude des transferts en milieux poreux par la thermodynamique du non équilibre - cas de la dissolution des gaz dans les sols, Thèse de doctorat, INSA Toulouse (1984).
10. Chammari A., Transfert gazeux dans les sols avec changement de phase - Application à quelques aspects de géotechnique environnementale, Thèse de doctorat, Université Montpellier 2 (2002).

11. Chammari A., Naon B., Cherblanc F., and Benet J.C., Transfert d'eau en sol aride avec changement de phase, C.R. Mecanique 331 (2003) 759-765.

Tunnels in Saturated Elasto-plastic Soils: Three-dimensional Validation of a Plane Simulation Procedure

Carlo Callari, Stefano Casini

Dipartimento di Ingegneria Civile
Università di Roma "Tor Vergata",
via del Politecnico, 1
00133 Roma, Italy

Abstract. This paper presents a study of tunneling in saturated poro-elastoplastic soils by means of two and three-dimensional coupled numerical analyses. To perform 2D simulations of the excavation stage, we employ an approach proposed in a previous paper as an extension to the saturated case of the so-called "convergence-confinement" method. To validate the plane procedure, the comparison with results of several 3D simulations of the excavation of shallow pervious tunnels is considered. By means of the hydro-mechanically coupled formulation employed in these comparisons, we investigate the face-advance rate influence on soil response to excavation. The proposed 2D method shows to be able to reproduce with satisfying approximation the effects of tunnel-face advancement on in-plane components of displacements and water flow.

1 Introduction

Near the excavation face, tunnelling is a fully three-dimensional problem. However, according to recent state-of-the-art reports [13,14,16], tunnels are still frequently analyzed considering two-dimensional schemes, simulating the excavation-face advancement by means of simplified approaches, such as the

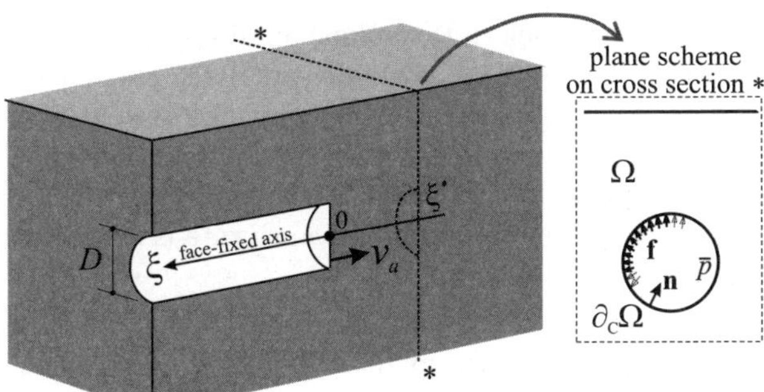

Fig. 1. The excavation of a pervious tunnel advancing with rate v_a in a saturated ground: schematic representation of the plane simulation procedure

so-called "convergence-confinement" method [19]. Even the traditional one-dimensional approach to ground-support interaction, assuming the tunnel problem as axially symmetric, is still considered as an effective tool for preliminary support design, as showed by some recent research works [1].

Therefore, it could be inferred that the increasing possibility to perform 3D numerical analyses is not leading to a total substitution of simplified and effective 2D calculations. This is probably due to the possibility of straightforward interpretation offered by plane results and by the still high computational cost of 3D simulations. The latter motivation is particularly strong when a number of extremely variable situations is to be investigated, as it is the case for tunnels in the urban environment interacting with pre-existing structures [15]. Furthermore, the consideration of many different configurations of the problem can be in general required by typical uncertainties on soil profile, in-situ stress state and physico-mechanical properties. In all these cases, at the same computational cost, it can be preferred to perform several plane analyses instead of few three-dimensional calculations.

The above considerations are particularly relevant for tunnels driven under the water table, where a crucial role is normally played by the interaction between the mechanical response of the solid skeleton and the interstitial fluid flow. In fact, the consideration of such a coupling in three-dimensional simulations of tunneling is not common at all, as demonstrated by the small number of fully coupled 3D analyses available in the literature [11,12,17,18,20].

As a matter of fact, also plane analyses of tunneling in saturated media are rarely performed by coupled numerical methods (less than 20% of the 135 case studies considered in [16]). On the contrary, the problem is typically treated by an uncoupled "limit" approach, assuming a "drained" or "undrained" response of the saturated ground to tunnel excavation. These common limit approaches are not effective in that cases where consolidation and excavation times are comparable. Furthermore, a hydro-mechanically coupled approach can simulate face-advance rate effects, which are frequently observed in saturated ground response to tunneling and that cannot be explained just as a consequence of the rate-dependent response of the solid skeleton.

However, it can be observed that even in that plane analyses of tunneling where a coupled approach is considered, this is normally restricted to the *post-excavation* response ("long-term" behaviour), i.e. the coupled formulation is applied assuming a large distance between the studied section and the tunnel face. The previous *excavation* phase ("short-term" behaviour), in fact, is typically assumed as instantaneous (i.e. an infinite advance rate is assumed), thus allowing for an undrained calculation [6,7,10].

As an accessory tool for the numerical study of strain localization presented in [3], we proposed a method for the plane simulation of tunneling in a coupled saturated ground with *finite* face-advance rate. In fact, to our knowledge, a similar procedure was not available in the literature and we formulated it as an extension of the aforementioned convergence-confinement method, accounting for the time-evolution of both mechanical unloading and drainage conditions induced by advancing excavation face.

In the present paper, results of several three-dimensional coupled analyses of shallow tunnelling in a poro-elastoplastic soil are employed to validate such a two-dimensional approach. An outline of the rest of the paper is then as follows. We briefly recall in Sect. 2 the aforementioned procedure. In Sects. 3 and 4, essential information is provided about the poro-plastic model and the finite-element formulation employed in numerical simulations. Three-dimensional results are analyzed with a view to the 2D procedure in Sect. 5.1, thus leading to a preliminary setting of the latter. The proposed plane approach is then validated in Sect. 5.2 by means of comparisons between 2D and 3D results. Finally, concluding remarks are reported in Sect. 6.

2   Procedure for plane simulation of tunneling in saturated ground

In a ground section Ω normal to a *pervious* tunnel axis (Fig. 1), three-dimensional effects of excavation under the water table can be approximated by reducing the so-called "excavation forces" [19]

$$\mathbf{f}(t) := [1 - \lambda(t)]\mathbf{f}_0 \quad \text{on} \quad \partial_c\Omega \tag{1}$$

and the "excavation pore pressures" [3]

$$\bar{p}(t) := [1 - \alpha(t)]\, p_0|_{\partial_c\Omega} \quad \text{on} \quad \partial_c\Omega \tag{2}$$

applied to the cavity boundary $\partial_c\Omega$ with unit normal $\mathbf{n}$. In (1) and (2), we denote by $\mathbf{f}_0 := (\boldsymbol{\sigma}_0\mathbf{n})|_{\partial_c\Omega}$ the surface actions equilibrating the pre-excavation state of stress $\boldsymbol{\sigma}_0$ and gravity body loads, and by p_0 the pre-excavation pore pressure field, respectively. Coefficients $\lambda(t)$ and $\alpha(t)$ describe the evolutions of excavation forces and drainage conditions, respectively, from initial time instants t_0 and t_{p0} to instants t_u and t_{pu} corresponding to total unloading of excavation forces and final pore pressures at the cavity boundary:

$$\lambda(t) = \begin{cases} 0 \text{ for } t \leq t_0 \\ 1 \text{ for } t > t_u \end{cases} \quad \text{and} \quad \alpha(t) = \begin{cases} 0 \text{ for } t \leq t_{p0} \\ 1 \text{ for } t > t_{pu} \end{cases} \tag{3}$$

The imposed excavation forces (1) and excavation pore pressures (2) are considered as that "equivalent" *fictitious* forces and drainage conditions able to reproduce in the two-dimensional scheme, the same in-plane components of displacement and flow fields, respectively, characterizing the three-dimensional excavation. As a consequence, the dependence of the unloading coefficients on time t is to be understood in the following sense:

$$\lambda(t) = \hat{\lambda}(\xi^*(t)) \qquad \alpha(t) = \hat{\alpha}(\xi^*(t)) \tag{4}$$

where $\xi^*(t)$ is the position of the studied section relative to the advancing excavation face (Fig. 1). Therefore, the "excavation" time instants t_0, t_{p0} (initial) and t_u, t_{pu} (final) correspond to the first and to the last appearance, respectively, of significant face-advance effects on displacement and flow field observed at the considered ground section. Such a definition can then be formulated in terms of time derivative $\dot{\xi}^*(t)$:

$$\begin{array}{ll} \text{in the 3D problem,} & \begin{cases} \text{for } t < t_0 \ \text{ or } t \geq t_u \ \Rightarrow \text{ no effect of } \dot{\xi}^*(t) \text{ on displ.} \\ \text{for } t < t_{p0} \text{ or } t \geq t_{pu} \Rightarrow \text{ no effect of } \dot{\xi}^*(t) \text{ on flow} \end{cases} \\ \text{at } \xi = \xi^*(t): & \end{array} \quad (5)$$

Denoting by $v_a(t)$ the magnitude of face-advancement rate (Fig. 1), it is $\dot{\xi}^*(t) = v_a(t)$. By integration, the following relations between the average face-advance rate $\bar{v}_a$ and the excavation time intervals are obtained:

$$T_u := t_u - t_0 = \frac{\xi_u^* - \xi_0^*}{\bar{v}_a} \quad \text{and} \quad T_p := t_{pu} - t_{p0} = \frac{\xi_{pu}^* - \xi_{p0}^*}{\bar{v}_a} \quad (6)$$

where:

$$\xi_0^* := \xi^*(t_0) \qquad \xi_u^* := \xi^*(t_u) \qquad \xi_{p0}^* := \xi^*(t_{p0}) \qquad \xi_{pu}^* := \xi^*(t_{pu}) \quad (7)$$

In summary, the application of the proposed two-dimensional simulation procedure requires the setting of excavation time instants t_0, t_u, t_{p0}, $t_p u$ and of the unloading laws $\lambda(t)$, $\alpha(t)$.

Denoting by D the characteristic size of a tunnel face advancing in a *dry* ground, the setting

$$\xi_0^* = -D \qquad \xi_u^* = 2D \quad (8)$$

is commonly assumed for the "extreme" positions of the considered section relative to tunnel face. Such an assumption is based on numerical results obtained for the case of an axial-symmetric cavity in an elastic ground [19].

For the case of tunneling under the water table, the relative positions defined by (7) in terms of the excavation time instants characterized in (5) are evaluated in Sect. 5.1, by means of three-dimensional numerical analyses of tunneling. In Sect. 5.2, the comparison between three-dimensional and plane numerical results is employed to evaluate the effectiveness of simple expressions for the unloading laws $\lambda(t)$, $\alpha(t)$.

Numerical simulations are performed by means of a finite element formulation of the coupled poro-elastoplastic theory described in the following.

3   The poro-elastoplastic model

In this Section, we recall the main equations of the considered poro-elastoplastic theory [2,9]. Denoting by $\mathbf{q}_w$ the fluid flow field in the saturated porous solid Ω, the fluid mass balance is expressed in the form:

$$\dot{M} = -\operatorname{div} \mathbf{q}_w \quad (9)$$

for the fluid content M, i.e. the variation of the fluid mass (referred to the initial value) per unit initial volume of the porous solid. The fluid flow is defined in terms of the pore pressure field p by means of the Darcy's law:

$$\mathbf{q}_w = -\rho_w \mathbf{k}(\nabla p - \rho_w \mathbf{g}) \tag{10}$$

for the density of fluid ρ_w, the permeability tensor of the porous solid $\mathbf{k}$ and the gravity acceleration vector $\mathbf{g}$. In this paper, tensile stresses $\boldsymbol{\sigma}$ and compressive pore pressures p are assumed as positive.

The considered poro-elastoplastic model is based on the additive decomposition of strains and fluid content:

$$\boldsymbol{\varepsilon} = \boldsymbol{\varepsilon}^e + \boldsymbol{\varepsilon}^p \qquad\qquad M = M^e + M^p \tag{11}$$

Denoting by $\mathbb{C}_{sk}$ the solid-skeleton ("drained") elastic tensor, the rate form of the constitutive relations is

$$\dot{\boldsymbol{\sigma}} = \dot{\boldsymbol{\sigma}}' - b\,\dot{p}\,\mathbf{1} \qquad \text{with} \qquad \dot{\boldsymbol{\sigma}}' = \mathbb{C}_{sk}\dot{\boldsymbol{\varepsilon}}^e \tag{12a}$$

$$\dot{p} = \frac{Q}{\rho_{wo}}\left[\dot{M}^e - \rho_{wo}b\,\dot{\boldsymbol{\varepsilon}}^e : \mathbf{1}\right] \tag{12b}$$

for the Biot's effective stress tensor $\boldsymbol{\sigma}'$, the Biot's modulus Q and the Biot's coefficient b. We assume an isotropic hardening characterized by a single scalar strain-like internal variable h with conjugated stress-like variable q. In the non-associated case considered herein, the evolution of the plastic internal variables is given by rate equations:

$$\dot{\boldsymbol{\varepsilon}}^p = \gamma\,\partial_{\boldsymbol{\sigma}} g_{sk} \qquad \dot{h} = \gamma\,\partial_q f_{sk} \qquad \dot{M}^p = \rho_{wo}\,b\,\dot{\varepsilon}_v^p \tag{13}$$

where the yield function $f_{sk}(\boldsymbol{\sigma}', q)$ and the plastic potential $g_{sk}(\boldsymbol{\sigma}')$ are defined in terms of effective stresses and $\varepsilon_v^p := \boldsymbol{\varepsilon}^p : \mathbf{1}$ is the volumetric plastic strain. The plastic multiplier γ is determined by the usual Kuhn-Tucker loading/unloading and consistency conditions.

In numerical simulations presented in Sect. 5, we adopt a constitutive model for the solid skeleton consisting of an isotropic linear elastic law and of the following Drucker-Prager yield function

$$f_{sk}\left(\boldsymbol{\sigma}', h\right) = \|\mathbf{s}\| + \beta\frac{1}{3}\,\boldsymbol{\sigma}' : \mathbf{1} + \sqrt{\frac{2}{3}}\,q(h) \tag{14}$$

and plastic potential

$$g_{sk}\left(\boldsymbol{\sigma}'\right) = \|\mathbf{s}\| + b_\psi\frac{1}{3}\,\boldsymbol{\sigma}' : \mathbf{1} + D \tag{15}$$

for the deviatoric part of stress tensor $\mathbf{s} := \mathrm{dev}(\boldsymbol{\sigma}')$, the pressure coefficient β, the dilatancy coefficient b_ψ and the dummy constant D. In (14), the linear softening law

$$q(h) = -\sigma_y - \mathcal{H}h \tag{16}$$

is considered, for the initial yield stress σ_y and a constant hardening modulus $\mathcal{H}$. An isotropic permeability $\mathbf{k} = k\,\mathbf{1}$ is assumed in (10).

4   Finite element formulation

In the following we briefly describe the finite element formulation of the coupled poro-elastoplastic model recalled in the previous Section. In a point $\mathbf{x}$ of a generic finite element Ω_e, the fields of displacements and pore pressures are approximated by the isoparametric interpolations

$$\mathbf{u}_e(\mathbf{x}) = \mathbf{N}_e(\mathbf{x})\,\mathbf{d}_e \qquad \text{and} \qquad p_e(\mathbf{x}) = \mathbf{N}_e^p(\mathbf{x})\,\mathbf{p}_e\ , \tag{17}$$

respectively, for the the nodal displacements $\mathbf{d}_e$, the nodal pore pressures $\mathbf{p}_e$ and the shape functions $\mathbf{N}_e(\mathbf{x})$, $\mathbf{N}_e^p(\mathbf{x})$. The approximations adopted for the fields of strain and fluid flow are

$$\boldsymbol{\varepsilon}_e = \bar{\mathbf{B}}_e \mathbf{d}_e \qquad \text{and} \qquad \mathbf{q}_w = -\rho_{wo}\mathbf{k}\left(\mathbf{B}_e^p \mathbf{p}_e - \rho_{wo}\mathbf{g}\right) \tag{18}$$

for an assumed strain operator $\bar{\mathbf{B}}_e$ and the standard gradient operator $\mathbf{B}_e^p := \nabla\mathbf{N}_e^p(\mathbf{x})$.

The finite element discretizations of the weak formulations of equilibrium and fluid mass balance lead to the equations governing the mechanical and the hydraulic problems, respectively. This non linear system is written in residual form and solved in the generic time step $[t_n, t_{n+1}]$ by the Newton-Raphson method, with the equilibrium equation calculated at instant t_{n+1} and the rate equation of mass balance approximated by a backward-Euler scheme. Denoting by $\Delta t = t_{n+1} - t_n$ the time interval and omitting index $n + 1$ to simplify the notation, the mechanical and the hydraulic equations can be written as

$$\begin{cases} \mathbf{R}_m = \mathbf{F}^{ext} - \overset{n_{elem}}{\underset{e=1}{\mathbf{A}}}\left[\int_{\Omega_e} \bar{\mathbf{B}}_e^T \boldsymbol{\sigma}\,d\Omega\right] = \mathbf{0} \\[2em] \mathbf{R}_f = \mathbf{W}^{ext} - \overset{n_{elem}}{\underset{e=1}{\mathbf{A}}}\left[\int_{\Omega_e} \mathbf{N}_e^{pT}\frac{M - M_n}{\Delta t}\,d\Omega + \int_{\Omega_e} \mathbf{B}_e^{pT}\mathbf{q}_w\,d\Omega\right] = \mathbf{0} \end{cases} \tag{19}$$

respectively, for the vectors of imposed external forces $\mathbf{F}^{ext}$ and flows $\mathbf{W}^{ext}$. The assembly of the contributions of all the n_{elem} finite elements is denoted by symbol $\mathbf{A}$. The total stresses $\boldsymbol{\sigma}$ are given by relation $(12\mathrm{a})_1$ in terms of the effective stresses $\boldsymbol{\sigma}'$ integrated through a return mapping algorithm.

At iteration $(k+1)$, the increments $\Delta(\cdot) = (\cdot)_{n+1}^{(k+1)} - (\cdot)_{n+1}^{(k)}$ are then obtained by the solution of the following linearized form of governing equations (19):

$$\begin{cases} \mathbf{R}_m^{(k)} = \overset{n_{elem}}{\underset{e=1}{\mathbf{A}}}\left(\mathbf{K}_e\,\Delta\mathbf{d}_e - \frac{1}{\rho_{wo}}\mathbf{Q}_e\,\Delta\mathbf{p}_e\right) \\[2em] \mathbf{R}_f^{(k)} = \overset{n_{elem}}{\underset{e=1}{\mathbf{A}}}\left[\frac{1}{\Delta t}\mathbf{Q}_e^T\,\Delta\mathbf{d}_e + \left(\mathbf{H}_e + \frac{1}{\Delta t}\mathbf{S}_e\right)\Delta\mathbf{p}_e\right] \end{cases} \tag{20}$$

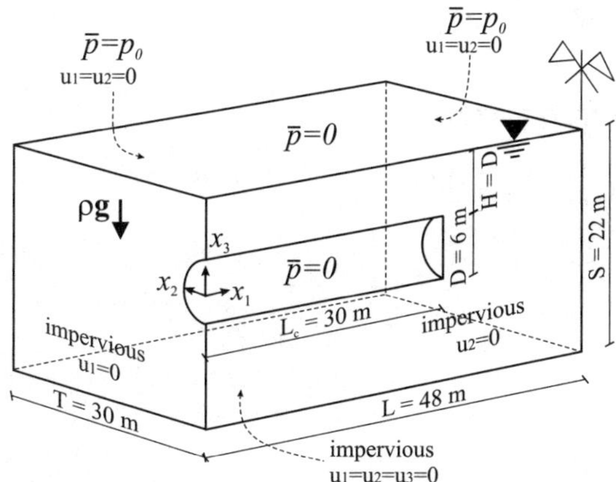

Fig. 2. Three-dimensional simulation of the excavation of a shallow pervious tunnel in a saturated ground: configuration of the problem with assumed boundary conditions on pore pressures ($\bar{p}$) and displacement components (u_1, u_2, u_3)

where all the matrices are evaluated at iteration (k) (indices omitted). In particular, the element matrices of stiffness, coupling, permeability and storage are given by

$$\mathbf{K}_e = \int_{\Omega_e} \bar{\mathbf{B}}_e^T \mathbb{C}_{sk} \, \bar{\mathbf{B}}_e \, d\Omega \qquad \mathbf{Q}_e = \int_{\Omega_e} \rho_{wo} b \, \bar{\mathbf{B}}_e^T \mathbf{1} \, \mathbf{N}_e^p \, d\Omega$$

$$\mathbf{H}_e = \int_{\Omega_e} \rho_{wo} \, \mathbf{B}_e^{pT} \mathbf{k} \, \mathbf{B}_e^p \, d\Omega \qquad \mathbf{S}_e = \int_{\Omega_e} \frac{\rho_{wo}}{Q} \, \mathbf{N}_e^{pT} \mathbf{N}_e^p \, d\Omega \tag{21}$$

respectively. The finite element formulation has been implemented in the general code FEAP [22].

5 Numerical simulations of tunneling in a saturated ground

We consider the full-face excavation of a shallow circular tunnel in a homogeneous soil layer with water table at the ground surface elevation. For the three-dimensional case, the configuration of the problem with applied loads and assumed boundary conditions is illustrated in Fig. 2. In the pervious and unsupported cavity, atmospheric pressure is assumed. The bedrock under the soil layer is considered as rigid and impervious. The ratio $T/D = 5$ is set between the width T of the considered domain cross section and the tunnel diameter D (Fig. 2). Numerical simulations are performed by discretizing the (pre-excavation) three-dimensional domain in 4176 eight-noded "brick" elements (i.e. 116 elements in the cross section and 36 divisions along the longitudinal axis x_1, Fig. 4). A mixed formulation with constant interpolation of spherical part of stress and linear interpolation of displacements is considered

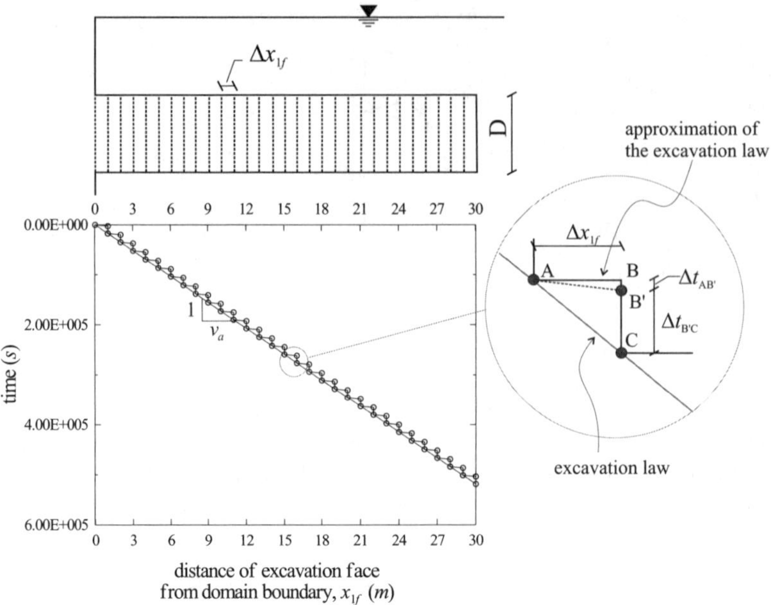

Fig. 3. Three-dimensional simulation of tunnel face advance with rate v_a

for the brick elements, in order to control locking effects near the undrained limit of the coupled response. Linear shape functions are employed also for the pore pressure field. Calculations are repeated for different values of tunnel diameter D (6.00, 8.00 and 10.00 m), keeping constant the ratio T/D as well as the number of finite elements. With this setting, negligible boundary effects on the cross-section response have been observed, consistently with the statistics reported in [16].

The excavation-face advance is simulated by progressive deactivation of 30 finite-element regions as thick as $D/6$. The deactivation procedure, illustrated in Fig. 3, is then the iteration of the following two-step sequence:

1. A practically instantaneous face advance is simulated by deactivation of a single region (segment AB': $\Delta x_{1f} > 0$ with $\Delta t_{AB'} \cong 0$). Free-drainage conditions (zero pore pressures) are then imposed at the cavity boundary, including the excavation face.
2. The tunnel-face position remains fixed (segment $B'C$: $\Delta x_{1f} = 0$) for the time interval $\Delta t_{B'C} = \Delta x_{1f}/v_a - \Delta t_{AB'} \cong \Delta x_{1f}/v_a$.

Therefore, a nearly undrained excavation is simulated in step 1. By means of preliminary calculations, we have estimated the minimum values of time step $\Delta t_{AB'}$ ensuring the absence of spatial oscillations of pore-pressure solution [21]. As expected, such a minimum time step depends on saturated solid parameters, on considered discretization and on dimension Δx_{1f} of the deactivated finite-element region.

The considered material parameters, reported in Table 1, can be realistically considered as characterizing a soil ($b = 1$) saturated with water (cf. Q and ρ_{wo} values). The assumed values of Drucker-Prager parameters β, σ_y and b_Ψ can be related to Mohr-Coulomb friction angle $\varphi = 18°$, drained cohesion $c = 30\ kPa$ and dilatancy angle $\psi = 9°$ by means of

$$\beta = \sqrt{\frac{2}{3}}\,\frac{6\sin\varphi}{3-\sin\varphi} \qquad \sigma_y = \frac{6\,c\cos\varphi}{3-\sin\varphi} \qquad b_\Psi = \sqrt{\frac{2}{3}}\,\frac{6\sin\psi}{3-\sin\psi} \tag{22}$$

respectively. Hydrostatic pore pressures and a "K_0-type" effective stress state are generated in the pre-excavation undeformed configuration. The initial horizontal stress ratio is $K_0 = 0.54$. The hydraulic permeability reported in Table 1 is calculated as $k_h = \gamma_w k$, for the unit weight of water $\gamma_w = 10\ kN/m^3$. The effects of tunneling on water table position are assumed as negligible.

Table 1. Material parameters considered in numerical simulations

saturated soil density	ρ	$2.0 \cdot 10^3$	kg/m^3
drained Young modulus	E_{sk}	20000	kPa
drained Poisson coefficient	ν_{sk}	0.35	
pressure coefficient	β	0.56	
initial yield stress	σ_y	63.41	kPa
hardening modulus	$\mathcal{H}$	200	kPa
dilatancy coefficient	b_Ψ	0.28	
Biot coefficient	b	1.0	
Biot modulus	Q	$3.33 \cdot 10^6$	kPa
permeability	k_h	$1.0 \cdot 10^{-7}$	m/s
fluid density	ρ_{wo}	$1.0 \cdot 10^3$	kg/m^3

With reference to the case $D = 6.0\ m$ and $v_a = 5.00\ m/day$, the calculated distributions of vertical displacements, pore pressures, piezometric heads as well as the deformed configuration of the mesh are reported in Fig. 4. However, we refer to [4] for a complete account of three-dimensional results, including the case of lining-supported tunnel, and their comparison with empirical relations and available monitoring data. In the present work, the consideration of 3D results is just finalized to the validation of the plane approach.

5.1   Analysis of 3D results with a view to the plane procedure

The plane approach is proposed for straightforward application to practical problems. Therefore, a main goal of present work is to assess if a sufficiently approximate estimate of "extreme" excavation face positions $(\xi_0^*; \xi_u^*)$ and $(\xi_{p0}^*; \xi_{pu}^*)$, defined by (5,7), can be based only on tunnel size D, as it is commonly assumed in the case of *dry* ground (cf. Eq. 8).

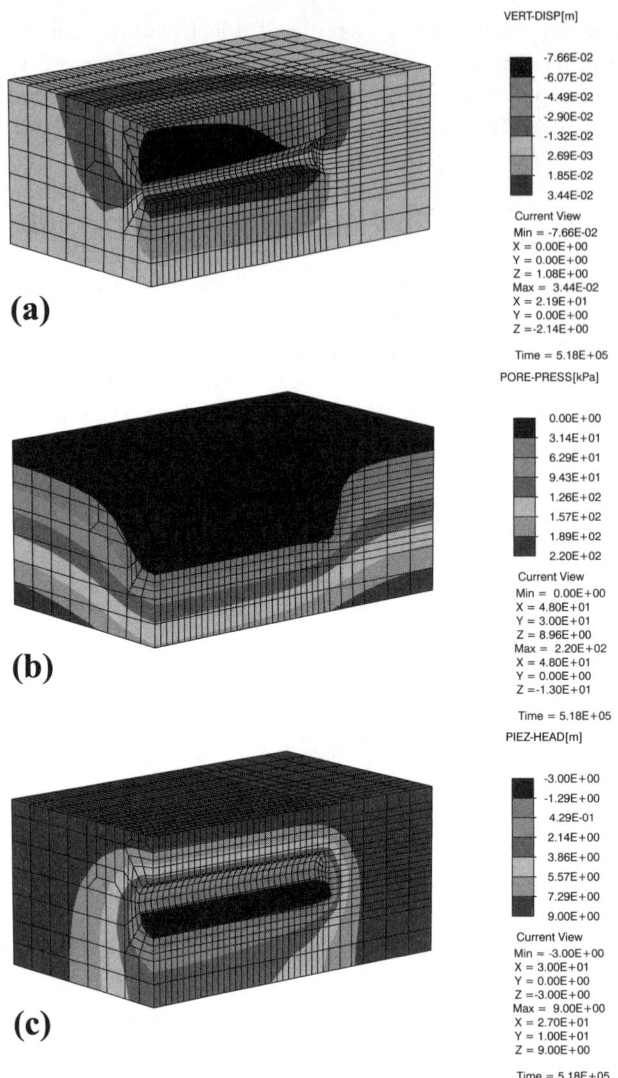

Fig. 4. Excavation of a shallow pervious tunnel in a saturated ground ($D = 6.0\ m$ and $v_a = 5.00\ m/day$): distributions of vertical displacements (a) and pore pressures (b) on the deformed mesh (displacement amplification factor $= 25$) and piezometric head field reported on the undeformed configuration (c)

In view of definition (5) and of the two-step sequence employed to simulate excavation (cf. Sect. 5), the excavation time instants $(t_0; t_u)$ and $(t_{p0}; t_{pu})$ can be evaluated from the analysis of three-dimensional results as shown in Figs. 5–6. In fact, in terms of the position ξ^* of a given cross section with reference to excavation face (Fig. 1), we have $\Delta\xi^* > 0$ in step 1 and $\Delta\xi^* = 0$ in step 2

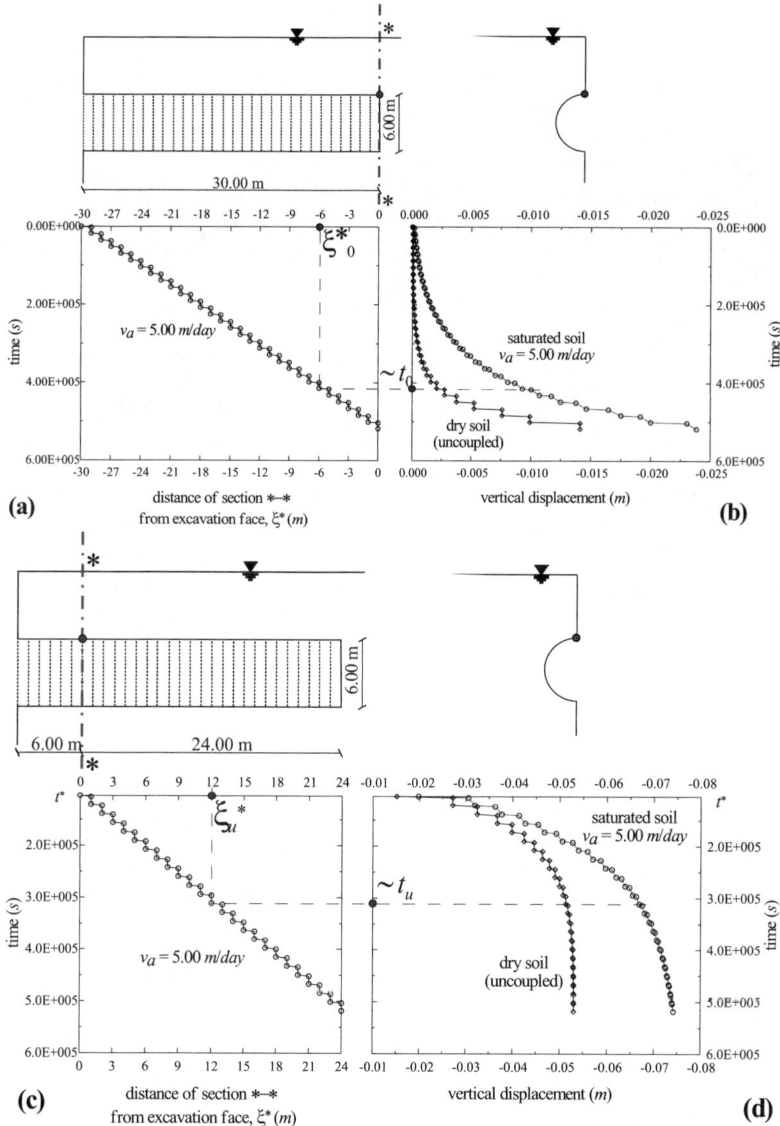

Fig. 5. Three-dimensional simulation of excavation rate 5.00 m/day with $D = 6.00\ m$: analysis of vertical displacements calculated at tunnel crown to evaluate t_0 (b), ξ^*_0 (a) and t_u (d), ξ^*_u (c)

of the aforementioned sequence. As a consequence, it is possible to separate effects of face advance (step 1) from effects of consolidation with space-fixed excavation face (step 2). For example, in Fig. 5b and d, the response calculated at tunnel crown is analyzed to evaluate t_0 and t_u as the instants of

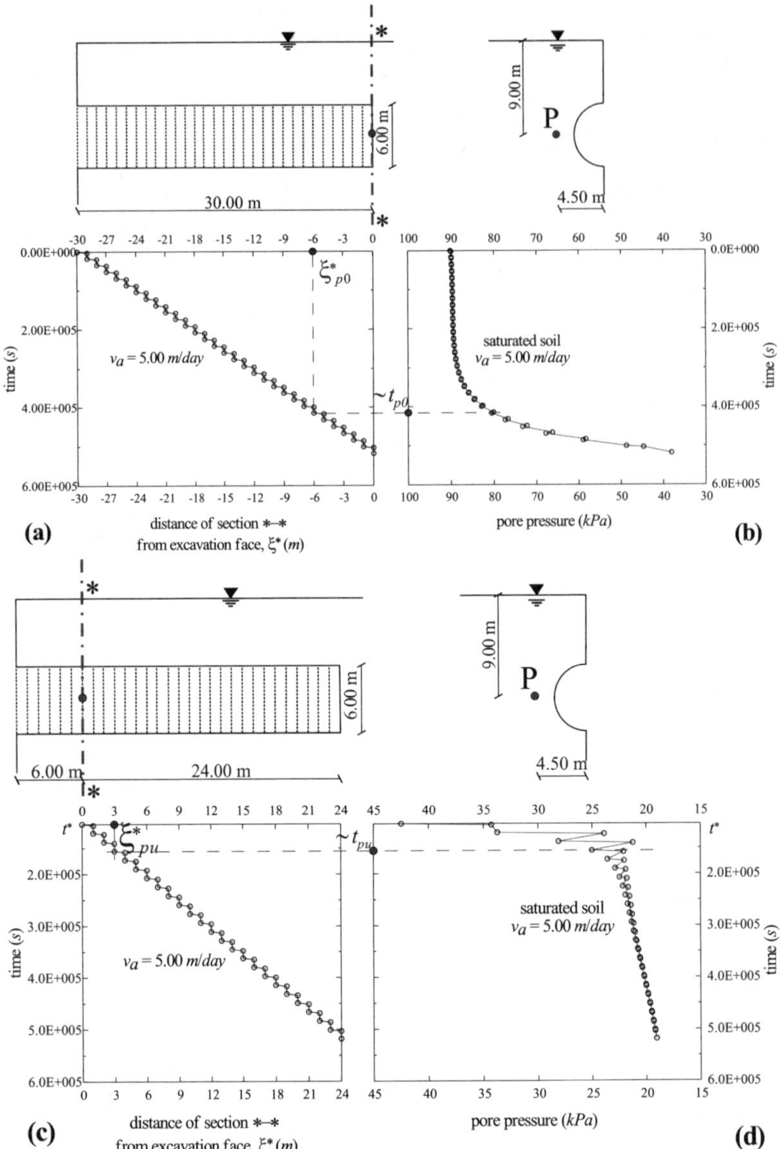

Fig. 6. Three-dimensional simulation of excavation rate 5.00 m/day with $D = 6.00\ m$: analysis of pore pressures calculated at point P near the tunnel shoulders to evaluate t_{p0} (b), ξ^*_{p0} (a) and t_{pu} (d), ξ^*_{pu} (c)

first and last appearance, respectively, of not negligible settlement *increments* induced by the (practically) instantaneous face advancements (steps 1). In other words, the chosen t_0 should correspond to the first observation of appreciable horizontal segments in the graph reported in Fig. 5b and t_u should approximately characterize the instant in which such segments become negligible in Fig. 5d. The corresponding excavation-face positions ξ_0^* and ξ_u^* are then evaluated as illustrated in Figs. 5a and c, respectively.

To estimate time instants t_{p0} and t_{pu}, the pore pressures calculated near the tunnel shoulders are analyzed with the same approach (e.g., cf. Fig. 6b and d, respectively). We remark that, to minimize the effects of mesh boundaries normal to longitudinal tunnel axis x_1, the sections at $x_1 = 30\ m$ and $x_1 = 6\ m$ have been considered to evaluate the effects of approaching (Figs. 5a–b, 6a–b) and leaving excavation face (Figs. 5c–d, 6c–d), respectively.

Numerical simulations with $D = 6.0\ m$ have been repeated for different values of face-advance rate v_a (0.10, 5.00 and 20 m/day). The analysis of corresponding results in terms of vertical displacements at tunnel axis (at crown and ground surface elevations) have indicated

$$\xi_0^* = \xi_{p0}^* = -D \qquad \xi_u^* = 2D \qquad \xi_{pu}^* = D/2 \tag{23}$$

as sufficiently approximate relations. The excavation time intervals are then calculated through equations (6) as:

$$T_u = 3\,\frac{D}{\bar{v}_a} \qquad \text{and} \qquad T_p = \frac{3}{2}\,\frac{D}{\bar{v}_a} = \frac{T_u}{2} \tag{24}$$

It can be observed that the setting $\xi_0^* = -D$ and $\xi_u^* = 2D$ is the same typically adopted for the case of tunneling in a dry ground (8). Therefore, in Fig. 5, we also report results obtained by a purely mechanical finite-element formulation, for a dry soil characterized by the same drained parameters reported in Table 1. The same value of density is set for the dry and the saturated grounds, thus leading to coincident pre-excavation total stress states.

The effectiveness of (23) has been confirmed also by results obtained for different assumptions on the dilatant response (non associated and associated flow rule) as well as for the case of impervious excavation face, supported by a uniform pressure (equal to in-situ horizontal total stress at face center elevation).

On the other hand, for increasing values of the tunnel diameter (8.00 and 10.00 m), a non-linear dependence of ξ_u^* on D is observed. For such diameters, in fact, large plastic strains are calculated, as shown by the comparison between the maximum vertical diametral convergences calculated for the dry soil in the assumptions of fully elastic (c_e) and elasto-plastic (c_{ep}) behaviour (for $D = 6.00$, 8.00 and 10.00 m, we have $c_e = 8$, 14 and 22 cm, $c_{ep} = 10$, 23 and 71 cm, respectively).

Relation $\xi_u^* = 2D$ is then effective if large and widespread plastic strains are not to be expected (e.g. for $c_{ep}/c_e \leq 1.5$). In other cases, the aforementioned expression of ξ_u^* could be corrected by means of simple approaches frequently employed for the dry case and based on the ratio c_{ep}/c_e [8], which is predictable by plane calculations.

5.2   Comparison between three-dimensional and plane results

The geometry of the considered two-dimensional domain and the imposed boundary conditions on displacements, tractions and pore pressures are depicted in Fig. 7a for the case $D = 6.00\ m$. Several preliminary calculations have been performed to investigate the influence of plane discretization refinement. In particular, for the case of tunneling in a dry elastic soil, we have checked the expected agreement between the responses calculated for zero excavation forces in the 2D scheme and in a 3D-domain cross section at large distance from the excavation face. For example, in terms of vertical diametral convergence, we have noted that values significantly smaller than 3D ones are obtained by means of a 2D mesh of about 2×100 three-noded triangles, i.e. with a refinement degree similar to the one adopted in the 3D-domain cross section (cf. Sect. 5). As a consequence of such a "stiffer" response of the 2D mesh, due to the imposed plane-strain constraint, a minimum number of about 2×700 linear triangles is required to attain a good agreement between plane and three-dimensional results. Further refinements of the 2D discretization do not lead to appreciable changes in the calculated solution. We remark that such a preliminary comparison does not involve the consideration of the procedure recalled in Sect. 2, since only *post*-excavation results of tunneling in dry ground are considered.

The performance of the proposed two-dimensional approach is assessed in the following, by comparison with three-dimensional analysis of *excavation*. In view of the preliminary results described above, a mesh of 724 three-noded triangle elements with linear interpolation of displacements and pore-pressures is employed in plane simulations (Fig. 7b). A mixed formulation with constant interpolation of volumetric stresses is considered again.

A possible expression for the unloading law $\hat{\lambda}(\xi^*)$ regulating excavation forces (1) is recovered from the simulations of tunneling in a dry soil. For example, a point-by-point combination of the curve: displacement versus λ,

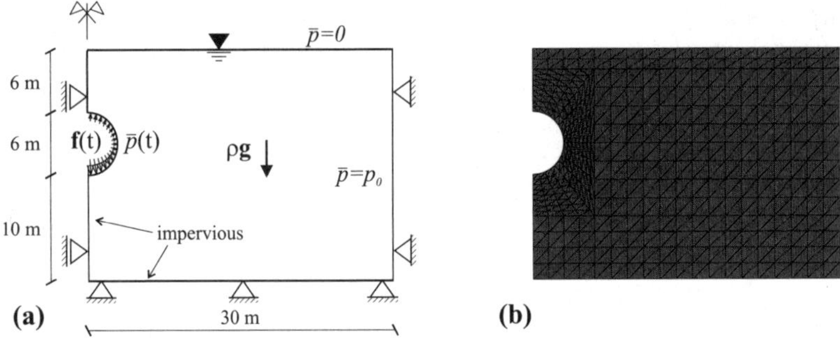

Fig. 7. Plane simulation of the excavation of a shallow pervious tunnel in a saturated ground: a) configuration of the problem with assumed boundary conditions on pore pressures ($\bar{p}$) and displacements; b) finite-element discretization

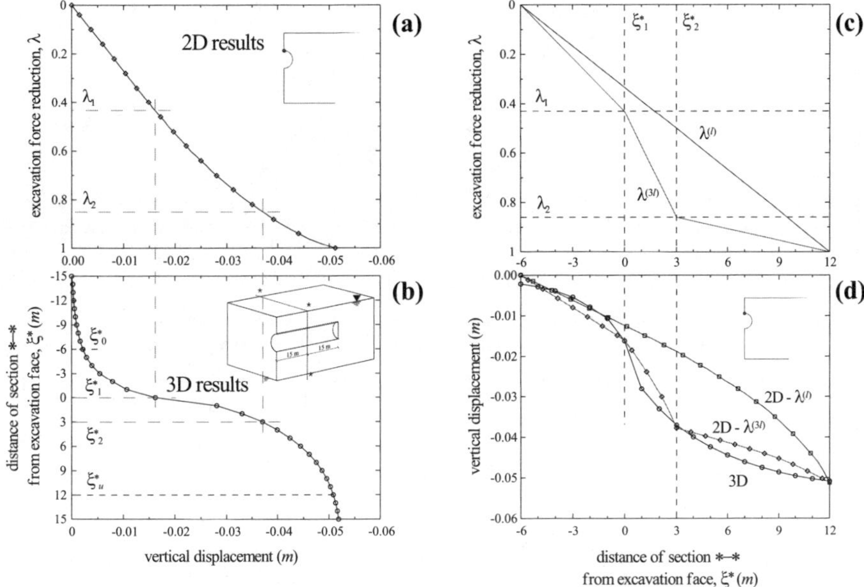

Fig. 8. Tunnel excavation in a dry ground ($D = 6.00\ m$). Vertical displacements at tunnel crown obtained from 2D analysis vs. unloading factor λ (a) and from 3D analysis vs. distance from excavation face (b). c) Linear ($\lambda^{(l)}$) and piece-wise linear ($\lambda^{(3l)}$) approximations for the unloading law $\hat{\lambda}(\xi^*)$. d) Vertical displacements at tunnel crown vs. distance from excavation face obtained from 2D and 3D simulations

calculated in the plane scheme (Fig. 8a) with the curve: displacement versus ξ^*, obtained in 3D analysis (Fig. 8b) leads to a bell-shaped graph for $\hat{\lambda}(\xi^*)$. An accurate interpolation of such a curve could be devised. However, a coarse approximation such as the piece-wise linear law defined in plane $\xi^* - \lambda$ by the points ($\lambda^{(3l)}$ in Fig. 8c)

$$\begin{cases} \xi_0^* = -D & \lambda_0 = 0 \\ \xi_1^* = 0 & \lambda_1 = 0.43 \\ \xi_2^* = D/2 & \lambda_2 = 0.86 \\ \xi_u^* = 2D & \lambda_u = 1 \end{cases} \tag{25}$$

leads to a satisfying agreement between three-dimensional and plane results (Fig. 8d). On the contrary, the agreement is poor if a fully linear unloading is assumed ($\lambda^{(l)}$ in Fig. 8d).

As regards the excavation pore pressures (2), for the unloading law $\hat{\alpha}(\xi^*)$, we consider the linear function identified by points (Fig. 9a):

$$\begin{cases} \xi_{p0}^* = -D & \alpha_0 = 0 \\ \xi_{pu}^* = D/2 & \alpha_u = 1 \end{cases} \tag{26}$$

Time-evolution relations $\lambda(t)$ and $\alpha(t)$ are then obtained by combination of the face-advance law $\xi^*(t)$ with $\hat{\lambda}(\xi^*)$ and $\hat{\alpha}(\xi^*)$, respectively. For the

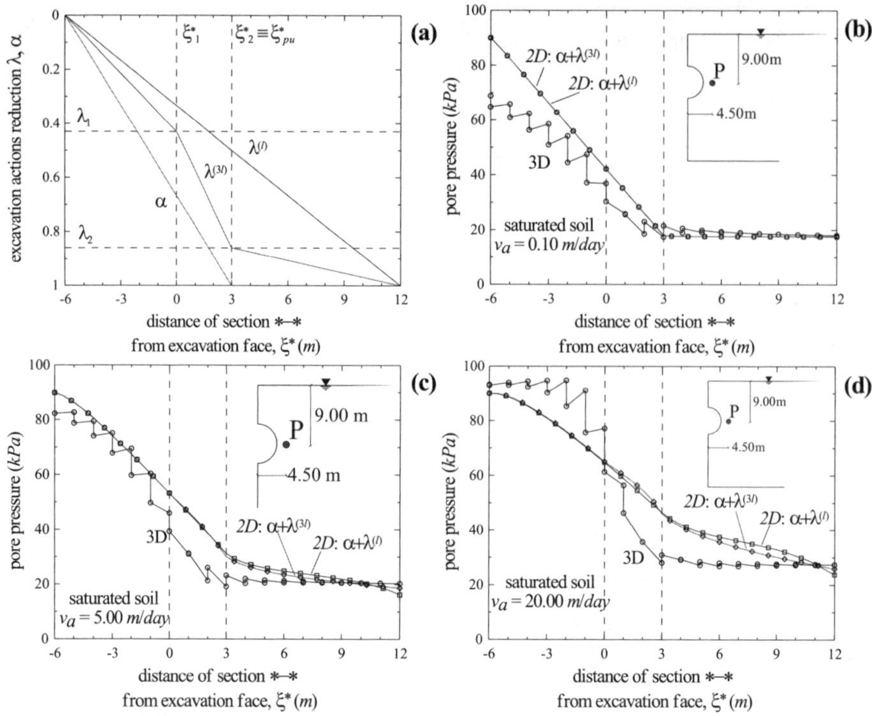

Fig. 9. Tunnel excavation in a saturated ground ($D = 6.00\ m$). a) Linear approximation for the unloading law $\hat{\alpha}(\xi^*)$, linear ($\lambda^{(l)}$) and piece-wise linear ($\lambda^{(3l)}$) approximations for the unloading law $\hat{\lambda}(\xi^*)$. b–d) Pore pressures at point P vs. distance from excavation face; results obtained from 2D and 3D simulations and for different advance rates v_a

considered case of constant face-advance rate (i.e. linear $\xi^*(t)$) and setting $t_0 = 0$, the considered approximations (25) and (26) leads to

$$\begin{cases} t_0 = 0 & \lambda_0 = 0 \\ t_1 = T_u/3 & \lambda_1 = 0.43 \\ t_2 = T_u/2 & \lambda_2 = 0.86 \\ t_u = T_u & \lambda_u = 1 \end{cases} \qquad \begin{cases} t_{p0} = 0 & \alpha_0 = 0 \\ t_{pu} = T_p & \alpha_u = 1 \end{cases} \qquad (27)$$

for the excavation time intervals T_u and T_p given by (24).

To assess the effectiveness of (27) in the plane simulation of tunneling in a saturated ground, several comparisons with three-dimensional results have been performed. For the three considered face-advance rates, a good agreement is observed between pore pressures calculated near the tunnel shoulders (Fig. 9b–d) and in terms of vertical displacements at tunnel crown (Fig. 10a,c,e) and at the ground surface (Fig. 10b,d,f). These comparisons in terms of displacements show that the piece-wise linear interpolation (25) of $\hat{\lambda}(\xi^*)$ is to be preferred to a linear approximation.

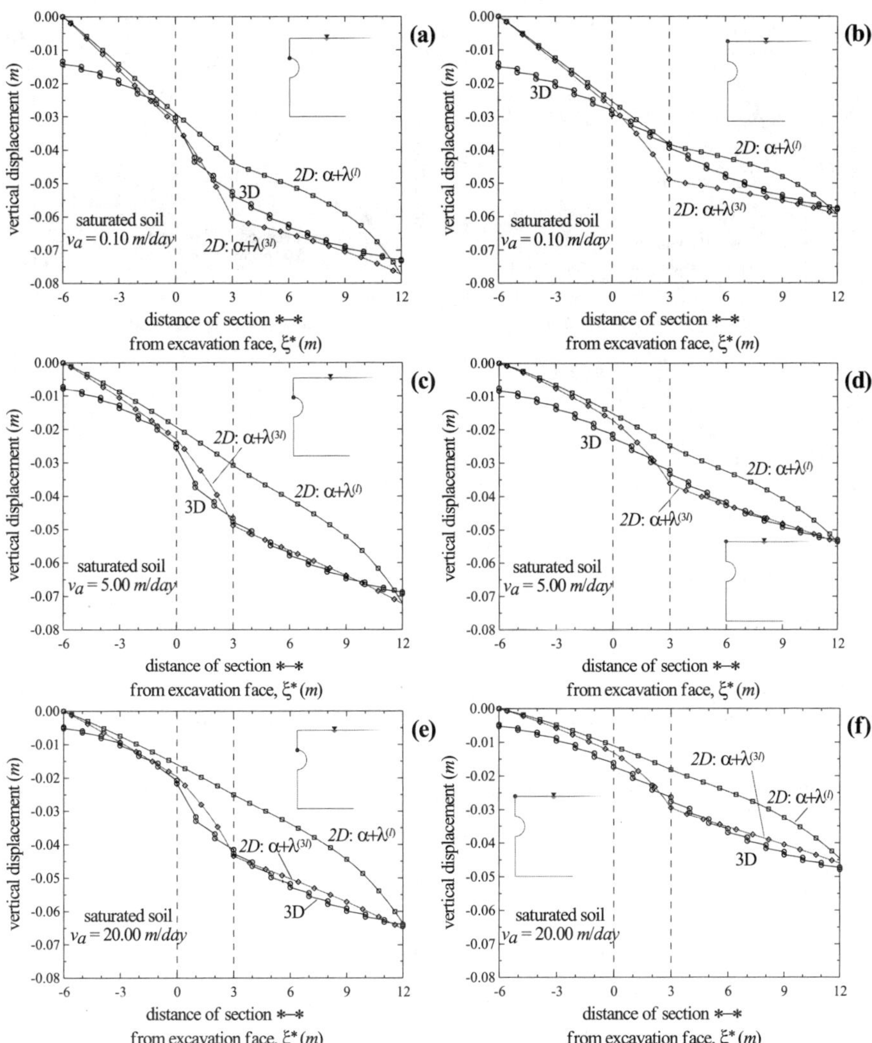

Fig. 10. Tunnel excavation in a saturated ground ($D = 6.00\ m$). Vertical displacements at tunnel crown (a,c,e) and at ground surface (b,d,f) vs. distance from excavation face. Results obtained from 2D and 3D simulations and for different advance rates v_a

Comparisons between subsidence troughs at ground surface calculated for $\xi^* = \xi_u^* = 2D$ show a very good agreement for the case of dry soil (Fig. 11a). For the case of tunneling under the water table, appreciable differences are observed between plane and three-dimensional results obtained for very slow excavation rates (Fig. 11a). The agreement is significantly better for

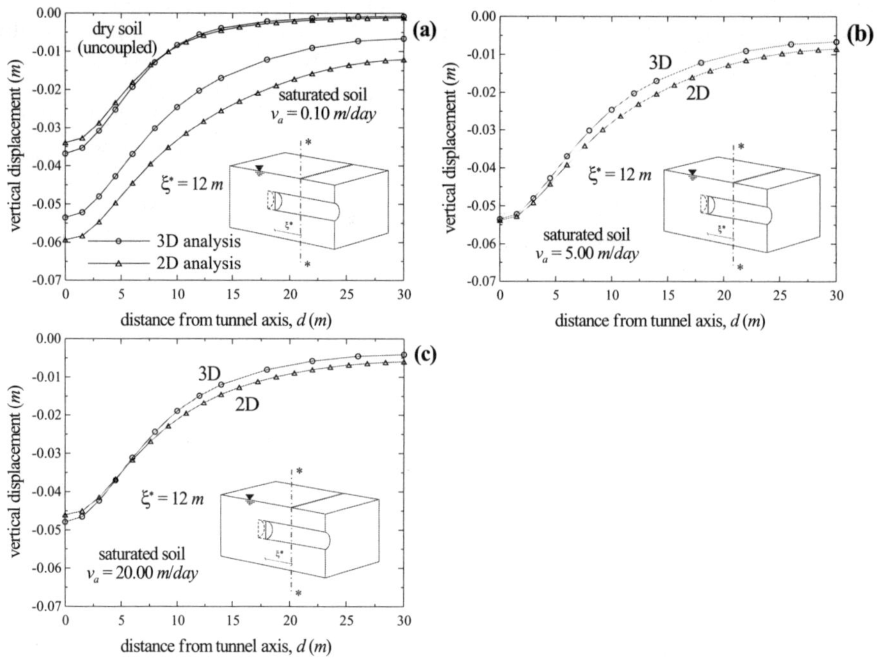

Fig. 11. Distributions of vertical displacements at ground surface ($D = 6.00\ m$): results obtained from 2D and 3D simulations and for different advance rates v_a

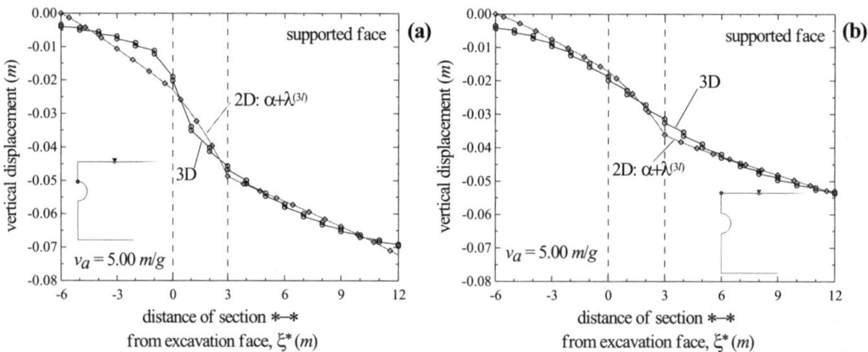

Fig. 12. Excavation in saturated ground with application of a supporting pressure at tunnel face ($D = 6.00\ m$). Vertical displacements at tunnel crown (a) and at ground surface (b) obtained from 2D and 3D simulations

intermediate and high rates, with 2D curves slightly wider than 3D ones (Fig. 11b–c).

According to the three-dimensional results, the effects of a supporting pressure applied to the tunnel face are significant in the region to be excavated (i.e. for $\xi^* < 0$), but negligible in the cavity (i.e. for $\xi^* > 0$), as shown

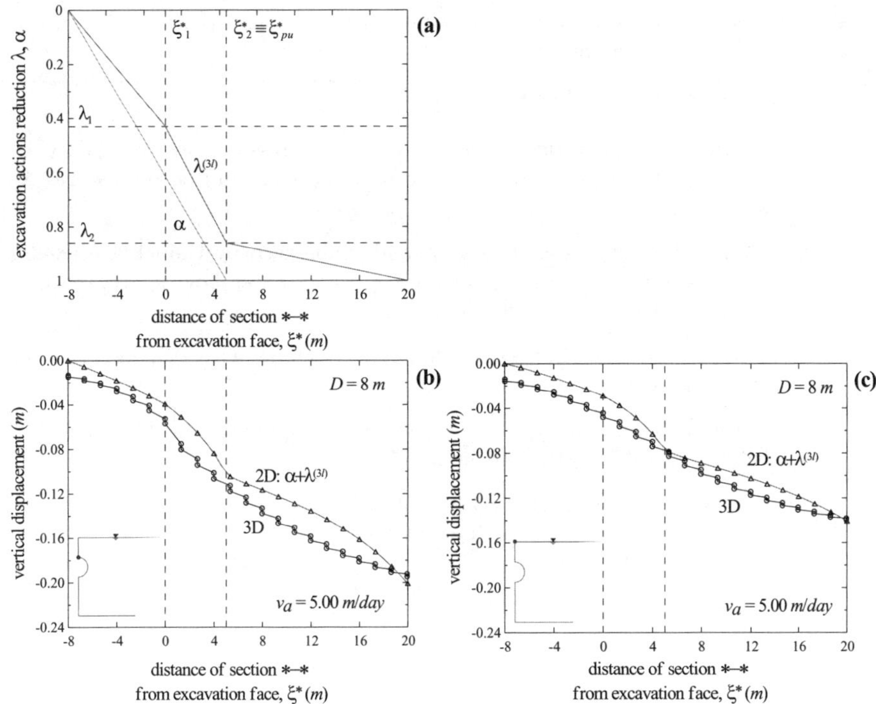

Fig. 13. Tunnel of diameter $D = 8\,m$ in a saturated ground. a) Piece-wise linear approximation ($\lambda^{(3l)}$) for the unloading law $\hat{\lambda}(\xi^*)$ and linear approximation for the unloading law $\hat{\alpha}(\xi^*)$. Vertical displacements at tunnel crown (b) and at ground surface (c) obtained from 2D and 3D simulations

by the comparison between the (3D) displacement curves obtained for the "supported" (Fig. 12a and b) and "unsupported" (Fig. 10c and d) cases. Therefore, the agreement of these curves with plane results ($\alpha + \lambda^{(3l)}$ in Fig. 12) is satisfying also in this case. However, further improvement of this fitting could be obtained by simple modifications of the adopted unloading law $\hat{\lambda}(\xi^*)$ to account for the tunnel-face supporting action [5].

We recall that a non-linear dependence of ξ_u^* on diameter D is observed when excavation induces a large plastic region around the cavity (cf. Sect. 5.1). For $D = 8.00\,m$, the analysis of 3D numerical results indicate $\xi_u^* = \delta \cdot (2D)$ with $\delta = 1.25$ as a proper setting. The same correction is applied to ξ_{pu}^* and associated time intervals T_u and T_{pu} employed in 2D unloading laws (Fig. 13a). A comparison with three-dimensional results in terms of vertical displacements is shown in Fig. 13b–c.

Table 2. Plane simulation of the excavation of a pervious tunnel in a saturated ground: main steps of the procedure

---

In the considered tunnel cross section,

1. compute pre-excavation stresses and pore pressures at the cavity boundary (to define initial "excavation" forces (1) and pore pressures (2));

2. estimate the longitudinal axis intervals influenced by tunnel-face advance, in terms of displacement $[\xi_0^*; \xi_u^*]$ and water flow $[\xi_{p0}^*; \xi_{pu}^*]$ (a possible setting is given in (23));

3. calculate corresponding excavation time intervals with (6), for given face-advance rate v_a;

4. set the unloading laws $\lambda(t)$ and $\alpha(t)$ appearing in (1) and (2), respectively (a possible form is given in (27));

5. perform tunneling simulation by unloading the excavation actions according to evolution laws defined at point 4

---

6   Concluding remarks

The comparison between the results of several 2D and 3D analyses of tunneling in coupled poro-plastic soil shows that the excavation-face influence on displacements and pore pressures can be effectively reproduced by the plane procedure proposed in [3] as an extension of the convergence-confinement method to the case of saturated ground.

Among the main results of this research, there is a consistent definition (6) of an "excavation" time T_u and its evaluation by $(24)_1$ in terms of face-advance rate v_a and tunnel diameter D. Therefore, a so-calculated T_u can be *a-priori* compared with an estimate of the consolidation "characteristic" time T_c to motivate a drained $(T_c/T_u \ll 1)$, undrained $(T_c/T_u \gg 1)$ or coupled $(T_c/T_u \cong 1)$ approach to tunneling analysis. Following a common position, T_c can be evaluated by extending the solution of one-dimensional elastic consolidation to the considered tunneling problem, thus obtaining the ratio:

$$\frac{T_c}{T_u} = \frac{\bar{v}_a}{c_v}\frac{L^2}{3D} \qquad \text{with} \quad \frac{1}{c_v} := \frac{1}{k}\left(\frac{b^2}{E_{oe}} + \frac{1}{Q}\right) \qquad (28)$$

where c_v is the consolidation coefficient [2], E_{oe} is the oedometric elastic modulus of the solid skeleton and L is a length characterizing the considered multi-dimensional consolidation problem. For example, it can be set $L = D/2$ for deep tunnels, or $L = H_w$ for shallow tunnels located at depth H_w under the water table.

If $T_c/T_u \cong 1$ and in all the uncertain situations, where the face-advance rate effects are to be taken explicitly into account, a two-dimensional analysis of tunneling can be performed following the steps outlined in Table 2.

Acknowledgements

This work has been carried out in the framework of the activities of European research group *Laboratoire Lagrange*.
Danilo Russo, currently at the Italian branch of Systra, was actively involved in the preliminary phase of research, during the preparation of his graduation thesis at University "Tor Vergata".

References

1. Alonso E., Alejano L.R., Varas F., Fdez-Manín G., Carranza-Torres C. (2003) Ground response curves for rock masses exhibiting strain-softening behaviour, Int. J. Num. Analyt. Meth. Geomech. **27**, 1153–1185.
2. Biot M.A. (1941) General theory of three-dimensional consolidation, J. Appl. Phys. **12**, 155–164.
3. Callari C. (2004) Coupled numerical analysis of strain localization induced by shallow tunnels in saturated soils, Computers and Geotechnics, **31**, 193–207.
4. Callari C., Casini S., Three-dimensional shallow tunneling in a hydro-mechanically coupled saturated soft ground, in preparation
5. Callari C., Marchetti P. (2003) Analisi di gallerie con il metodo convergenza-confinamento, Manuale di Ingegneria Civile e Ambientale, **1** with CD-ROM, fourth edition, Zanichelli/Esac.
6. Carter J.P., Booker J.R. (1982) The analysis of consolidation and creep around a deep circular tunnel in clay, Proc. 4th Int. Conf. on Numerical Methods in Geomechanics, Edmonton, 537–544.
7. Cividini A., Gioda G., Barla G. (1985) A numerical analysis of tunnels in saturated two-phase media, Proc. 11th ICSMFE, San Francisco, **2**, 729–732.
8. Corbetta F., Bernaud D., Nguyen Minh D. (1991) Contribution à la méthode convergence-confinement par la principe de la similitude, Revue Française de Géotechnique, **54**, 5–12.
9. Coussy O. (1995) Mechanics of porous continua, Chichester, Wiley.
10. Giraud A., Picard J.M., Rousset G. (1993) Time-dependent behaviour of tunnels excavated in porous mass, Int. J. Rock Mech. Min. Sci. Geomech. Abstr., **30**, 1453–1459.
11. Kasper T., Meschke G., (2003), Modeling aspects and numerical simulation of shield tunnelling in soft soils, Proc. 7th International Conference on Computational Plasticity (Complas VII), Owen D.R.J., Oñate E., Hinton E., editors, CIMNE, Barcelona.
12. Komiya K., Soga K., Akagi H., Hagiwara T., Bolton M.D. (1999) Finite element modelling of excavation and advancement processes of a shield tunnelling machine, Soils and Foundations, **39**, 37–52.
13. Leca E., Leblais Y., Kuhnhenn K. (2000) Underground works in soils and soft rock tunneling, GeoEng2000 Conf. Melbourne, Key note lecture.

14. Mair R.J., Taylor R.N. (1997) Bored tunnelling in the urban environment, State of the Art Report, Proc. 14th ICSMFE, Hamburg. Balkema, **4**.

15. Mair R.J., Taylor R. N., Burland J. B. (1996) Prediction of ground movements and assessment of risk of building damage due to bored tunneling, Proc. Int. Symp. Geotechnical Aspects of Underground Construction in Soft Ground, Mair R.J., Taylor R.N., editors, London, Balkema, 713–718.

16. Mestat Ph., Bourgeois E., Riou Y. (2004) Numerical modelling of embankments and underground works, Computers and Geotechnics, **31**, 227–236.

17. Ng C.W.W., Lee G.T.K. (2002) A three-dimensional parametric study of the use of soil nails for stabilising tunnel faces, Computers and Geotechnics, **29**, 673–697.

18. Ohtsu H., Ohnishi Y., Taki H., Kamemura K. (1999) A study on problems associated with finite element excavation analysis by the stress-flow coupled method, Int. J. Num. Analyt. Meth. Geomech. **23**, 1473–1492.

19. Panet M., Guellec P. (1974) Contribution à l'étude du soutenement d'un tunnel à l'arrière du front de taille, Proc. Third ISRM Congress, Denver.

20. Swoboda G., Abu-Krisha A. (1999) Three-dimensional numerical modelling for TBM tunnelling in consolidated clay, Tunneling and Underground Space Technology, **14**, 327–333.

21. Thomas H. R., Zhou Z. (1997) Minimum time step size for diffusion problem in FEM analysis. Int. J. Numer. Methods Engrg., **40**, 3865–3880.

22. Zienkiewicz O.C., Taylor R.L., The finite element method, McGraw Hill, New York, **I**, 1989, **II**, 1991, **III**, 2000.

A Plasticity Model and Hysteresis Cycles

Nelly Point[1,2], Denise Vial[1]

[1] Conservatoire National des Arts et Métiers,
   Département de Mathématiques ,
   292 rue Saint Martin,
   F-75141 Paris Cedex 03, France
[2] Ecole Nationale des Ponts et Chaussées,
   Laboratoire Analyse des Matériaux et Identification,
   6-8 avenue Blaise Pascal,
   Cité Descartes,
   Champs-sur-Marne,
   F-77455 Marne la Vallée Cedex 2, France

Abstract. Physical considerations on the micromechanisms involved in plasticity phenomenon are used to derive a four parameters plasticity model. The choice of the more appropriate set of parameters is important for this highly non linear model. Once these parameters, characteristic of the material, are identified, the response to any series of loading and unloading is easily computed. An uniaxial tensile test or experimental hysteresis cycles permit to identify the parameters through a nonlinear least square method. The study of experimental hysteresis cycles for important loads permits an easy evaluation of these characteristics parameters.

1   Introduction

Increasing capabilities are demanded to materials and structures. Most of them are submitted to cyclic loading and unloading which bring them into post-elastic domain and their response exhibits hysteritic behaviour. To optimize the maintenance and to estimate the lifetime of the structure, a good understanding of plastic evolution is necessary. The model must be as simple as possible but at the same time accurate enough to take into account all the observed behaviors.

Plastic evolution involves complex mechanisms, usually described by internal variables. Physical observations have led to consider, in a linear framework, two micromechanisms of plasticity coupled by a single yield condition. This model only needs four parameters plus the Young modulus to describe any typical tensile curve observed for metallic materials. These parameters are linked to physical properties and once they are identified, the evolution of the material and the hysteresis cycles are easily computed for any series of loading and unloading, it is only necessary to know exactly the initial state of the material and the history of the loading.

2   The plasticity model

The considered model is of the standard generalized class. It is linked to the model proposed by Zarka and Casier and called MASSI (Modèle d'Analyse Simplifiée des Structures Inélastiques) ([1],[2],[3]). The main feature of this model is to separate the plastic mechanisms into two coupled families :

- α_1 is the local plastic strain
- α_2 takes into account free dislocations of the metallic polycristalline material

The corresponding local stresses are denoted by σ_1 and σ_2 . In general these quantities are tensors.

The relation between (α_1, α_2) , (σ_1, σ_2) and the global deviatoric stress S is assumed to be linear and is written as:

$$\begin{aligned} \sigma_{1,ij} &= S_{ij} - b_{11}\alpha_{1,ij} - b_{12}\alpha_{2,ij} \\ \sigma_{2,ij} &= \qquad -b_{21}\alpha_{1,ij} - b_{22}\alpha_{2,ij} \end{aligned} \tag{1}$$

with $i, j = 1, 2, 3$ and where $B = (b_{k,l})$,

The matrix B is called the work hardening tensor of the material. Due to physical properties, B is positive and symmetric and is assumed here to be strictly positive.

The two micromechanisms are coupled through a single plastic criterion : the local microstresses σ_1 and σ_2 are assumed to remain in a fixed yield domain defined by :

$$\sigma^T \sigma = \sum_{l=1}^{2} \sigma_{l,ij}\sigma_{l,ij} \leq S_0^2 \tag{2}$$

where S_0 is a positive scalar.

The material plastic properties are defined by S_0 and three coefficients for the matrix B .

The identification of these four parameters can be performed using a uniaxial tensile test.

3   Uniaxial tensile test

Classically, the total strain ε^{tot} is broken down into two parts : the elastic strain ε^e and the plastic strain ε^p . Thus, using the relation

$$\varepsilon^p = \frac{\Sigma}{E} - \varepsilon^{tot} \tag{3}$$

where E is the Young modulus, the experimental curve representing the evolution of the stress Σ as a function of plastic strain ε^p is obtained (cf. Fig. 1)
.

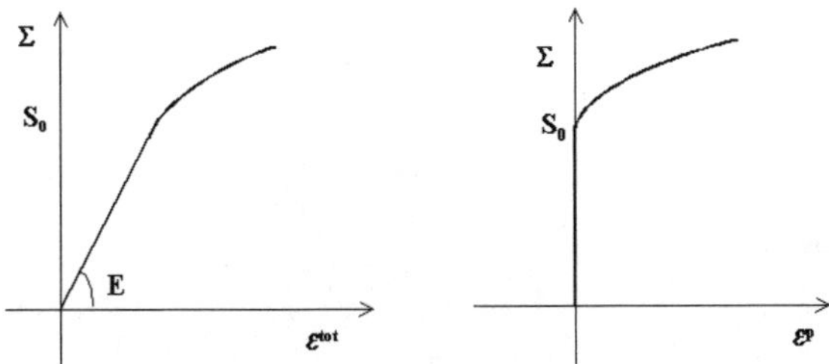

Fig. 1. Tensile test. (**a**) Stress-strain curve. (**b**) Plastic stress-strain curve

In the isotropic case and for uniaxial tensile test, the microstresses and the micromechanisms are described by scalar quantities σ_1 , σ_2 and α_1 , α_2 .

During a uniaxial tensile test, as long as the stress Σ remains smaller than the limit quantity S_0 , the variation of the stress is linear with respect to the total strain ε^{tot} , and $da_1 = da_2 = 0$.

For Σ greater than S_0 , the phenomenon ceases to be linear: plasticity appears and the point which represents the vector $\sigma = (\sigma_1, \sigma_2)$ moves from A to B on the yield circle defined by $\sigma_1^2 + \sigma_2^2 = S_0^2$ (Fig. 2). As the stress Σ increases, the local stress σ tends to an asymptotic limit denoted by σ_{as} ([4] [5]).

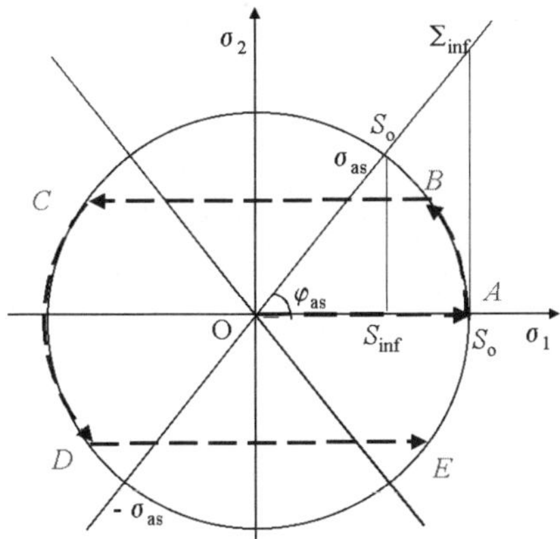

Fig. 2. Evolution in the yield domain

On the figure 3 representing Σ versus ε^p , the asymptotic limit is a straight line with slope p_{inf} , passing by $(0, \Sigma_{\text{inf}})$.

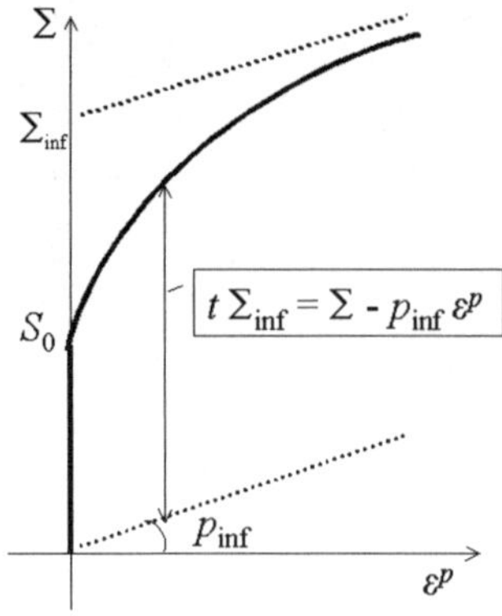

Fig. 3. The global stress versus plastic-strain during a tensile test

4   Cyclic loading and unloading

If the stress Σ ceases to increase, and begins to decrease, then the phenomenon is linear again, $d\sigma_2 = 0$ and $d\sigma_1 = d\Sigma$, the local stress σ_2 remains constant, whereas σ_1 decreases with Σ until the phenomenon starts again to be plastic. If the global stress Σ goes on decreasing, the evolution remains plastic until Σ stops to decrease.

On figure 2, this means that the point which represents $\sigma = (\sigma_1, \sigma_2)$ moves on the segment BC during the elastic evolution and then, on the circle, from C towards an asymptotic limit corresponding to $-\sigma_{as}$. When Σ ceases to decrease to increase again, then the point representing σ stops in D and moves on the segment DE . And so on ...

On figure 4, the corresponding evolution is plotted in the (ε^p, Σ)-plan.

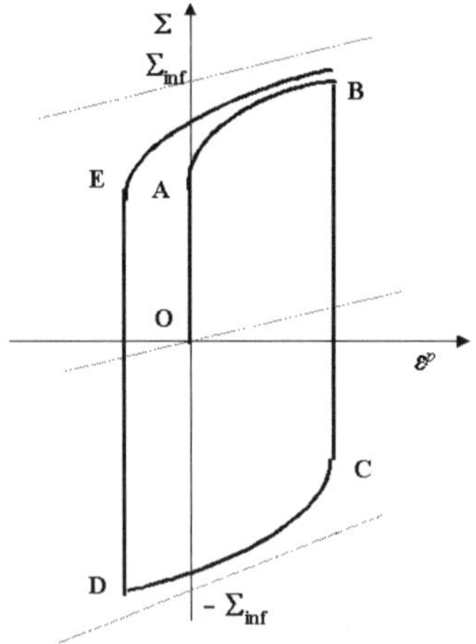

Fig. 4. Hysteresis cycles

The length of the segment BC on Fig. 2 and on Fig. 4 is the same since the variation of σ_1 is equal to the variation of Σ in case of linear evolution.

5   Description of the hardening tensor

The choice of the set of parameters used to describe the hardening tensor B is very important.

After different attempts, it appeared that the more relevant way to describe B was to write it as :

$$B = \begin{bmatrix} p_{\text{inf}} + r^2 b & -rb \\ -rb & b \end{bmatrix} \tag{4}$$

where p_{inf} and b are stresses and r is a ratio coupling the two micromechanisms of plasticity.

When $r = 0$, the two micromechanisms are uncoupled and the corresponding plasticity model is the bilinear one.

More precisely, the quantity $b_{11} = p_{\text{inf}} + r^2 b$ is equal to the slope of the curve Σ versus ε^p when $\varepsilon^p = 0$, p_{inf} is the slope of the asymptotic direction of this curve and b represents the intensity of the second micromechanism.

The parameter r is the tangent of the angle φ_{as} of σ_{as} with σ_1 which is the direction of the global stress Σ . If Σ_{inf} denotes the ordinate of the intersection of this asymptotic line with the vertical axis, it has been shown that :

$$\Sigma_{\text{inf}} = \frac{S_0}{\cos \varphi_{as}} = S_0 \sqrt{1 + r^2} \tag{5}$$

where S_0 is the yield parameter.

Hence, the plastic evolution can be described using these 4 following parameters: $\Sigma_{\text{inf}}, \ p_{\text{inf}}, \ r, \ b$.

6   Analysis of the plastic evolution

A parametric description of the curve Σ versus ε^p has been established in [5] and an implicit equation has been derived.

The plastic strain ε^p can be expressed as a function of the quantity t represented on Fig. 3 and defined by :

$$t = \frac{\Sigma - p_{\text{inf}} \ \varepsilon^p}{\Sigma_{\text{inf}}} \tag{6}$$

To be specific :

$$\varepsilon^p = \varepsilon^p_{in} + \frac{\Sigma_{\text{inf}}}{b(1 + r^2)^2} \left(h(t; r) - h(t_{in}; r) \right) \tag{7}$$

where ε^p_{in} and t_{in} are the initial values of ε^p and t and where $h(.;r)$ is defined by :

$$h(t;r) = \frac{1}{2}\ln\left(\frac{1+t}{1-t}\right) \pm r\sqrt{1-t^2} + (r^2-1)\,t \tag{8}$$

The sign in h changes with the sign of the time's derivative of ε^p .

The formulae (6,7,8) have been used in [4] to identify the four parameters Σ_{inf}, p_{inf}, r, b, through a non-linear leastsquare method for different data $(\varepsilon^p_i, \Sigma_i)$ corresponding to different metallic materials and obtained from experimental tensile tests.

7   Computation of hysteritic responses

Conversely, knowing the parameters Σ_{inf}, p_{inf}, r, b for a given material, it is now possible to simulate its response to any series of loading and unloading, periodic or not.

This computation has been performed, starting from an initial state at rest, that is without any plastic strain in the material. The first load has been taken equal to $S_0 + \Delta\Sigma$, and the following loads have been increased each time by the quantity $\Delta\Sigma$ which has been chosen arbitrarily. The unloadings have been increased in the same way (Fig. 5 and 6).

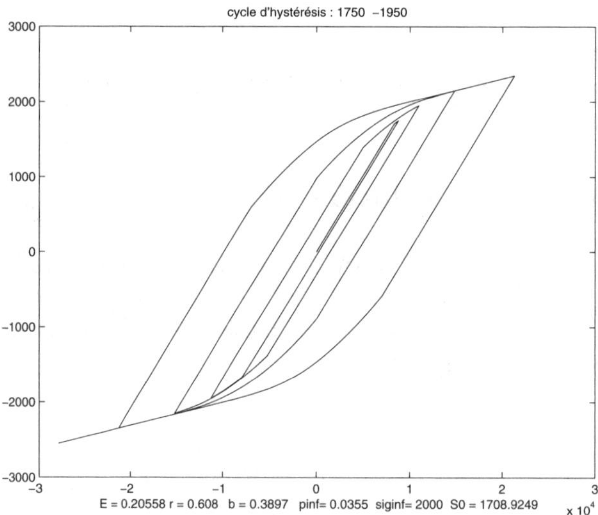

Fig. 5. Progressive cyclic loading in the stress-strain plan

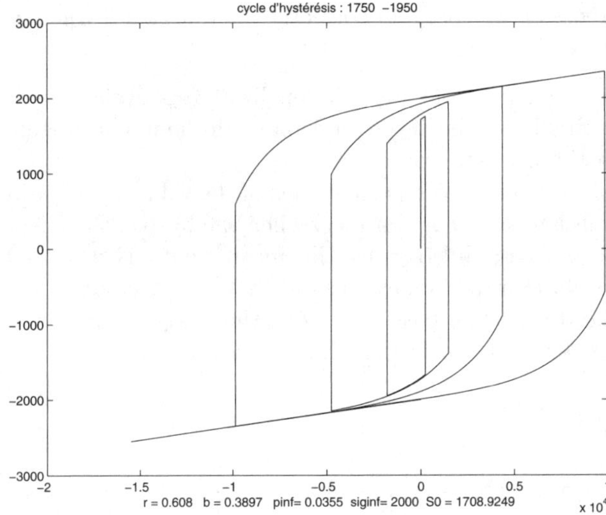

Fig. 6. Progressive cyclic loading in the stress-plastic strain plan

8 Identification through hysteresis cycles

Figure 2 shows that the minimal length of the segment BC , corresponding to elastic unloading, is equal to $2S_{\mathrm{inf}}$ with:

$$S_{\mathrm{inf}} = S_0 \cos \varphi_{as} = \frac{S_0}{\sqrt{1 + r^2}} \tag{9}$$

The relations (9) and (5) permit to link the parameter r to a very simple aspect of the plastic strain-stress hysteresis cycles , namely :

$$r^2 = \frac{\Sigma_{\mathrm{inf}} - S_{\mathrm{inf}}}{S_{\mathrm{inf}}} \tag{10}$$

This provides a quite simple way to evaluate the parameter r .

During series of loading and unloading, the length of the segment BC (resp. DE) is equal to the absolute value of the variation of the global stress . When this quantity increased, the successive length of the segments $B^{(k)}C^{(k)}$ decrease and tend to $2S_{\mathrm{inf}}$.

Moreover, knowing the length of the segment BC, it is possible to estimate the value of the internal variable σ in C and hence to know the initial state of the material at the beginning of the plastic evolution from C to D . Whereas, when a tensile test is used, there is no evidence that the initial state corresponds really to internal microstresses equal to 0.

The only problem is that, in practice, it is not possible to increase the loads infinitely for, at least, two reasons :

- the phenomenon is no more plastic for too important loads, damage and even breakdown may happen.

- numerically, it is impossible to compute correctly the function h when Σ tends to $\pm\infty$ since t tends to ±1 .

But any way, the use of data from well chosen hysteresis cycles seems more efficient than the simple tensile test to estimate the four parameters needed to define the plastic properties of a material.

Besides, the aspect of the cycles gives a coarse idea of the value of the ratio coupling the two micromechanisms. For instance, when the hysteresis cycles, for important load, are close to parallelograms, this means that r is close to 0 and that the plasticity is close to the bilinear plasticity. More precisely, when C on Fig. 4 is under the dotted line passing by O , then $\Sigma_{\text{inf}} \leq 2S_{\text{inf}}$ and $r \leq 1$, on the contrary, when C is above, then $\Sigma_{\text{inf}} \geq 2S_{\text{inf}}$ and $r \geq 1$.

9   Conclusion

Starting from physical considerations, a model of plasticity has been constructed that permits to predict the behavior of a material under cyclic loading, periodic or not, as soon as characteristics parameters are known and the initial state defined.

An identification procedure has been constructed which uses the experimental data of a classical tensile test. But, it has appeared that the evaluation of the asymptote is not easy.

The above study of hysteresis cycles permits to understand that the evaluation of the different parameters is easier using the data of a complete hysteresis cycle for a large variation of the applied global stress. The asymptote appears more clearly and successive cycles with increasing loads give directly a reliable estimation of r .

The use of data from well chosen hysteresis cycles seems more efficient to estimate the four parameters needed to define the plastic properties of a material. This permits to have a more robust prediction of these coefficients.

References

1. Zarka J., Casier J. (1979) Elastic plastic response of a structure to cyclic loadings: practical rules. Mechanics Today, **6**, Ed. Nemat-Nasser, Pergamon Press
2. Khabou M.L., Castex L., Inglebert G. (1985) Eur. J. Mech., A/Solids, **9**, **6**, 537-549.
3. Halphen B., Nguyen Q.S., (1975) Sur les matériaux standards généralisés. J. de Mécanique, **14**, 1 , 39-63
4. Inglebert G., Vial D., Point N. (1997) Identification of a four parameters plasticity model. Mathematical Modelling & Scientific Computing, Washington, D.C., **8**
5. Inglebert G., Vial D., Point N. (1999) Modèle micromécanique à quatre paramètres pour le comportement élastoplastique. Groupe pour l'Avancement de la Mécanique Industrielle, **52**, march 1999

Computational Analysis
of Isotropic Plasticity Models

Nunziante Valoroso[1], Luciano Rosati[2]

[1] Istituto per le Tecnologie della Costruzione
   Consiglio Nazionale delle Ricerche
   viale Marx, 15
   00137 Roma, Italy
[2] Dipartimento di Scienza delle Costruzioni
   Università di Napoli "Federico II"
   Via Claudio, 21
   80125 Napoli, Italy

Abstract. The implementation of plasticity models with general isotropic yield surfaces is discussed. The computation of the return map solution and the algorithm linearization are carried out by suitably exploiting the isotropic character of the elastic constitution and of the yield function; the consistent tangent tensor is given an intrinsic explicit representation and the relevant coefficients are provided. It is also addressed the implementation of the constitutive algorithm in the subspace defined by the plane stress condition. This is obtained only by specializing the three-dimensional formulation to a two-dimensional ambient space, with the result that the structure of the return mapping scheme and the formal expression of the consistent tangent tensor are preserved. The effectiveness of the approach is demonstrated by means of representative numerical simulations.

1 Introduction

Computational plasticity has been characterized by significant advancements in the last two decades mainly after the work of J.C. Simo, whose seminal contributions in the field have been collected in two recent monographs [1,2].

It is now well understood that accurate and stable algorithms for integrating the rate constitutive equations are of major importance for carrying out efficient stress computation schemes; furthermore, the paramount role of the consistent tangent both at the local and at the global level has been put forward by several authors. Nonetheless, in many cases the exact algorithm linearization and the consistent tangent may be extremely demanding or computationally expensive to obtain. This is actually the case for complex yield criteria such as three-invariant models for concrete or frictional materials or when highly nonlinear hardening/softening rules come into play.

As explicitly remarked in [1, sec. 3.6.2, p.147], a first source of difficulty in obtaining consistent tangent tensors lies in the evaluation of the gradient of the plastic flow i.e., for standard material models, in computing the second derivatives of the yield function. This is however only a preliminary task to accomplish for obtaining the final result, since this requires the inversion of the jacobian associated with the local stress computation scheme.

The general framework here outlined addresses isotropic elasto(visco-)plasticity, i.e. material models whose elastic and (visco)plastic behaviours are described through an isotropic stored energy and an elastic domain given as the level set of an isotropic yield function.

In [3] the authors have presented a quite general solution strategy allowing the treatment of J_3-dependent plasticity models in a way to take proper advantage of the isotropic properties. In particular, they have shown how to obtain an intrinsic representation of the consistent tangent tensor and have provided its explicit expression with no use of matrix operations.

In the paper [4] the above ideas have been further elaborated and extended to the principal axis formulation of isotropic plasticity. This has been obtained by illustrating the relationships existing between the first and second derivatives of isotropic scalar functions of a symmetric rank-two tensor argument, assigned either as function of the eigenvalues or of the invariants, and by proving some basic results concerning the differentiation of the eigenprojectors of a symmetric rank-two tensor and the principal space representation and inversion of rank-four positive-definite symmetric tensors.

Starting from the above mentioned contributions, the present work aims to provide further insights into the implementation of the Newton closest-point projection scheme and its consistent specialization to the significant class of plane stress problems.

This paper is ordered as follows. In section 2 it is presented a summary of the three-dimensional continuum problem formulation; the time-discrete problem and its algorithmic implementation are then discussed in section 3. Section 4 focuses on the algorithm linearization, and in particular on the construction of the formal inverse of the elastoplastic compliance tensor entering the local stress computation scheme and the expression of the algorithmic tangent. A novel expression of the consistent tangent tensor, more compact and effective with respect to other existing expressions, is then provided in section 5 along with an explicit representation formula, whose coefficients are given in the appendix. In section 6 is then discussed the implementation of the general return mapping scheme in the subspace defined by the plane stress condition; the consistent linearization and the explicit expression of the algorithmic tangent for the plane stress case are then given in section 7. Finally, a numerical example is presented in section 8 to illustrate the performances of the devised solution schemes.

2   Problem set up. Continuum formulation

Let us consider an elastoplastic structural model undergoing a quasi-static loading process whose events are ordered by a scalar parameter t that will be referred to in the sequel as *time*.

Under the assumption of small transformations the kinematics of deformation at each point is characterized by the infinitesimal strain measure and by the additive decomposition:

$$\varepsilon = \mathbf{e} + \mathbf{p} \tag{1}$$

$\mathbf{e}$ and $\mathbf{p}$ respectively denoting the elastic and plastic strain tensors.

Addressing the purely mechanical case, the stored energy function can be given in fully decoupled form as [5]:

$$\psi(\mathbf{e}, \boldsymbol{\alpha}) = \psi_{el}(\mathbf{e}) + \psi_h(\boldsymbol{\alpha}) \tag{2}$$

where ψ_{el} and ψ_h are isotropic functions, both assumed to be twice differentiable with positive definite Hessian. They represent in turn the elastic energy and the hardening potential, the latter characterizing the inelastic response in terms of a strain-like variable $\boldsymbol{\alpha}$ that, for the developments that follow, will be partitioned into a tensorial variable $\boldsymbol{\eta}$ and a scalar variable ζ respectively accounting for the kinematic and isotropic hardening mechanisms.

The constitutive relationships for the stress-like variables follow from the standard thermodynamic argument [6]. In particular, for linear elasticity one has the constitutive law for the Cauchy stress:

$$\boldsymbol{\sigma} = \mathbb{E}(\varepsilon - \mathbf{p}) \tag{3}$$

$\mathbb{E}$ being the elastic tensor:

$$\mathbb{E} = 2G\,\mathbb{I} + \lambda(\mathbf{1} \otimes \mathbf{1}) = 2G\,(\mathbf{1} \boxtimes \mathbf{1}) + \lambda(\mathbf{1} \otimes \mathbf{1}) \tag{4}$$

where G and λ are the Lamé's moduli. The symbols $\mathbb{I}$ and $\mathbf{1}$ in (4) respectively denote the rank-four and rank-two identity tensors that are related each other through the so-called square tensor product [7]; its definition along with the relevant composition rules are postponed to Appendix A.

Kinematic and isotropic hardening of the model are assumed to be governed by the relationships:

$$\boldsymbol{\beta} = \mathbb{H}_{kin}\boldsymbol{\eta} \tag{5a}$$

$$\vartheta = h_{iso}(\zeta) \tag{5b}$$

where $\mathbb{H}_{kin} = h_{kin}\mathbb{I}$, h_{kin} denotes the kinematic hardening modulus and h_{iso} is a nonlinear isotropic hardening function.

The pair of internal variables $\boldsymbol{\beta}$ and ϑ, usually referred to as the *back stress* and the *drag stress* [8], account for the evolution in stress space of the yield locus, defined as the level set of a scalar function that will be assumed convex and smooth. In particular, in the sequel we shall make reference to a general isotropic yield function given as:

$$\tilde{\phi}(\boldsymbol{\sigma}, \boldsymbol{\beta}, \vartheta) = \phi(\boldsymbol{\tau}, \vartheta) = \varphi(I_1, J_2, J_3) - \vartheta - Y_o \tag{6}$$

where Y_o depends upon the initial yield limits of the material,

$$I_1 = \operatorname{tr}(\boldsymbol{\tau}); \qquad J_2 = \frac{1}{2}\operatorname{tr}\mathbf{S}^2; \qquad J_3 = \frac{1}{3}\operatorname{tr}\mathbf{S}^3 \tag{7}$$

are stress invariants and $\mathbf{S}$ is the deviator of the relative stress $\boldsymbol{\tau} = \boldsymbol{\sigma} - \boldsymbol{\beta}$.

The definition of the model is completed by providing the evolutionary equations for the strain-like variables; for standard materials they follow from the principle of maximum plastic dissipation as:

$$\begin{cases} \dot{\mathbf{p}} = \ \ \dot{\gamma}\,\mathrm{d}_{\boldsymbol{\sigma}}\tilde{\phi}(\boldsymbol{\sigma},\boldsymbol{\beta},\vartheta) \\[2mm] \dot{\boldsymbol{\eta}} = -\dot{\gamma}\,\mathrm{d}_{\boldsymbol{\beta}}\tilde{\phi}(\boldsymbol{\sigma},\boldsymbol{\beta},\vartheta) \\[2mm] \dot{\zeta} = -\dot{\gamma}\,\mathrm{d}_{\vartheta}\tilde{\phi}(\boldsymbol{\sigma},\boldsymbol{\beta},\vartheta) \end{cases} \tag{8}$$

where $\dot{\gamma}$ is the plastic consistency parameter that, for rate–independent plasticity, is characterized as a Lagrange multiplier subject to the Karush-Kuhn-Tucker (KKT) conditions [9]:

$$\tilde{\phi} \leq 0; \qquad \dot{\gamma} \geq 0; \qquad \dot{\gamma}\tilde{\phi} = 0 \tag{9}$$

It is worth noting that, due to the assumed form of the yield function, the rate of the internal variable $\boldsymbol{\eta}$ equals that of the plastic strain $\mathbf{p}$. Assuming, without loss of generality, that prior to any loading these are both identically zero, one then infers the equality $\boldsymbol{\eta} = \mathbf{p}$, whence the constitutive equation for the back stress tensor:

$$\boldsymbol{\beta} = \mathbb{H}_{kin}\mathbf{p} \tag{10}$$

currently addressed in the literature [8]. Accordingly, for the problem at hand the variables $\boldsymbol{\eta}$ and $\boldsymbol{\beta}$ can be dropped out from the formulation and the stress computation can be carried out by making reference only to the relative stress tensor:

$$\boldsymbol{\tau} = \mathbb{E}\boldsymbol{\varepsilon} - \mathbb{E}_H\mathbf{p} = (\mathbb{E} - \mathbb{E}_H)\boldsymbol{\varepsilon} + \mathbb{E}_H\mathbf{e} \tag{11}$$

$\mathbb{E}_H$ being the elasto-hardening tensor:

$$\mathbb{E}_H = \mathbb{E} + \mathbb{H}_{kin} = (2G + h_{kin})\mathbb{I} + \lambda(\mathbf{1} \otimes \mathbf{1}) \tag{12}$$

Moreover, in order to put forward the basic structure of the evolution problem, the flow equations are recast in the form:

$$\begin{cases} \dot{\mathbf{e}} = \dot{\boldsymbol{\varepsilon}} - \dot{\gamma}\mathbf{n}_H \\[2mm] \dot{\zeta} = \dot{\gamma} \end{cases} \tag{13}$$

where $\mathbf{n}_H$ is the gradient of the yield function with respect to $\boldsymbol{\tau}$:

$$\begin{aligned} \mathbf{n}_H = \mathrm{d}_{\boldsymbol{\tau}}\phi(\boldsymbol{\tau},\vartheta) &= \frac{\partial\varphi}{\partial I_1}\mathbf{1} + \frac{\partial\varphi}{\partial J_2}\mathbf{S} + \frac{\partial\varphi}{\partial J_3}\Big(\mathbf{S}^2 - \frac{2}{3}J_2\mathbf{1}\Big) = \\ &= (\,\mathrm{d}_1\varphi - \tfrac{2}{3}J_2\,\mathrm{d}_3\varphi)\mathbf{1} + \mathrm{d}_2\varphi\mathbf{S} + \mathrm{d}_3\varphi\mathbf{S}^2 = \\ &= n_1\mathbf{1} + n_2\mathbf{S} + n_3\mathbf{S}^2 \end{aligned} \tag{14}$$

3   Discrete formulation and return map. 3D case

As a consequence of the additive character of (13), an update algorithm can be defined based on an elastic-plastic split by first integrating its elastic part to get the elastic prediction; this last one is taken in turn as the initial condition for the plastic correction, whereby the elastically predicted stress state is relaxed onto the updated yield surface [10].

In particular, the decomposition of the equations (13) into elastic and plastic parts is as follows. The former is defined as:

$$\begin{cases} \dot{\mathbf{e}} = \dot{\boldsymbol{\varepsilon}} \\ \dot{\zeta} = 0 \end{cases} \tag{15}$$

while the latter is given by:

$$\begin{cases} \dot{\mathbf{e}} = -\dot{\gamma}\mathbf{n}_H \\ \dot{\zeta} = \dot{\gamma} \end{cases} \tag{16}$$

for which the KKT conditions act as a constraint.

Equations (15) along with the initial conditions:

$$\mathbf{e}(t_n) = \boldsymbol{\varepsilon}(t_n) - \mathbf{p}(t_n) = \mathbf{e}_o; \qquad \zeta(t_n) = \zeta_o \tag{17}$$

are amenable to exact solution:

$$\begin{cases} \mathbf{e} = \mathbf{e}^{tr} = \mathbf{e}_o + \Delta\boldsymbol{\varepsilon} \\ \zeta = \zeta^{tr} = \zeta_o \end{cases} \tag{18}$$

where unsuffixed quantities denote the unknown values of the state variables at the end of the typical time step $[t_n, t_{n+1}]$.

The corresponding values of the stress-like variables are then:

$$\begin{cases} \boldsymbol{\tau}^{tr} = (\mathbb{E} - \mathbb{E}_H)\boldsymbol{\varepsilon} + \mathbb{E}_H\mathbf{e}^{tr} \\ \vartheta^{tr} = h_{iso}\left(\zeta^{tr}\right) \end{cases} \tag{19}$$

and define the trial stress state to be checked for plastic consistency.

Owing to the positive definiteness of the Hessian of the strain energy and to the convexity of the yield surface the algorithmic statement of the loading/unloading conditions can be formulated exclusively in terms of the trial state; indeed, the nature of the response can be determined based only on the trial value of the yield function:

$$\phi^{tr} = \phi\left(\boldsymbol{\tau}^{tr}, \vartheta^{tr}\right) \tag{20}$$

since the above hypotheses ensure that (20) does never underestimate the value of the yield function at the end of the time step [1].

Accordingly, if $\phi^{tr} \leq 0$ equations (16) do not need to be integrated and the trial state equals the one at solution; otherwise, plastic consistency has to be restored by solving the plastic equations for which (18) provide the initial conditions. In this last case, upon time integration of the flow equations (16) by means of the fully implicit (backward Euler) scheme, one obtains the nonlinear system:

$$\begin{cases} \mathbf{e} - \mathbf{e}^{tr} = -\gamma \mathbf{n}_H \\ \zeta - \zeta^{tr} = \gamma \end{cases} \tag{21}$$

where γ is the algorithmic plastic consistency parameter obeying the discrete consistency conditions in KKT form:

$$\phi(\boldsymbol{\tau}, \vartheta) \leq 0; \qquad \gamma \geq 0; \qquad \gamma\phi(\boldsymbol{\tau}, \vartheta) = 0 \tag{22}$$

For plastic loading ($\phi^{tr} > 0$) one has then the residual equations:

$$\begin{cases} \mathbf{r_e} = \mathbb{E}_H^{-1}(\boldsymbol{\tau} - \boldsymbol{\tau}^{tr}) + \gamma \mathbf{n}_H = \mathbf{0} \\ \mathbf{r}_\zeta = H_{iso}^{-1}(\vartheta - \vartheta^{tr}) - \gamma = 0 \\ \mathbf{r}_\phi = \phi(\boldsymbol{\tau}, \vartheta) = 0 \end{cases} \tag{23}$$

whose linearization around the $k-$th estimate of the solution yields:

$$\begin{bmatrix} \mathbf{r}_e^{(k)} \\ \mathbf{r}_\zeta^{(k)} \\ \mathbf{r}_\phi^{(k)} \end{bmatrix} + \begin{bmatrix} \mathbb{G}_H^{(k)} & \mathbf{0} & \mathbf{n}_H^{(k)} \\ \mathbf{0} & \left(H_{iso}^{(k)}\right)^{-1} & -1 \\ \left(\mathbf{n}_H^{(k)}\right)^T & -1 & 0 \end{bmatrix} \begin{bmatrix} \delta\boldsymbol{\tau}_{(k)}^{(k+1)} \\ \delta\vartheta_{(k)}^{(k+1)} \\ \delta\gamma_{(k)}^{(k+1)} \end{bmatrix} = \mathbf{0} \tag{24}$$

where $H_{iso}^{(k)} = h'_{iso}(\zeta^{(k)})$ is the isotropic hardening tangent modulus and $\mathbb{G}_H^{(k)}$ is the rank-four tensor:

$$\mathbb{G}_H^{(k)} = \mathbb{E}_H^{-1} + \gamma^{(k)} \, \mathrm{d}_{\tau\tau}^2 \phi^{(k)} \tag{25}$$

Note that $\mathbb{G}_H^{(k)}$ is positive-definite owing to the positive definiteness of the tensor $\mathbb{E}_H^{-1}$ and the convexity of the yield function. Hence, solving for the increment of the plastic parameter we get:

$$\delta\gamma_{(k)}^{(k+1)} = \frac{\mathbf{r}_\phi^{(k)} - \mathbb{G}_H^{(k)-1} \mathbf{r}_e^{(k)} \cdot \mathbf{n}_H^{(k)} + H_{iso}^{(k)} \mathbf{r}_\zeta^{(k)}}{\mathbb{G}_H^{(k)-1} \mathbf{n}_H^{(k)} \cdot \mathbf{n}_H^{(k)} + H_{iso}^{(k)}} \tag{26}$$

from which the updating formulas for the whole set of state variables follow.

Remark 1 As put forward by (26), the solution of the local return mapping algorithm is here accomplished by computing the inverse of the rank-four tensor $\mathbb{G}_H$ although this operation is not strictly required for computing the increments of the state variables, that could also have been obtained by directly solving the linear system (24). The reason underlying this choice is the fact that, as shown in the sequel, the tensor $\mathbb{G}_H^{-1}$ has to be computed at the end of the constitutive iterations in order to build up the expression of the algorithmic tangent operator, that is needed for the consistent linearization of the elastoplastic boundary value problem [11,12]. $\qquad\square$

4   The $\mathbb{G}_H$ tensor and its inverse

The standard approach exploited in the literature, see e.g. [1,13], for the computation of the inverse of the rank-four tensor $\mathbb{G}_H$ amounts to represent it as a 6x6 (4x4) matrix for 3D (plane strain or axisymmetric) problems and to perform the numerical inversion. However, as shown in [3], a more efficient approach can be exploited by providing suitable representation formulas for the tensor $\mathbb{G}_H$ and its inverse. In this respect, the following considerations are in order.

Setting:

$$\mathbb{E}_H = h_1 \mathbb{I} + h_2 (\mathbf{1} \otimes \mathbf{1}) \tag{27}$$

the elasto-hardening compliance tensor is computed as:

$$\mathbb{E}_H^{-1} = \frac{1}{h_1}\mathbb{I} - \frac{h_2}{h_1(h_1 + 3h_2)}(\mathbf{1} \otimes \mathbf{1}) \tag{28}$$

via the application of Sherman-Morrison-Woodbury's formula [14]. On account of (14) the second derivative of the yield function is given by [15]:

$$
\begin{aligned}
\mathrm{d}^2_{\tau\tau}\phi &= n_2\,\mathrm{d}_\tau \mathbf{S} + n_3\,\mathrm{d}_\tau \mathbf{S}^2 + \mathbf{1} \otimes \mathrm{d}_\tau n_1 + \mathbf{S} \otimes \mathrm{d}_\tau n_2 + \mathbf{S}^2 \otimes \mathrm{d}_\tau n_3 = \\
&= e_1 \mathbb{I} + e_2(\mathbf{S} \boxtimes \mathbf{1} + \mathbf{1} \boxtimes \mathbf{S}) + \\
&\quad + e_3(\mathbf{1} \otimes \mathbf{1}) + e_4(\mathbf{S} \otimes \mathbf{1} + \mathbf{1} \otimes \mathbf{S}) + e_5(\mathbf{S}^2 \otimes \mathbf{1} + \mathbf{1} \otimes \mathbf{S}^2) + \\
&\quad + e_6(\mathbf{S} \otimes \mathbf{S}) + e_7(\mathbf{S}^2 \otimes \mathbf{S} + \mathbf{S} \otimes \mathbf{S}^2) + e_8(\mathbf{S}^2 \otimes \mathbf{S}^2)
\end{aligned} \tag{29}
$$

where the coefficients of the expansion read:

$$
\begin{aligned}
&e_1 = \mathrm{d}_2\varphi = n_2; \qquad e_2 = \mathrm{d}_3\varphi = n_3 \\
&e_3 = \mathrm{d}_{11}\varphi - \tfrac{1}{3}\mathrm{d}_2\varphi - \tfrac{4}{3}J_2\,\mathrm{d}_{13}\varphi + \tfrac{4}{9}J_2^2\,\mathrm{d}_{33}\varphi; \\
&e_4 = \mathrm{d}_{12}\varphi - \tfrac{2}{3}\mathrm{d}_3\varphi - \tfrac{2}{3}J_2\,\mathrm{d}_{23}\varphi; \qquad e_5 = \mathrm{d}_{13}\varphi - \tfrac{2}{3}J_2\,\mathrm{d}_{33}\varphi; \\
&e_6 = \mathrm{d}_{22}\varphi; \qquad e_7 = \mathrm{d}_{23}\varphi, \qquad e_8 = \mathrm{d}_{33}\varphi
\end{aligned} \tag{30}
$$

$\mathrm{d}_i\varphi$ and $\mathrm{d}^2_{ij}\varphi$ being the first and second derivatives of the yield function with respect to the generic (pair of) invariant(s).

Accordingly, one has representation formula for $\mathbb{G}_H$:

$$
\begin{aligned}
\mathbb{G}_H &= g_1 \mathbb{I} + g_2(\mathbf{S} \boxtimes \mathbf{1} + \mathbf{1} \boxtimes \mathbf{S}) + \\
&\quad + g_3(\mathbf{1} \otimes \mathbf{1}) + g_4(\mathbf{S} \otimes \mathbf{1} + \mathbf{1} \otimes \mathbf{S}) + g_5(\mathbf{S}^2 \otimes \mathbf{1} + \mathbf{1} \otimes \mathbf{S}^2) + \\
&\quad + g_6(\mathbf{S} \otimes \mathbf{S}) + g_7(\mathbf{S}^2 \otimes \mathbf{S} + \mathbf{S} \otimes \mathbf{S}^2) + g_8(\mathbf{S}^2 \otimes \mathbf{S}^2)
\end{aligned} \tag{31}
$$

for:

$$
g_1 = \frac{1}{h_1} + \gamma e_1; \qquad g_2 = \gamma e_2; \qquad g_3 = -\frac{h_2}{h_1(h_1+3h_2)} + \gamma e_3;
$$

$$
g_4 = \gamma e_4; \qquad g_5 = \gamma e_5; \qquad g_6 = \gamma e_6; \qquad g_7 = \gamma e_7; \qquad g_8 = \gamma e_8
$$

The isotropic character of the tensor $\mathbb{G}_H$ follows from the isotropy of the elastic relationship and that of the yield function. Moreover, we note that it does possess the structure:

$$\mathbb{G}_H = \sum_{\alpha=0}^{2}\sum_{\beta=0}^{2}\left[p_{\alpha\beta}(\mathbf{S}^\alpha \otimes \mathbf{S}^\beta + \mathbf{S}^\beta \otimes \mathbf{S}^\alpha)\right]+$$
$$+\sum_{\alpha=0}^{1}\sum_{\beta=0}^{1}\left[q_{\alpha\beta}(\mathbf{S}^\alpha \boxtimes \mathbf{S}^\beta + \mathbf{S}^\beta \boxtimes \mathbf{S}^\alpha)\right] \tag{32}$$

where $\mathbf{S}^0 = \mathbf{1}$ while $p_{\alpha\beta}$ and $q_{\alpha\beta}$ are isotropic scalar functions.

In order to establish a representation formula for the fourth-order tensor $\mathbb{G}_H^{-1}$ consider the tensor equation:

$$\mathbb{G}_H\mathbf{X} = \mathbf{H} \tag{33}$$

in the unknown $\mathbf{X}$. Clearly, $\mathbf{X}$ is function of $\mathbf{H}$ and of the tensor $\mathbf{S}$ entering the definition (32) of $\mathbb{G}_H$:

$$\mathbf{X} = \mathbf{F}(\mathbf{S},\mathbf{H}) \tag{34}$$

It is not difficult to prove, by exploiting the composition rules between tensor products referred to in Appendix A, that $\mathbf{F}$ is an isotropic function of $\mathbf{S}$ and $\mathbf{H}$ simultaneoulsy. In this respect one has to show that:

$$\mathbf{Q}\mathbf{F}(\mathbf{S},\mathbf{H})\mathbf{Q}^T = \mathbf{F}(\mathbf{Q}\mathbf{S}\mathbf{Q}^T, \mathbf{Q}\mathbf{H}\mathbf{Q}^T) \qquad \forall \mathbf{Q} \in \text{Orth} \tag{35}$$

i.e. that $\mathbf{Q}\mathbf{X}\mathbf{Q}^T$ is solution of (33) provided that $\mathbf{S}$ and $\mathbf{H}$ are replaced by $\mathbf{Q}\mathbf{S}\mathbf{Q}^T$ and $\mathbf{Q}\mathbf{H}\mathbf{Q}^T$.

To provide a concise proof of the final result we further assume $\mathbf{H}$ to be symmetric; the more general proof can be obtained by following the same path of reasoning illustrated in [16] with reference to the simpler case of the tensor equation $\mathbf{A}\mathbf{X} + \mathbf{X}\mathbf{A} = \mathbf{H}$.

Being $\mathbf{F}$ linear in $\mathbf{H}$, the representation theorem for isotropic tensorial functions of two symmetric tensor arguments [6] yields:

$$\mathbf{X} = \mathbb{G}_H^{-1}\mathbf{H} = = a\mathbf{H} + b(\mathbf{S}\mathbf{H} + \mathbf{H}\mathbf{S}) + c(\mathbf{S}^2\mathbf{H} + \mathbf{H}\mathbf{S}^2)+$$
$$+d_{00}(\operatorname{tr}\mathbf{H})\mathbf{1} + d_{01}(\operatorname{tr}\mathbf{H})\mathbf{S} + d_{02}(\operatorname{tr}\mathbf{H})\mathbf{S}^2+$$
$$+d_{10}(\operatorname{tr}\mathbf{H}\mathbf{S})\mathbf{1} + d_{11}(\operatorname{tr}\mathbf{H}\mathbf{S})\mathbf{S} + d_{12}(\operatorname{tr}\mathbf{H}\mathbf{S})\mathbf{S}^2+ \tag{36}$$
$$+d_{20}(\operatorname{tr}\mathbf{H}\mathbf{S}^2)\mathbf{1} + d_{21}(\operatorname{tr}\mathbf{H}\mathbf{S}^2)\mathbf{S} + d_{22}(\operatorname{tr}\mathbf{H}\mathbf{S}^2)\mathbf{S}^2$$

whereby, owing to the symmetry of $\mathbf{H}$ and $\mathbf{S}$, one has:

$$\mathbb{G}_H^{-1} = a\mathbb{I} + b(\mathbf{S}\boxtimes\mathbf{1} + \mathbf{1}\boxtimes\mathbf{S}) + c(\mathbf{S}^2\boxtimes\mathbf{1} + \mathbf{1}\boxtimes\mathbf{S}^2)+$$
$$+d(\mathbf{1}\otimes\mathbf{1}) + e(\mathbf{S}\otimes\mathbf{1} + \mathbf{1}\otimes\mathbf{S}) + f(\mathbf{S}^2\otimes\mathbf{1} + \mathbf{1}\otimes\mathbf{S}^2)+ \tag{37}$$
$$+g(\mathbf{S}\otimes\mathbf{S}) + h(\mathbf{S}^2\otimes\mathbf{S} + \mathbf{S}\otimes\mathbf{S}^2) + i(\mathbf{S}^2\otimes\mathbf{S}^2)$$

Using the first Rivlin's identity for tensor polynomials [17], see Appendix A, we finally get the representation:

$$\mathbb{G}_H^{-1} = i_1\mathbb{I} + i_2(\mathbf{S} \boxtimes \mathbf{1} + \mathbf{1} \boxtimes \mathbf{S}) + i_3(\mathbf{S} \boxtimes \mathbf{S}) +$$

$$+ i_4(\mathbf{1} \otimes \mathbf{1}) + i_5(\mathbf{S} \otimes \mathbf{1} + \mathbf{1} \otimes \mathbf{S}) + i_6(\mathbf{S}^2 \otimes \mathbf{1} + \mathbf{1} \otimes \mathbf{S}^2) + \tag{38}$$

$$+ i_7(\mathbf{S} \otimes \mathbf{S}) + i_8(\mathbf{S}^2 \otimes \mathbf{S} + \mathbf{S} \otimes \mathbf{S}^2) + i_9(\mathbf{S}^2 \otimes \mathbf{S}^2)$$

whose coefficients $i_1, \ldots, i_9$ can be obtained by enforcing one of the two conditions:

$$\mathbb{G}_H\mathbb{G}_H^{-1} = \mathbb{G}_H^{-1}\mathbb{G}_H = \mathbb{I} \tag{39}$$

Remark 2 The assumed representation formula (38) is obviously not unique; actually, other equivalent expressions can be obtained from it via Rivlin's identities. However, the expression (38) is retained since, as shown in detail in [3], it allows a considerable simplification in the practical computation of the unknown coefficients $i_1, \ldots, i_9$, that are obtained by equating the coefficients of the tensors on both sides of the expressions stemming from the application of (39) and using the composition rules between tensor products along with the Cayley-Hamilton theorem:

$$\mathbf{A}^3 - I_\mathbf{A}\mathbf{A}^2 + II_\mathbf{A}\mathbf{A} - III_\mathbf{A}\mathbf{1} = \mathbf{0} \qquad \forall \mathbf{A} \in \mathrm{Lin} \tag{40}$$

It is also worth noting that use of (39) leads to non-symmetric tensor expressions; however, the linear system expressing the condition that the coefficient of the term $\mathbf{1} \boxtimes \mathbf{1}$ equals 1 while the remaining ones have to vanish does possess a particular structure. Actually, this system consists of three blocks of equations having the same coefficient matrix and sharing only one unknown so that, in practice, one has to solve three linear systems whose order decreases progressively from three to one (a single scalar equation).

This results in a very compact and efficient programing of the solution, see also [3] where the listing of the Fortran code for the 3D inverse computation is reported. $\qquad\qquad\square$

5   Consistent tangent. 3D case

The notion of consistent tangent tensor, introduced in the celebrated paper by Simo and Taylor [12], represents an essential ingredient for the effective solution of the elastoplastic boundary value problem via full Newton-Raphson linearization. The expression of the consistent tangent presented in [18]:

$$\mathbb{E}_{tan} = \mathbb{E} - \mathbb{E}\mathbb{F}_H^{-1}\mathbb{E} - \frac{(\mathbb{E} - \mathbb{E}\mathbb{F}_H^{-1}\mathbb{E}_H)\mathbf{n}_H \otimes (\mathbb{E} - \mathbb{E}\mathbb{F}_H^{-1}\mathbb{E}_H)\mathbf{n}_H}{(\mathbb{E}_H - \mathbb{E}_H\mathbb{F}_H^{-1}\mathbb{E}_H)\mathbf{n}_H \cdot \mathbf{n}_H + H} \tag{41}$$

where:

$$\mathbb{F}_H^{-1} = (\gamma\, \mathrm{d}_{\boldsymbol{\tau}\boldsymbol{\tau}}^2\phi) - (\gamma\, \mathrm{d}_{\boldsymbol{\tau}\boldsymbol{\tau}}^2\phi)\mathbb{G}_H^{-1}(\gamma\, \mathrm{d}_{\boldsymbol{\tau}\boldsymbol{\tau}}^2\phi) \tag{42}$$

generalizes the one originally contributed for J_2 plasticity in [12] since it applies to general isotropic elastoplastic models endowed with linear hardening.

The above relationship is however not optimal in view of actual implementation since it is obtained by considering the problem of the construction of the consistent tangent tensor completely disjoint from that of the stress computation. Indeed, relationship (41) fails to exploit the fact that all what is needed for implementing the consistent tangent can be readily obtained from the linearized form of the residual equations, thus resulting in an expression which is quite complicated and, in addition, not well-suited for multisurface plasticity and not prone to an immediate specialization to problems with a reduced number of stress components (as in the plane stress case).

As also shown in [19], a more direct derivation of the expression of the consistent tangent can be obtained by using the following arguments.

Since in the return mapping algorithm reference is made to the relative stress tensor $\boldsymbol{\tau}$, we start by considering the linearizations of (3) and (11):

$$d_\varepsilon\boldsymbol{\sigma} = \mathbb{E} - \mathbb{E}\,d_\varepsilon\mathbf{p} \tag{43a}$$

$$d_\varepsilon\boldsymbol{\tau} = \mathbb{E} - \mathbb{E}_H\,d_\varepsilon\mathbf{p} \tag{43b}$$

whose comparison yields the relationship existing between the consistent tangent $\mathbb{E}_{tan} = d_\varepsilon\boldsymbol{\sigma}$ and the rank-four tensor $d_\varepsilon\boldsymbol{\tau}$ as:

$$d_\varepsilon\boldsymbol{\sigma} = \mathbb{E}_{tan} = \mathbb{E} - \mathbb{E}\mathbb{E}_H^{-1}\mathbb{E} + \mathbb{E}\mathbb{E}_H^{-1}\,d_\varepsilon\boldsymbol{\tau} \tag{44}$$

The previous equation can be effectively exploited for computing the consistent tangent since the derivative $d_\varepsilon\boldsymbol{\tau}$ is easily obtained from the linearization of the residual equations at the local converged state, i.e. from the system:

$$\begin{bmatrix} -\mathbb{E}_H^{-1}\mathbb{E}\delta\varepsilon^* \\[2mm] 0 \\[2mm] 0 \end{bmatrix} + \begin{bmatrix} \mathbb{G}_H & \mathbf{0} & \mathbf{n}_H \\[2mm] \mathbf{0} & H_{iso}^{-1} & -1 \\[2mm] (\mathbf{n}_H)^T & -1 & 0 \end{bmatrix} \begin{bmatrix} d_\varepsilon\boldsymbol{\tau}\,\delta\varepsilon^* \\[2mm] d_\varepsilon\vartheta\cdot\delta\varepsilon^* \\[2mm] d_\varepsilon\gamma\cdot\delta\varepsilon^* \end{bmatrix} = \mathbf{0} \quad \forall\delta\varepsilon^* \tag{45}$$

embodying the full linearization of (11) and (10) along with Prager's consistency condition [5].

Solving for $d_\varepsilon\boldsymbol{\tau}$ one has then:

$$d_\varepsilon\boldsymbol{\tau} = \mathbb{G}_H^{-1}\mathbb{E}_H^{-1}\mathbb{E} - \frac{\mathbb{G}_H^{-1}\mathbf{n}_H \otimes \mathbb{G}_H^{-1}\mathbf{n}_H}{\mathbb{G}_H^{-1}\mathbf{n}_H\cdot\mathbf{n}_H + H_{iso}}\mathbb{E}_H^{-1}\mathbb{E} = \mathbb{D}_{tan}\mathbb{E}_H^{-1}\mathbb{E} \tag{46}$$

whence, on account of (44), the expression of the consistent tangent is obtained as:

$$\mathbb{E}_{tan} = \mathbb{E} - \mathbb{E}\mathbb{E}_H^{-1}\mathbb{E} + (\mathbb{E}\mathbb{E}_H^{-1} \boxtimes \mathbb{E}\mathbb{E}_H^{-1})\,\mathbb{D}_{tan} \tag{47}$$

and the definition of the square product between rank-four tensors follows from the generalization of that between rank-two tensors, see Appendix A.

We note in passing that in absence of any hardening the previous formula collapses to the well-known expression of the consistent tangent of ideal plasticity:

$$\mathbb{E}_{tan} = \mathbb{G}^{-1} - \frac{\mathbb{G}^{-1}\mathbf{n} \otimes \mathbb{G}^{-1}\mathbf{n}}{\mathbb{G}^{-1}\mathbf{n} \cdot \mathbf{n}} \tag{48}$$

$\mathbb{G}$ and $\mathbf{n}$ being the obvious specializations of $\mathbb{G}_H$ and $\mathbf{n}_H$ to this last case.

In the light of the previous considerations it is apparent that, once the tensor $\mathbb{G}_H^{-1}$ is known, the expression of the consistent tangent can be arrived at via elementary algebraic operations since one basically needs to compute the symmetric tensor $\mathbb{E}\mathbb{E}_H^{-1}$, whose expression reads:

$$\mathbb{E}\mathbb{E}_H^{-1} = \mathbb{E}_H^{-1}\mathbb{E} = m_1\mathbb{I} + m_2(\mathbf{1} \otimes \mathbf{1}) \tag{49}$$

with coefficients m_1 and m_2 given by:

$$\begin{aligned} m_1 &= \frac{2G}{2G + h_{kin}} \\ m_2 &= \frac{m_1 \lambda h_{kin}}{2G(2G + h_{kin} + 3\lambda)} \end{aligned} \tag{50}$$

Combining the previous relationships with (14) and (38), one has then the final expression of the consistent tangent:

$$\begin{aligned} \mathbb{E}_{tan} = {}& t_1\mathbb{I} + t_2(\mathbf{S} \boxtimes \mathbf{1} + \mathbf{1} \boxtimes \mathbf{S}) + t_3(\mathbf{S} \boxtimes \mathbf{S}) + \\ & + t_4(\mathbf{1} \otimes \mathbf{1}) + t_5(\mathbf{S} \otimes \mathbf{1} + \mathbf{1} \otimes \mathbf{S}) + t_6(\mathbf{S}^2 \otimes \mathbf{1} + \mathbf{1} \otimes \mathbf{S}^2) + \\ & + t_7(\mathbf{S} \otimes \mathbf{S}) + t_8(\mathbf{S}^2 \otimes \mathbf{S} + \mathbf{S} \otimes \mathbf{S}^2) + t_9(\mathbf{S}^2 \otimes \mathbf{S}^2) \end{aligned} \tag{51}$$

whose coefficients $t_1 \ldots t_9$ are polynomial functions of J_2, J_3, $i_1 \ldots i_9$, h_1, h_2 and of the Lamé's constants.

For the sake of completeness the explicit form of the coefficients of the consistent tangent (51) are reported in Appendix B.

Remark 3 The effectiveness of the approach developed in the previous sections is further emphasized in the case of isotropic multisurface plasticity models. Indeed, in this case use of the generalized Koiter's flow rule [20] leads to an expression for the consistent tangent analogous to (47) in which the term $\mathbb{D}_{tan}$ reads:

$$\mathbb{D}_{tan} = \mathbb{G}_H^{-1} - \sum_{\alpha,\beta \in \mathcal{S}_A} \left(\frac{\mathbb{G}_H^{-1}\mathbf{n}_\alpha \otimes \mathbb{G}_H^{-1}\mathbf{n}_\beta}{\sum\limits_{\alpha,\beta \in \mathcal{S}_A} \mathbb{G}_H^{-1}\mathbf{n}_\alpha \cdot \mathbb{G}_H^{-1}\mathbf{n}_\beta + H_{iso}} \right) \tag{52}$$

where the tensor $\mathbb{G}_H$ is given as:

$$\mathbb{G}_H = \mathbb{E}_H^{-1} + \sum_{\alpha \in \mathcal{S}_A} \gamma_\alpha \, \mathrm{d}^2_{\tau\tau}\phi_\alpha \tag{53}$$

$\mathcal{S}_A$ being the set of active constraints, see also [1].

The developed tensor algebra allows one to obtain the tangent tensor in entirely explicit form so that only the matrix representation is needed for implementing the consistent tangent. On the contrary, lacking the theoretical background put forward in the previous sections, the expression of the consistent tangent to be assembled into the global structural tangent matrix can be arrived at only via lengthy matrix operations.   □

Remark 4 A useful result stemming from the assumption of isotropy is that the return mapping algorithm takes place at fixed principal axis; accordingly, only the principal values of the state variables need to be iterated upon and the local stress update algorithm formulated in the principal frame simply requires the inversion of the 3x3 principal tangent compliance matrix [21].

It is then quite natural to wonder if the computation of the formal inverse of the rank-four tensor $\mathbb{G}_H$ does entail any significant implication in a principal space implementation.

In [4] is shown that this inversion is fully equivalent to the inversion of the rank-four compliance tangent tensor in the global coordinate system, a result that is arrived at by investigating the properties of isotropic functions of a symmetric tensor argument. As a consequence, the usual procedure of expressing the tangent tensor in the principal reference frame and then transforming it to the given reference frame via matrix manipulations can be by-passed since the final expression of the consistent tangent in the global coordinate system can be directly constructed, with the additional advantage that all the terms entering it can be clearly identified. Moreover, even in the principal axis formulation the yield function can be assigned either in terms of eigenvalues or of stress invariants without affecting the derivation of the tensor quantities entering the return mapping and the tangent operator.   □

6   The plane stress problem. Formulation and return map solution

Let us now turn to discuss the plane stress problem, i.e. the three-dimensional problem for which the plane stress condition:

$$\sigma_{13} = \sigma_{23} = \sigma_{33} = 0$$

is enforced.

At first sight, the elastoplastic constitutive problem directly formulated in the plane stress subspace would require an algorithmic treatment specifically designed for the problem at hand [22]. In particular, apart from the obvious modification of the elastic relationship, one should seemingly need to re-define the yield function in terms of the non-vanishing stress components and re-compute the relevant linearizations, an approach that, besides being very time consuming, turns out to be fairly involved for yield criteria including the cubic stress invariant J_3, see e.g. [23].

This is however unnecessary since, as shown in [19,24], the solution procedure for plane stress elastoplastic models endowed with general isotropic yield criteria can be consistently derived from the corresponding 3D one and, in addition, one can still make reference to the expression of the yield function given in terms of the 3D stress invariants (7).

The rationale for providing a consistent derivation of the plane stress constitutive algorithm relies upon the splitting of the equations governing the three-dimensional problem into two parts accounting for the in-plane and out-of-plane model response. To this end we define the rank-four projection operators:

$$\mathbb{P} = \mathbb{P}^T = (\mathbb{P})^n \qquad n > 0 \tag{54a}$$

$$\mathbb{P}_c = \mathbb{I} - \mathbb{P} \tag{54b}$$

respectively mapping a generic symmetric second-order tensor $\mathbf{A}$ onto the complementary subspaces $\mathcal{S}_p$ and $\mathcal{S}_c$ of Lin defined by its in-plane and out-of-plane components.

In what follows we shall use the notation $\hat{\mathbf{A}}$ ($\check{\mathbf{A}}$) to denote the 2D tensor collecting the in-plane (out-of-plane) components of $\mathbf{A}$ defined through:

$$\mathbb{P}\,\mathbf{A} \rightleftharpoons \hat{\mathbf{A}} \tag{55a}$$

$$\mathbb{P}_c\mathbf{A} \rightleftharpoons \check{\mathbf{A}} \tag{55b}$$

where the symbol $\rightleftharpoons$ is used to emphasize that the 3D tensors on the left are isometrically mapped to the 2D tensors on the right-hand side.

A decomposition analogous to the one introduced above for rank-two tensors can be given for a generic symmetric rank-four 3D tensor $\mathbb{A}$ as:

$$\mathbb{A} = (\mathbb{P} \boxtimes \mathbb{P})\mathbb{A} + (\mathbb{P} \boxtimes \mathbb{P}_c)\mathbb{A} + (\mathbb{P}_c \boxtimes \mathbb{P})\mathbb{A} + (\mathbb{P}_c \boxtimes \mathbb{P}_c)\mathbb{A} =$$
$$= \mathbb{P}\mathbb{A}\mathbb{P}^T + \mathbb{P}\mathbb{A}\mathbb{P}_c^T + \mathbb{P}_c\mathbb{A}\mathbb{P}^T + \mathbb{P}_c\mathbb{A}\mathbb{P}_c^T \tag{56}$$

by means of the rank-eight tensors $\mathbb{P}\boxtimes\mathbb{P}$, $\mathbb{P}\boxtimes\mathbb{P}_c$, $\mathbb{P}_c\boxtimes\mathbb{P}$ and $\mathbb{P}_c\boxtimes\mathbb{P}$. Hence, by appealing to the matrix representation induced by the component ordering:

$$[\mathbb{A}] = \big[[\hat{\mathbb{A}}] \quad [\check{\mathbb{A}}]\big]^T \tag{57}$$

the matrix form corresponding to the decomposition (56) reads:

$$[\mathbb{A}] = \begin{bmatrix} [\hat{\mathbb{A}}] & [\tilde{\mathbb{A}}] \\ [\tilde{\mathbb{A}}] & [\check{\mathbb{A}}] \end{bmatrix} \tag{58}$$

where the following isometric mappings:

$$\mathbb{P}\mathbb{A}\mathbb{P}^T \rightleftharpoons \hat{\mathbb{A}}; \qquad \mathbb{P}\mathbb{A}\mathbb{P}_c^T \rightleftharpoons \tilde{\mathbb{A}}$$
$$\mathbb{P}_c\mathbb{A}\mathbb{P}^T \rightleftharpoons \tilde{\mathbb{A}}; \qquad \mathbb{P}_c\mathbb{A}\mathbb{P}_c^T \rightleftharpoons \check{\mathbb{A}} \tag{59}$$

are in order.

Following the previous ideas it is not difficult to show that one can infer the plane stress elastic constitutive equation:

$$\hat{\boldsymbol{\sigma}} = \hat{\mathbb{E}}(\hat{\boldsymbol{\varepsilon}} - \hat{\mathbf{p}}) \tag{60}$$

from the three-dimensional relationship (3), where the in-plane elastic tensor is given by:

$$\hat{\mathbb{E}} = 2G\hat{\mathbb{I}} + \frac{2G\lambda}{(2G + \lambda)}(\hat{\mathbf{1}} \otimes \hat{\mathbf{1}}) = \left(\widehat{\mathbb{P}\mathbb{E}^{-1}\mathbb{P}^T}\right)^{-1} \tag{61}$$

Moreover, the in-plane/out-of-plane splitting of the time-discrete flow equations (21) is obtained as:

$$\hat{\mathbf{e}} - \hat{\mathbf{e}}^{tr} = -\hat{\mathbf{p}} + \hat{\mathbf{p}}_o = -\gamma\, \mathrm{d}_{\hat{\tau}}\phi \tag{62a}$$

$$\check{\mathbf{e}} - \check{\mathbf{e}}^{tr} = -\check{\mathbf{p}} + \check{\mathbf{p}}_o = -\gamma\, \mathrm{d}_{\check{\tau}}\phi \tag{62b}$$

where $\mathrm{d}_{\hat{\tau}}\phi$ and $\mathrm{d}_{\check{\tau}}\phi$ denote the rank-two 2D tensors representing the derivatives of the yield function ϕ with respect to the in-plane and out-of-plane relative stresses:

$$\hat{\boldsymbol{\tau}} = \hat{\mathbb{E}}\hat{\boldsymbol{\varepsilon}} - \hat{\mathbb{E}}_H\hat{\mathbf{p}} \tag{63a}$$

$$\check{\boldsymbol{\tau}} = \qquad - \check{\mathbb{H}}_{kin}\check{\mathbf{p}} \tag{63b}$$

having denoted by $\hat{\mathbb{E}}_H$ and $\check{\mathbb{H}}_{kin}$ the in-plane elasto-hardening tensor and the out-of-plane kinematic hardening operator respectively.

Accordingly, the residual equations can be reformulated in staggered form as:

$$\mathbf{r}_{\check{\mathbf{e}}} = \check{\mathbb{H}}_{kin}^{-1}\left(\check{\boldsymbol{\tau}} - \check{\boldsymbol{\tau}}^{tr}\right) + \gamma\, \mathrm{d}_{\check{\tau}}\phi(\hat{\boldsymbol{\tau}}, \check{\boldsymbol{\tau}}, \vartheta) = \mathbf{0} \tag{64}$$

$$\begin{cases} \mathbf{r}_{\hat{\mathbf{e}}} = \hat{\mathbb{E}}_H^{-1}\left(\hat{\boldsymbol{\tau}} - \hat{\boldsymbol{\tau}}^{tr}\right) + \gamma\, \mathrm{d}_{\hat{\tau}}\phi(\hat{\boldsymbol{\tau}}, \check{\boldsymbol{\tau}}, \vartheta) = \mathbf{0} \\[2mm] \mathbf{r}_{\zeta} = H_{iso}^{-1}\left(\vartheta - \vartheta^{tr}\right) - \gamma = 0 \\[2mm] \mathbf{r}_{\phi} = \phi(\hat{\boldsymbol{\tau}}, \check{\boldsymbol{\tau}}, \vartheta) = 0 \end{cases} \tag{65}$$

where the trial stresses are given by:

$$\hat{\boldsymbol{\tau}}^{tr} = \hat{\mathbb{E}}\hat{\boldsymbol{\varepsilon}} - \hat{\mathbb{E}}_H\hat{\mathbf{p}}_o; \qquad \check{\boldsymbol{\tau}}^{tr} = -\check{\mathbb{H}}_{kin}\check{\mathbf{p}}_o \tag{66}$$

and the dependence of ϕ upon the in-plane $\hat{\boldsymbol{\tau}}$ and out-of-plane $\check{\boldsymbol{\tau}}$ parts of $\boldsymbol{\tau}$ has been emphasized by writing $\phi(\hat{\boldsymbol{\tau}}, \check{\boldsymbol{\tau}}, \vartheta)$.

Note that the variables in (64) interact with those in (65) through the tensor $\check{\boldsymbol{\tau}}$ and that (62b) establishes a nonlinear relation between the out-of-plane plastic strain and the stresses $\hat{\boldsymbol{\tau}}$ and $\check{\boldsymbol{\tau}}$. Accordingly, one has:

$$\mathrm{d}_{\hat{\tau}}\check{\mathbf{p}} = \gamma\, \mathrm{d}_{\check{\tau}\check{\tau}}^2\phi + \gamma(\mathrm{d}_{\check{\tau}\hat{\tau}}^2\phi)\mathrm{d}_{\hat{\tau}}\check{\boldsymbol{\tau}} + \mathrm{d}_{\check{\tau}}\phi \otimes \mathrm{d}_{\hat{\tau}}\gamma \tag{67}$$

that, once combined with the derivative of (63b) with respect to $\hat{\boldsymbol{\tau}}$, provides:

$$\mathrm{d}_{\hat{\tau}}\check{\boldsymbol{\tau}} = -\check{\mathbb{A}}_H^{-1}(\gamma\, \mathrm{d}_{\check{\tau}\hat{\tau}}^2\phi + \mathrm{d}_{\check{\tau}}\phi \otimes \mathrm{d}_{\hat{\tau}}\gamma) \tag{68}$$

where $\check{\mathbb{A}}_H$ is the rank-four two-dimensional tensor:

$$\check{\mathbb{A}}_H = \check{\mathbb{H}}_{kin}^{-1} + \gamma\, \mathrm{d}_{\check{\boldsymbol{\tau}}\check{\boldsymbol{\tau}}}^2 \phi \tag{69}$$

and the short-hand notation:

$$\mathrm{d}_{\alpha\beta}^2 \phi \qquad \{\alpha, \beta\} \in \{\hat{\boldsymbol{\tau}}, \check{\boldsymbol{\tau}}\} \tag{70}$$

has been used to denote the rank-four two-dimensional tensors obtained from the Hessian of the yield function via the mappings (59).

The perfect analogy between the plane stress return mapping and the three-dimensional stress computation scheme (24) is self-evident by considering the linearized equations:

$$\begin{bmatrix} \mathbf{r}_{\hat{\mathbf{e}}}^{(k)} \\[6pt] \mathbf{r}_{\zeta}^{(k)} \\[6pt] \mathbf{r}_{\phi}^{(k)} \end{bmatrix} + \begin{bmatrix} \hat{\mathbb{G}}_H^{(k)} & \mathbf{0} & \hat{\mathbf{n}}_H^{(k)} \\[6pt] \mathbf{0} & \left(H_{iso}^{(k)}\right)^{-1} & -1 \\[6pt] \left(\hat{\mathbf{n}}_H^{(k)}\right)^T & -1 & -\Theta^{(k)} \end{bmatrix} \begin{bmatrix} \delta\hat{\boldsymbol{\tau}}_{(k)}^{(k+1)} \\[6pt] \delta\vartheta_{(k)}^{(k+1)} \\[6pt] \delta\gamma_{(k)}^{(k+1)} \end{bmatrix} = \begin{bmatrix} \mathbf{0} \\[6pt] 0 \\[6pt] 0 \end{bmatrix} \tag{71}$$

where it has been set:

$$\hat{\mathbf{n}}_H^{(k)} = \mathrm{d}_{\hat{\boldsymbol{\tau}}}\phi^{(k)} - (\gamma^{(k)}\, \mathrm{d}_{\check{\boldsymbol{\tau}}\hat{\boldsymbol{\tau}}}^2 \phi^{(k)})(\check{\mathbb{A}}_H^{(k)})^{-1}\, \mathrm{d}_{\check{\boldsymbol{\tau}}}\phi^{(k)} \tag{72}$$

$$\Theta^{(k)} = (\check{\mathbb{A}}_H^{(k)})^{-1}\, \mathrm{d}_{\check{\boldsymbol{\tau}}}\phi^{(k)} \cdot \mathrm{d}_{\check{\boldsymbol{\tau}}}\phi^{(k)} \tag{73}$$

and $\hat{\mathbb{G}}_H$ is the rank-four two-dimensional tensor:

$$\hat{\mathbb{G}}_H^{(k)} = \hat{\mathbb{E}}_H^{-1} + \gamma^{(k)}\, \mathrm{d}_{\hat{\boldsymbol{\tau}}\hat{\boldsymbol{\tau}}}^2 \phi^{(k)} - (\gamma^{(k)}\, \mathrm{d}_{\hat{\boldsymbol{\tau}}\check{\boldsymbol{\tau}}}^2 \phi^{(k)})(\check{\mathbb{A}}_H^{(k)})^{-1}(\gamma^{(k)}\, \mathrm{d}_{\check{\boldsymbol{\tau}}\hat{\boldsymbol{\tau}}}^2 \phi^{(k)}) \tag{74}$$

As shown in [19], $\hat{\mathbb{G}}_H$ and $\check{\mathbb{A}}_H$ are positive-definite; hence, being also $H_{iso}^{(k)} > 0$, one can solve for the increment $\delta\gamma_{(k)}^{(k+1)}$ to get:

$$\delta\gamma_{(k)}^{(k+1)} = \frac{\mathbf{r}_{\phi}^{(k)} - \hat{\mathbb{G}}_H^{(k)\,-1}\mathbf{r}_{\hat{\mathbf{e}}}^{(k)} \cdot \hat{\mathbf{n}}_H^{(k)} + H_{iso}^{(k)}\mathbf{r}_{\zeta}^{(k)}}{\hat{\mathbb{G}}_H^{(k)\,-1}\hat{\mathbf{n}}_H^{(k)} \cdot \hat{\mathbf{n}}_H^{(k)} + \Theta^{(k)} + H_{iso}^{(k)}} \tag{75}$$

Remark 5 It is worth noting that, once the increments $\delta\hat{\boldsymbol{\tau}}_{(k)}^{(k+1)}$ and $\delta\vartheta_{(k)}^{(k+1)}$ have been recovered through (71), it is necessary to update the out-of-plane stress tensor $\check{\boldsymbol{\tau}}$.

This can be done via the updating formula:

$$\delta\check{\boldsymbol{\tau}}_{(k)}^{(k+1)} = -(\check{\mathbb{A}}_H^{(k)})^{-1}(\mathbf{r}_{\check{\mathbf{e}}}^{(k)} + \gamma^{(k)}\, \mathrm{d}_{\check{\boldsymbol{\tau}}\hat{\boldsymbol{\tau}}}^2 \phi^{(k)}\delta\hat{\boldsymbol{\tau}}_{(k)}^{(k+1)} + \mathrm{d}_{\check{\boldsymbol{\tau}}}\phi^{(k)}\delta\gamma_{(k)}^{(k+1)}) \tag{76}$$

that follows from the linearization of (64). $\square$

Remark 6 The solution of the plane stress constitutive algorithm could have been obtained via the simultaneous linearization of (64) and (65). As in the 3D case we have chosen to to proceed as shown in (71) and (75) mainly in view of the actual implementation of the consistent tangent, whose derivation is detailed in the next section. $\square$

7   Consistent tangent. Plane stress case

Objective of this section is that of evaluating the consistent tangent tensor [12], i.e. the the rank-four two-dimensional tensor:

$$\hat{\mathbb{E}}_{tan} = \mathrm{d}_{\hat{\varepsilon}}\hat{\boldsymbol{\sigma}} \tag{77}$$

consistent with the stress computation algorithm discussed in the previous section. To this end we shall follow the approach detailed in section 3 and start from considering the linearization of the elastic law (60) and of the constitutive equation for the in-plane relative stress (63a):

$$\mathrm{d}_{\hat{\varepsilon}}\hat{\boldsymbol{\sigma}} = \hat{\mathbb{E}} - \hat{\mathbb{E}}\,\mathrm{d}_{\hat{\varepsilon}}\hat{\mathbf{p}} \tag{78a}$$

$$\mathrm{d}_{\hat{\varepsilon}}\hat{\boldsymbol{\tau}} = \hat{\mathbb{E}} - \hat{\mathbb{E}}_H\,\mathrm{d}_{\hat{\varepsilon}}\hat{\mathbf{p}}, \tag{78b}$$

to get the relationship:

$$\hat{\mathbb{E}}_{tan} = \hat{\mathbb{E}} - \hat{\mathbb{E}}\hat{\mathbb{E}}_H^{-1}\hat{\mathbb{E}} + \hat{\mathbb{E}}\hat{\mathbb{E}}_H^{-1}\,\mathrm{d}_{\hat{\varepsilon}}\hat{\boldsymbol{\tau}} \tag{79}$$

In perfect analogy with the three-dimensional case, the expression of the derivative $\mathrm{d}_{\hat{\varepsilon}}\hat{\boldsymbol{\tau}}$ can be obtained from the linearization with respect to the driving variable $\hat{\varepsilon}$ of the residual equations at the local converged state as:

$$\mathrm{d}_{\hat{\varepsilon}}\hat{\boldsymbol{\tau}} = \hat{\mathbb{G}}_H^{-1}\,\hat{\mathbb{E}}_H^{-1}\hat{\mathbb{E}} - \frac{\hat{\mathbb{G}}_H^{-1}\hat{\mathbf{n}}_H \otimes \hat{\mathbb{G}}_H^{-1}\hat{\mathbf{n}}_H}{\hat{\mathbb{G}}_H^{-1}\hat{\mathbf{n}}_H \cdot \hat{\mathbf{n}}_H + \Theta + H_{iso}}\,\hat{\mathbb{E}}_H^{-1}\hat{\mathbb{E}} = \hat{\mathbb{D}}_{tan}\,\hat{\mathbb{E}}_H^{-1}\hat{\mathbb{E}} \tag{80}$$

whence the expression of the plane stress consistent tangent is obtained as:

$$\hat{\mathbb{E}}_{tan} = \hat{\mathbb{E}} - \hat{\mathbb{E}}\hat{\mathbb{E}}_H^{-1}\hat{\mathbb{E}} + (\hat{\mathbb{E}}\hat{\mathbb{E}}_H^{-1} \boxtimes \hat{\mathbb{E}}\hat{\mathbb{E}}_H^{-1})\,\hat{\mathbb{D}}_{tan} \tag{81}$$

where:

$$\hat{\mathbb{E}}\hat{\mathbb{E}}_H^{-1} = \hat{\mathbb{E}}_H^{-1}\hat{\mathbb{E}} = \hat{m}_1\hat{\mathbb{I}} + \hat{m}_2(\hat{\mathbf{1}} \otimes \hat{\mathbf{1}}) \tag{82}$$

$\hat{m}_1$ and $\hat{m}_2$ being given by:

$$\hat{m}_1 = \frac{2G}{2G + h_{kin}} = m_1$$

$$\hat{m}_2 = \frac{m_1 \lambda h_{kin}}{(2G + h_{kin})(2G + \lambda) + 4G\lambda} \tag{83}$$

Remark 7 Using the matrix representation, the evaluation of $\hat{\mathbb{G}}_H^{-1}$ simply requires the inversion of a 3x3 matrix.

However, following a path of reasoning similar to the one used in the three-dimensional case it can be proved that $\hat{\mathbb{G}}_H^{-1}$ is amenable to the following representation [24]

$$\hat{\mathbb{G}}_H^{-1} = \hat{\imath}_1(\hat{\mathbf{1}} \boxtimes \hat{\mathbf{1}}) + \hat{\imath}_2(\hat{\mathbf{1}} \otimes \hat{\mathbf{1}}) + \hat{\imath}_3(\hat{\mathbf{1}} \otimes \hat{\mathbf{S}} + \hat{\mathbf{S}} \otimes \hat{\mathbf{1}}) + \hat{\imath}_4(\hat{\mathbf{S}} \otimes \hat{\mathbf{S}}) \tag{84}$$

where the unknown coefficients can be determined via the condition:

$$\hat{\mathbb{G}}_H \hat{\mathbb{G}}_H^{-1} = \hat{\mathbb{G}}_H^{-1} \hat{\mathbb{G}}_H = \hat{\mathbb{I}} \tag{85}$$

that basically requires the solution of a linear system of order two.

Accordingly, no matrix operations are needed for obtaining the consistent tangent $\hat{\mathbb{E}}_{tan}$ that can be given an entirely explicit (inverse-free) representation as:

$$\hat{\mathbb{E}}_{tan} = \hat{t}_1(\hat{\mathbf{1}} \boxtimes \hat{\mathbf{1}}) + \hat{t}_2(\hat{\mathbf{1}} \otimes \hat{\mathbf{1}}) + \hat{t}_3(\hat{\mathbf{1}} \otimes \hat{\mathbf{S}} + \hat{\mathbf{S}} \otimes \hat{\mathbf{1}}) + \hat{t}_4(\hat{\mathbf{S}} \otimes \hat{\mathbf{S}}) \tag{86}$$

The coefficients $\hat{t}_1, \ldots, \hat{t}_4$ of the above representation formula, that are polynomial functions of $\hat{\imath}_1 \ldots \hat{\imath}_4$, the Lamé's constants, the hardening moduli and the invariants of the two-dimensional stress deviator $\hat{\mathbf{S}}$, are explicitly given in Appendix C. $\qquad\qquad\square$

8 Numerical example

To demonstrate the performance of the solution algorithms discussed in the previous sections we present hereafter a numerical example; the simulation refers to an isotropic yield function obtained as a pressure-insensitive version of the celebrated five parameter concrete model originally proposed by Willam and Warnke in [25]:

$$\varphi(J_2, J_3) = \frac{[2J_2]^{1/2}}{r(\theta, e)} - \sqrt{\frac{2}{3}} f'_c \tag{87}$$

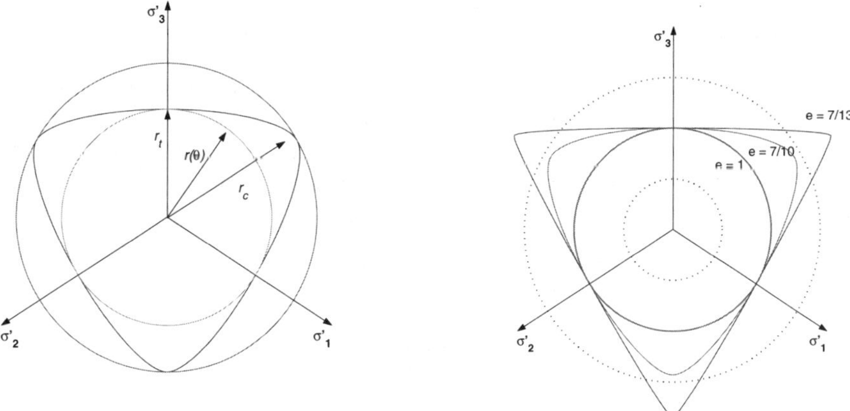

Fig. 1. Deviatoric sections of the yield locus.

The two model parameters f'_c and e respectively represent the magnitude of the limit stress in uniaxial compression and the eccentricity describing the out-of-roundness of the deviatoric section; this last one is defined through an elliptic interpolation between the tensile ($\theta = 0$) and compressive ($\theta = \pi/3$) meridians:

$$r(\theta, e) = \frac{2(1 - e^2)\cos\theta - (1 - 2e)[4(1 - e^2)\cos^2\theta + 5e^2 - 4e]^{1/2}}{4(1 - e^2)\cos^2\theta + (1 - 2e)^2} \tag{88}$$

θ being the Lode angle [26]:

$$\theta = \frac{1}{3}\arccos\left(\frac{3\sqrt{3}}{2}\frac{J_3}{J_2^{3/2}}\right); \qquad \theta \in \left[0, \frac{\pi}{3}\right] \tag{89}$$

The failure surface (87) can be shown to possess convexity within the range $\frac{1}{2} \leq e \leq 1$; the minimum shape factor $e = \frac{1}{2}$ corresponds to a triangular shape of the deviatoric section:

$$\varphi(J_2, J_3) = [2J_2]^{1/2}\, 2\cos\theta - \sqrt{\frac{2}{3}}f_c' \tag{90}$$

whereas for $e = 1$ the influence of the J_3 invariant through the Lode angle θ is dropped out and the failure surface collapses to the Von Mises cylinder. The deviatoric section of the yield locus (87) is depicted in figure 1.

8.1   Strip with circular hole

The problem studied, already considered in [22], consists of a tension strip with a central circular hole under plane stress conditions.

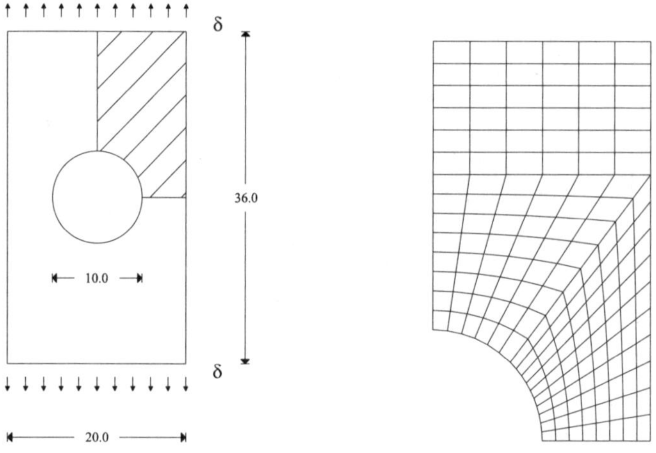

Fig. 2. Strip with circular hole. Model problem and FE mesh.

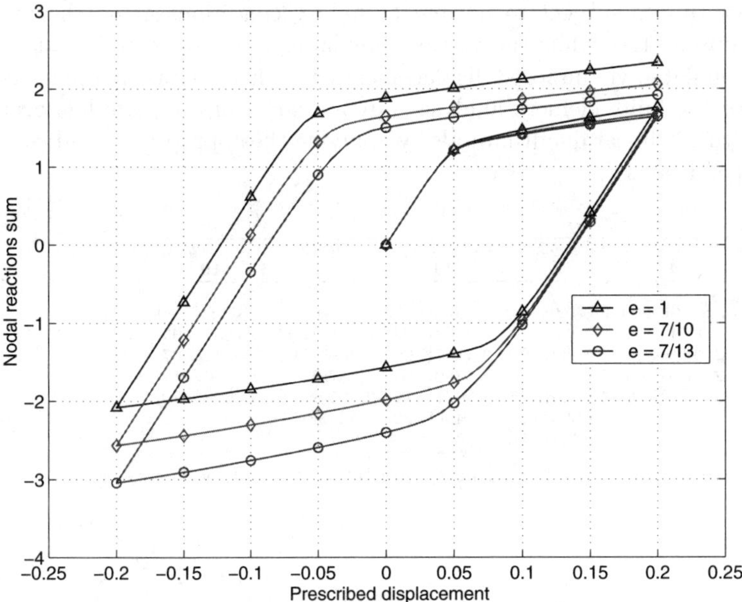

Fig. 3. Strip with circular hole. Load-displacement curves. $h_{iso} > 0$, $h_{kin} > 0$.

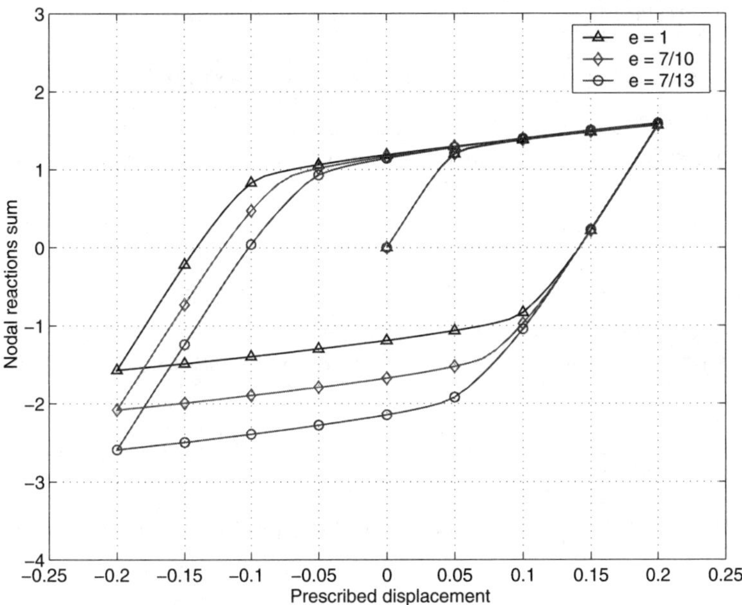

Fig. 4. Strip with circular hole. Load-displacement curves. $h_{iso} = 0$, $h_{kin} > 0$.

The structure is subject to normal boundary conditions along the two horizontal outer edges while the vertical boudaries are allowed to translate; loading is simulated via imposed displacements δ in the vertical direction, see figure 2. For the finite element analysis only a quarter of the plate has been modeled with 196 linear quadrilaterals by applying the appropriate symmetry boundary conditions.

displacement	*0.20*	*0.05*	*-0.15*	*-0.05*	*0.15*
load step	**4**	**7**	**11**	**15**	**19**
iteration					
1	1.8156E+00	1.7703E+00	1.8165E+00	1.7554E+00	1.8173E+00
2	1.3008E-04	6.5036E-03	1.0263E-04	2.5887E-03	5.1837E-05
3	2.3039E-06	1.5533E-03	2.4197E-06	1.2089E-03	3.3941E-07
4	2.8418E-09	5.6110E-06	2.6289E-09	3.5875E-06	1.6906E-12
5	5.0066E-16	1.0626E-07	5.6027E-15	2.3654E-09	5.5608E-20
6		6.0895E-11	5.5496E-22	1.5250E-17	
7		5.3231E-18			

Table 1. Strip with circular hole. Energy norms for typical load steps. $e = 1$.

displacement	*0.20*	*0.05*	*-0.15*	*-0.05*	*0.15*
load step	**4**	**7**	**11**	**15**	**19**
iteration					
1	1.8167E+00	1.7660E+00	1.8158E+00	1.7554E+00	1.8178E+00
2	1.7971E-04	4.0344E-03	7.9058E-05	1.7817E-03	8.4952E-05
3	4.2700E-06	1.6727E-03	6.2009E-07	2.1400E-04	9.9971E-07
4	1.8112E-09	2.2596E-06	2.3362E-10	5.7226E-06	4.2744E-11
5	3.5620E-16	4.5921E-09	1.5468E-16	4.7606E-08	1.0084E-18
6		2.6854E-15		2.6781E-09	
7		3.2383E-23		1.0659E-16	

Table 2. Strip with circular hole. Energy norms for typical load steps. $e = 7/10$.

displacement	*0.20*	*0.05*	*-0.15*	*-0.05*	*0.15*
load step	**4**	**7**	**11**	**15**	**19**
iteration					
1	1.8172E+00	1.7600E+00	1.8150E+00	1.7557E+00	1.8179E+00
2	1.4076E-04	5.1218E-03	1.4578E-04	8.8406E-04	5.6313E-05
3	9.6048E-06	3.4933E-03	4.1706E-06	1.7812E-05	1.7700E-06
4	3.2383E-08	4.6049E-06	8.2463E-09	4.0173E-07	3.3188E-11
5	4.2629E-13	5.7902E-08	1.3648E-13	8.4559E-10	8.7360E-19
6	6.2405E-21	2.1003E-12	1.0555E-18	4.7268E-18	
7		1.4990E-19			

Table 3. Strip with circular hole. Energy norms for typical load steps. $e = 7/13$.

The loading history is of cyclic type and has been assigned using four different sets of time increments: $\Delta t = 0.01, 0.02, 0.04, 0.05$ that correspond to equal prescribed increments of the vertical displacement on the upper boundary of the specimen.

Elastoplastic behaviour with linear isotropic and kinematic hardening has been considered; in particular, the elastic constants and the hardening moduli have been taken as:

$$E = 70.0; \qquad \nu = 0.2; \qquad h_{iso} = 2.24; \qquad h_{kin} = \frac{2}{3}(2.24)$$

with the following data set for the initial uniaxial yield limits:

$i)$ $f'_t = f'_c = 0.243;$
$ii)$ $f'_t = 0.243; f'_c = 0.347;$
$iii)$ $f'_t = 0.243; f'_c = 0.451$

that correspond to an eccentricity factor for the yield surface equal to 1, 7/10 and 7/13 respectively, see also figure 1. The relevant load-deflection curves are reported in figure 3 and refer to the solution obtained for $\Delta t = 0.05$. The solution obtained for pure kinematic hardening is also reported in figure 4 for comparison purposes.

Computations have been carried out by using a customized version of the finite element code FEAP rel. 7.4 [27]; all of them have been successfully completed in 100, 50, 25 and 20 load steps by using a linear line search algorithm both at the local and at the global level. In the numerical examples described below use has been made of a global termination criterion expressed in terms of the incremental energy norm [28] as:

$$E^{(i)} \leq \mu E^{(0)}$$

with a tolerance $\mu = 10^{-16}$.

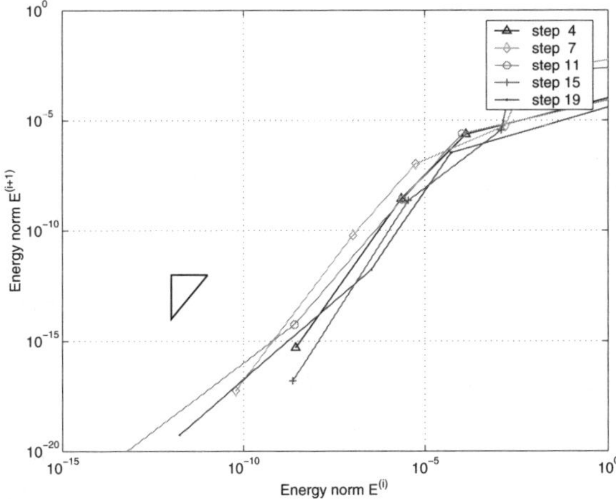

Fig. 5. Strip with circular hole. Convergence behaviour. $e = 1$.

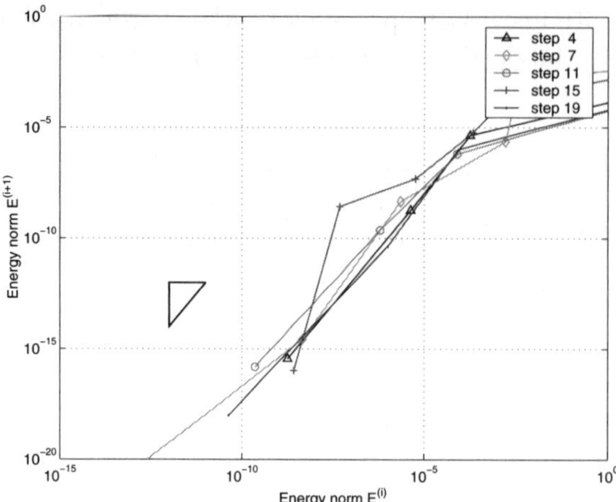

Fig. 6. Strip with circular hole. Convergence behaviour. $e = 7/10$.

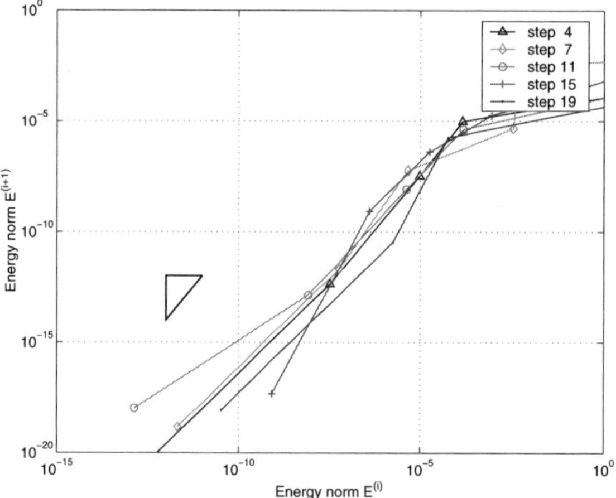

Fig. 7. Strip with circular hole. Convergence behaviour. $e = 7/13$.

The convergence behaviour is illustrated in tables 1-3 in terms of the energy norm for some typical load steps and refer to the solution computed for $\Delta t = 0.05$. The convergence rate exhibited by the Newton-Raphson method during this analysis is almost asymptotically quadratic as in all the other examined cases not documented here for brevity. This fact can be appreciated also in figures 5-7 by comparing the slope of the curves, which represent the incremental energy norm $E^{(i+1)}$ plotted vs. the energy norm at the previous iteration $E^{(i)}$, with the slope 2/1 of the triangle which, in the logarithmic scale, correspond to the quadratic convergence rate.

9   Summary and conclusions

The formulation and implementation of general three-invariant isotropic plasticity models have been discussed. In particular, the developments carried out in the paper have shown how the isotropic properties can be effectively exploited to obtain the tangent tensors required for the solution of the nonlinear equations arising in the framework of the popular Newton-like solution scheme. A novel expression of the consistent tangent, by far simpler with respect to other existing ones, has also been presented and its explicit form has been provided in terms of the coefficients of a representation formula by exploiting classical representation theorems for isotropic tensor functions.

The specialization of the proposed approach to the significant class of plane stress problems has also been addressed. In particular, it has been been shown that the consistent reduction of the problem originally formulated in a three-dimensional ambient space to a two-dimensional context allows for a natural derivation of the Newton-closest-point solution algorithm and the relevant closed-form linearization for the plane stress case and, more generally, for other plasticity problems formulated in terms of any reduced set of stress components.

The numerical simulations presented have demonstrated the efficiency of the devised strategy in practical computations.

Acknowledgements

The Authors acknowledge the Italian National Research Council (CNR) and the Italian Ministry of Education, University and Research (MIUR) for the financial support. This work has been partly carried out within the framework of the activities of the *Laboratoire Lagrange*, an European research group gathering CNR, CNRS, Università di Roma "Tor Vergata", Université de Montpellier II, ENPC, LCPC.

References

1. J.C. Simo and T.J.R. Hughes. *Computational Inelasticity*. Springer, Berlin, 1998.
2. J.C. Simo. Numerical analysis and simulation of plasticity. In P.G. Ciarlet and J.L. Lions, editors, *Handbook of Numerical Analysis*, number VI. Elsevier Science, Amsterdam, 1998.
3. V. Palazzo, L. Rosati, and N. Valoroso. Solution procedures for J_3 plasticity and viscoplasticity. *Computer Methods in Applied Mechanics and Engineering*, 191(8):903–939, 2001.
4. L. Rosati and N. Valoroso. A return map algorithm for general isotropic elasto/visco- plastic materials in principal space. *International Journal for Numerical Methods in Engineering*, 60(2):461–498, 2004.

5. J. Lubliner. *Plasticity Theory*. Macmillan, New York, 1990.
6. C. Truesdell and W. Noll. The non-linear field theories of mechanics. In S. Flügge, editor, *Handbuch der Physik Band III/3*. Springer, Berlin, 1965.
7. G. Del Piero. Some properties of the set of fourth-order tensors with application to elasticity. *Journal of Elasticity*, 3:245–261, 1979.
8. J. Lemaitre and J.L. Chaboche. *Mechanics of Solids Materials*. Cambridge University Press, Cambridge, 1990.
9. D.P. Bertsekas. *Constrained Optimization and Lagrange Multiplier Methods*. Academic Press, New York, 1982.
10. J.C. Simo and M. Ortiz. An analysis of a new class of integration algorithms for elastoplastic constitutive relations. *International Journal for Numerical Methods in Engineering*, 23:353–366, 1986.
11. J. Nagtegaal. On the implementation of inelastic constitutive equations with special reference to large deformation problems. *Computer Methods in Applied Mechanics and Engineering*, 33:469–484, 1982.
12. J.C. Simo and R.L. Taylor. Consistent tangent operators for rate-independent elastoplasticity. *Computer Methods in Applied Mechanics and Engineering*, 48:101–118, 1985.
13. M.A. Crisfield. *Non-Linear Finite Element Analysis of Solids and Structures*, volume I. John Wiley & Sons, Chichester, UK, 1991.
14. G.H. Golub and C.F. Van Loan. *Matrix Computations*. The John Hopkins University Press, Baltimore, 1989.
15. A. Matzenmiller and R.L. Taylor. A return mapping for isotropic elastoplasticity. *International Journal for Numerical Methods in Engineering*, 37:1994, 813-826.
16. L. Rosati. A novel approach to the solution of the tensor equation $\mathbf{AX} + \mathbf{XA} = \mathbf{H}$. *International Journal of Solids and Structures*, 37:3457–3477, 2000.
17. R.S. Rivlin. Further remarks on the stress-deformation relations for isotropic materials. *Journal of Rational Mechanics and Analysis*, 4:681–701, 1955.
18. G. Alfano and L. Rosati. A general approach to the evaluation of consistent tangent operators for rate-independent elastoplasticity. *Computer Methods in Applied Mechanics and Engineering*, 167(1):75–89, 1998.
19. N. Valoroso and L. Rosati. Consistent derivation of the constitutive algorithm for plane-stress isotropic plasticity. Part I: Theoretical formulation. 2004. Submitted for publication.
20. W.T. Koiter. General theorems for elastic-plastic solids. In I.N. Sneddon and R. Hill, editors, *Progress in Solid Mechanics*, number 6, pages 167–221. North-Holland, Amsterdam, 1960.
21. J.C. Simo. Algorithms for static and dynamic multiplicative plasticity that preserve the classical return mapping schemes of the infinitesimal theory. *Computer Methods in Applied Mechanics and Engineering*, 99:61–112, 1992.
22. J.C. Simo and R.L. Taylor. A return mapping algorithm for plane stress elastoplasticity. *International Journal for Numerical Methods in Engineering*, 22:649–670, 1986.
23. P. Fuschi, M. Dutko, D. Peric, and D.R.J. Owen. On numerical integration of the five parameter model for concrete. *Computers & Structures*, 53:825–838, 1994.
24. N. Valoroso and L. Rosati. Consistent derivation of the constitutive algorithm for plane-stress isotropic plasticity. Part II: Computational issues. 2004. Submitted for publication.

25. K.J. Willam and E.P. Warnke. Constitutive models for triaxial behaviour of concrete. International Association for Bridge and Structural Engineering: seminar on concrete structures subject to triaxial stresses, 1974. Paper III-01.
26. W.F. Chen and A.F. Saleeb. *Constitutive Equations for Engineering Materials.* Wiley, New York, 1982.
27. R.L. Taylor. *FEAP - User Manual.* University of California at Berkekey, *http://www.ce.berkeley.edu/~rlt.*
28. O.C. Zienkiewicz and R.L. Taylor. *The Finite Element Method, Vol. II: Solid Mechanics.* Butterworth-Heinemann, Oxford, 2000.

A   Some tensor algebra

Let $\mathcal{V}$ be a three-dimensional inner product space over the real field and denote by Lin ($\mathbb{L}$in) the space of all second- (fourth-) order tensors on $\mathcal{V}$.

The dyadic product of two elements $\mathbf{A}, \mathbf{B} \in$ Lin is defined as [14]:

$$(\mathbf{A} \otimes \mathbf{B})\mathbf{C} = (\mathbf{B} \cdot \mathbf{C})\mathbf{A} = \mathrm{tr}\,(\mathbf{B}^T\mathbf{C})\mathbf{A} \qquad \forall\, \mathbf{C} \in \mathrm{Lin} \tag{91}$$

where tr denotes the trace operator.

According to the definition given by Del Piero [7], the so-called square tensor product is introduced as:

$$(\mathbf{A} \boxtimes \mathbf{B})\mathbf{C} = \mathbf{A}\mathbf{C}\mathbf{B}^T \qquad \forall\, \mathbf{C} \in \mathrm{Lin} \tag{92}$$

Hence, denoting by $\mathbb{I}$ the identity tensor in $\mathbb{L}$in, it turns out to be:

$$\mathbb{I} = \mathbf{1} \boxtimes \mathbf{1}$$

where $\mathbf{1}$ is the identity tensor in Lin.

The component form of dyadic and square tensor products in a Cartesian frame are provided by the following relationships:

$$(\mathbf{A} \otimes \mathbf{B})_{ijkl} = \mathbf{A}_{ij}\mathbf{B}_{kl} \qquad \forall\, \mathbf{A},\, \mathbf{B} \in \mathrm{Lin} \tag{93}$$

$$(\mathbf{A} \boxtimes \mathbf{B})_{ijkl} = \mathbf{A}_{ik}\mathbf{B}_{jl} \qquad \forall\, \mathbf{A},\, \mathbf{B} \in \mathrm{Lin} \tag{94}$$

Given $\mathbf{A}, \mathbf{B}, \mathbf{C}, \mathbf{D} \in$ Lin and invoking (91) and (92), the following composition rules hold:

$$(\mathbf{A} \boxtimes \mathbf{B})(\mathbf{C} \boxtimes \mathbf{D}) = (\mathbf{A}\mathbf{C}) \boxtimes (\mathbf{B}\mathbf{D})$$

$$(\mathbf{A} \boxtimes \mathbf{B})(\mathbf{C} \otimes \mathbf{D}) = (\mathbf{A}\mathbf{C}\mathbf{B}^T) \otimes \mathbf{D} \tag{95}$$

$$(\mathbf{A} \otimes \mathbf{B})(\mathbf{C} \boxtimes \mathbf{D}) = \mathbf{A} \otimes (\mathbf{C}^T\mathbf{B}\mathbf{D})$$

We remind that in a three-dimensional setting dyadic and square tensor products can be related by means of three identities due to Rivlin [17]; in the case of a symmetric tensor $\mathbf{A}$ they can be specialized as follows:

$$\mathbf{A}^2 \boxtimes \mathbf{1} + \mathbf{1} \boxtimes \mathbf{A}^2 = -\mathbf{A} \boxtimes \mathbf{A} + (\mathbf{A}^2 \otimes \mathbf{1} + \mathbf{1} \otimes \mathbf{A}^2)+$$

$$+ I_{\mathbf{A}}(\mathbf{A} \boxtimes \mathbf{1} + \mathbf{1} \boxtimes \mathbf{A}) - I_{\mathbf{A}}(\mathbf{A} \otimes \mathbf{1} + \mathbf{1} \otimes \mathbf{A})+ \tag{96}$$

$$+ \mathbf{A} \otimes \mathbf{A} + II_{\mathbf{A}}(\mathbf{1} \otimes \mathbf{1}) - II_{\mathbf{A}}(\mathbf{1} \boxtimes \mathbf{1})$$

$$\mathbf{A}^2 \boxtimes \mathbf{A} + \mathbf{A} \boxtimes \mathbf{A}^2 = \ I_\mathbf{A}(\mathbf{A} \boxtimes \mathbf{A}) + (\mathbf{A}^2 \otimes \mathbf{A} + \mathbf{A} \otimes \mathbf{A}^2) +$$
$$- I_\mathbf{A}(\mathbf{A} \otimes \mathbf{A}) + III_\mathbf{A}(\mathbf{1} \otimes \mathbf{1}) - III_\mathbf{A}(\mathbf{1} \boxtimes \mathbf{1}) \tag{97}$$

$$\mathbf{A}^2 \boxtimes \mathbf{A}^2 = \ II_\mathbf{A}(\mathbf{A} \boxtimes \mathbf{A}) + (\mathbf{A}^2 \otimes \mathbf{A}^2) - III_\mathbf{A}(\mathbf{A} \boxtimes \mathbf{1} + \mathbf{1} \boxtimes \mathbf{A}) +$$
$$- II_\mathbf{A}(\mathbf{A} \otimes \mathbf{A}) + III_\mathbf{A}(\mathbf{A} \otimes \mathbf{1} + \mathbf{1} \otimes \mathbf{A}) \tag{98}$$

$I_\mathbf{A}, II_\mathbf{A}, III_\mathbf{A}$ being the principal invariants of $\mathbf{A}$:

$$I_\mathbf{A} = \operatorname{tr}(\mathbf{A}); \qquad II_\mathbf{A} = \frac{1}{2}\left[\operatorname{tr}(\mathbf{A})^2 - \operatorname{tr}(\mathbf{A}^2)\right]; \qquad III_\mathbf{A} = \det(\mathbf{A}) \tag{99}$$

In the two-dimensional case the previous relationships collapse to the following one:

$$\mathbf{A} \boxtimes \mathbf{1} + \mathbf{1} \boxtimes \mathbf{A} = (\mathbf{A} \otimes \mathbf{1} + \mathbf{1} \otimes \mathbf{A}) + I_\mathbf{A}(\mathbf{1} \boxtimes \mathbf{1}) - I_\mathbf{A}(\mathbf{1} \otimes \mathbf{1}) \tag{100}$$

where both $\mathbf{A}$ and $\mathbf{1}$ are understood as two-dimensional tensors.

B   Coefficients for the 3D consistent tangent

The explicit expressions of the coefficients entering the 3D consistent tangent tensor (51) are given as follows:

$$t_1 = 2G\left(1 - m_1\right) + i_1 m_1^2$$

$$t_2 = i_2 m_1^2 \tag{101}$$

$$t_3 = i_3 m_1^2$$

$$t_4 = \lambda(1 - m_1) - (2G + 3\lambda)m_2 +$$
$$+ m_2(3m_2 + 2m_1)i_1 + 2J_2 m_2^2 i_3 + (m_1 + 3m_2)^2 i_4 + \tag{102}$$
$$+ 4J_2 m_2(m_1 + 3m_2)i_6 + (2J_2)^2 m_2^2 i_9 - \frac{q_1^2}{den}$$

$$t_5 = 2m_1 m_2 i_2 + m_1(m_1 + 3m_2)i_5 + 2J_2 m_1 m_2 i_8 - m_1 \frac{q_1 q_2}{den} \tag{103}$$

$$t_6 = \ m_1 m_2 i_3 + m_1(m_1 + 3m_2)i_6 + 2J_2 m_1 m_2 i_9 - m_1 \frac{q_1 q_3}{den} \tag{104}$$

$$t_7 = m_1^2 i_7 - m_1^2 \frac{q_2^2}{den} \tag{105}$$

$$t_8 = m_1^2 i_8 - m_1^2 \frac{q_2 q_3}{den} \tag{106}$$

$$t_9 = m_1^2 i_9 - m_1^2 \frac{q_3^2}{den} \tag{107}$$

den being the denominator:

$$
\begin{aligned}
den =\ & [3i_1 + 9i_4 + (2i_3 + 12i_6 + 4i_9 J_2)J_2]n_1^2 + \\
& + [(8i_2 + 12i_5 + 8i_8 J_2)J_2 + (6i_3 + 18i_6 + 12i_9 J_2)J_3]n_1 n_2 + \\
& + [(2i_1 + 2i_3 J_2 + 4i_7 J_2)J_2 + (6i_2 + 12i_8 J_2 + 9i_9 J_3)J_3]n_2^2 + \\
& + 2\{[2J_2(i_1 + 3i_4 + J_2(i_3 + 5i_6 + 2i_9 J_2)) + \\
& + 3(2i_2 + 3i_5 + 2i_8 J_2)J_3]n_1 + [4J_2^2(i_2 + i_5 + i_8 J_2) + 3i_1 J_3 + \\
& + J_2(5i_3 + 6(i_6 + i_7 + i_9 J_2))J_3 + 9i_8 J_3^2]n_2\}n_3 + \\
& + \{2J_2^2[i_1 + 2i_4 + J_2(i_3 + 4i_6 + 2i_9 J_2)] + \\
& + 2J_2(5i_2 + 6i_5 + 6i_8 J_2)J_3 + 3(i_3 + 3i_7)J_3^2\}n_3^2 + H
\end{aligned}
\tag{108}
$$

The intermediate coefficients $q_1, \ldots, q_3$ are listed below.

$$
\begin{aligned}
q_1 =\ & [(m_1 + 3m_2)i_1 + 2J_2 m_2 i_3 + 3(m_1 + 3m_2)i_4 + \\
& + 2J_2(m_1 + 6m_2)i_6 + 4J_2^2 m_2 i_9]n_1 + \\[4pt]
& [4J_2 m_2 i_2 + J_3(m_1 + 3m_2)i_3 + 2J_2(m_1 + 3m_2)i_5 + \\
& + 3J_3(m_1 + 3m_2)i_6 + 4J_2^2 m_2 i_8 + 6J_2 J_3 m_2 i_9]n_2 + \\[4pt]
& [2J_2 m_2 i_1 + 2J_3(m_1 + 3m_2)i_2 + 2J_2^2 m_2 i_3 + \\
& + 2J_2(m_1 + 3m_2)i_4 + 3J_3(m_1 + 3m_2)i_5 + \\
& + 2J_2^2(m_1 + 5m_2)i_6 + 6J_2 J_3 m_2 i_8 + 4J_2^3 m_2 i_9]n_3
\end{aligned}
\tag{109}
$$

$$
\begin{aligned}
q_2 =\ & [2i_2 + 3i_5 + 2J_2 i_8]n_1 + [i_1 + J_2 i_3 + 2J_2 i_7 + 3J_3 i_8]n_2 + \\
& [2J_2 i_2 + J_3 i_3 + 2J_2 i_5 + 3J_3 i_7 + 2J_2^2 i_8]n_3
\end{aligned}
\tag{110}
$$

$$
\begin{aligned}
q_3 =\ & [i_3 + 3i_6 + 2J_2 i_9]n_1 + [2i_2 + 2J_2 i_8 + 3J_3 i_9]n_2 + \\
& [i_1 + J_2 i_3 + 2J_2 i_6 + 3J_3 i_8 + 2J_2^2 i_9]n_3
\end{aligned}
\tag{111}
$$

C   Coefficients for the 2D consistent tangent

The explicit expressions of the coefficients entering the plane stress consistent tangent tensor (86) are given as follows:

$$\hat{t}_1 = 2G\,(1 - m_1) + \hat{\imath}_1 m_1^2 \tag{112}$$

$$\hat{t}_2 = \frac{2G\lambda}{(2G + \lambda)}(1 - m_1) - (2G + \frac{4G\lambda}{(2G + \lambda)})\hat{m}_2 +$$
$$+ \hat{m}_2(2\hat{m}_2 + 2m_1)\hat{\imath}_1 + (m_1 + 2\hat{m}_2)^2\hat{\imath}_2 + \tag{113}$$
$$+ 2\hat{I}_1\hat{m}_2(m_1 + 2\hat{m}_2)\hat{\imath}_3 + \hat{I}_1^2\hat{m}_2^2\hat{\imath}_4 - \frac{\hat{q}_1^2}{den}$$

$$\hat{t}_3 = m_1(m_1 + 2\hat{m}_2)\hat{\imath}_3 + \hat{I}_1 m_1\hat{m}_2\hat{\imath}_4 - m_1\frac{\hat{q}_1\hat{q}_2}{den} \tag{114}$$

$$\hat{t}_4 = m_1^2\hat{\imath}_4 - m_1^2\frac{\hat{q}_2^2}{den} \tag{115}$$

den being the denominator:

$$den = (2\hat{\imath}_1 + 4\hat{\imath}_2 + 4\hat{I}_1\hat{\imath}_3 + \hat{I}_1^2\hat{\imath}_4)\hat{n}_1^2 +$$
$$+ (2\hat{I}_1\hat{\imath}_1 + 4\hat{I}_1\hat{\imath}_2 + (4\hat{J}_2 + 2\hat{I}_1^2)\hat{\imath}_3 + 2\hat{I}_1\hat{J}_2\hat{\imath}_4)\hat{n}_1\hat{n}_2 + \tag{116}$$
$$+ (\hat{J}_2\hat{\imath}_1 + \hat{I}_1^2\hat{\imath}_2 + 2\hat{I}_1\hat{J}_2\hat{\imath}_3 + \hat{J}_2^2\hat{\imath}_4)\hat{n}_2^2 + H$$

while $\hat{I}_1$ and $\hat{J}_2$ are invariants of the in-plane relative stress deviator $\hat{\mathbf{S}}$:

$$\hat{I}_1 = \hat{\mathbf{1}} \cdot \hat{\mathbf{S}} = \mathrm{tr}\,(\hat{\mathbf{S}}); \qquad \hat{J}_2 = \hat{\mathbf{S}} \cdot \hat{\mathbf{S}} = \mathrm{tr}\,(\hat{\mathbf{S}}^2) \tag{117}$$

The intermediate coefficients $\hat{q}_1, \hat{q}_2$ are given below.

$$\hat{q}_1 = \{[2\hat{\imath}_1 + 4\hat{\imath}_2 + 2\hat{I}_1\hat{\imath}_3 + \hat{I}_1(2\hat{\imath}_3 + \hat{I}_1\hat{\imath}_4)]m_2 +$$
$$+ (\hat{\imath}_1 + 2\hat{\imath}_2 + \hat{I}_1\hat{\imath}_3)m_1\}\hat{n}_1 +$$
$$+ \{[2\hat{I}_1\hat{\imath}_2 + 2\hat{J}_2\hat{\imath}_3 + \hat{I}_1(\hat{\imath}_1 + \hat{I}_1\hat{\imath}_3 + \hat{J}_2\hat{\imath}_4)]m_2 + \tag{118}$$
$$+ (\hat{I}_1\hat{\imath}_2 + \hat{J}_2\hat{\imath}_3)m_1\}\hat{n}_2$$

$$\hat{q}_2 = (2\hat{\imath}_3 + \hat{I}_1\hat{\imath}_4)\hat{n}_1 + (\hat{\imath}_1 + \hat{I}_1\hat{\imath}_3 + \hat{J}_2\hat{\imath}_4)\hat{n}_2 \tag{119}$$

A Non-linear Hardening Model Based on Two Coupled Internal Hardening Variables: Formulation and Implementation

Nelly Point[1,2], Silvano Erlicher[1,3]

[1] Ecole Nationale des Ponts et Chaussées,
Laboratoire d'Analyse des Matériaux et Identification,
6-8 avenue Blaise Pascal,
Cité Descartes, Champs-sur-Marne,
F-77455 Marne la Vallée Cedex 2, France
[2] Conservatoire National des Arts et Métiers,
Département de Mathématiques,
292 rue Saint Martin,
F-75141 Paris Cedex 03, France
[3] Università di Trento,
Dipartimento di Ingegneria Meccanica e Strutturale
Via Mesiano 77, 38050, Trento, Italy

Abstract. An elasto-plasticity model with coupled hardening variables of strain type is presented. In the theoretical framework of generalized associativity, the formulation of this model is based on the introduction of two hardening variables with a coupled evolution. Even if the corresponding hardening rules are linear, the stress-strain hardening evolution is non-linear. The numerical implementation by a standard return mapping algorithm is discussed and some numerical simulations of cyclic behaviour in the univariate case are presented.

1 Introduction

Starting from the analysis of the dislocation phenomenon in metallic materials, Zarka and Casier [1] and Kabhou et al. [2] proposed an elasto-plasticity model ("four-parameter model") where, in addition to the usual kinematic hardening internal variable, a second strain-like internal variable was introduced. It plays a role in a modified definition of the von Mises criterion and its evolution, defined by linear flow rules, is coupled with the one of the kinematic hardening variable. The resulting elasto-plastic model depends only on four parameters. Its non-linear hardening behavior was studied in [4] and a parameter identification method using essentially a cyclic uniaxial test was presented in [5] . In this note, the thermodynamic formulation of the classical elasto-plastic model with linear kinematic and isotropic hardening is first recalled. Then, by using the same theoretical framework, a generalization of the four-parameter model is suggested, relying on the introduction of an additional isotropic hardening variable. Finally, a return mapping implementation of the generalized model is presented and some numerical simulations are briefly discussed.

2 Thermodynamic formulation of a plasticity model with linear kinematic/isotropic hardening

Under the assumption of isothermal infinitesimal transformations and of isotropic material, the hydrostatic and the deviatoric responses can be treated separately (see, among others, [7]). Hence, the free energy density Ψ can be split into its spherical part Ψ_h and its deviatoric part Ψ_d. To obtain *linear* state equations, Ψ_h and Ψ_d are assumed quadratic. Moreover, experimental results for metals show that permanent strain is only due to deviatoric slip. Hence, an elastic spherical behaviour is assumed, leading to the following definition :

$$\Psi_h = \frac{1}{2}\left(\lambda + \frac{2\mu}{3}\right) tr\,(\varepsilon)^2 = \frac{1}{2}K tr\,(\varepsilon)^2 \tag{1}$$

where ε is the (small) strain tensor, λ and μ are the Lamé constants and K is the bulk modulus. Under the same assumptions, the deviatoric potential Ψ_d must depend only on deviatoric state variables. The plastic flow is associated to the plastic strain ε^p, while the kinematic/isotropic hardening behaviour is introduced by the tensorial internal variable α and by the scalar variable p :

$$\Psi_d = \Psi_d\,(\varepsilon_d, \varepsilon^p, \alpha, p) = \frac{2\mu}{2}\,(\varepsilon_d - \varepsilon^p) : (\varepsilon_d - \varepsilon^p) + \frac{B}{2}\alpha : \alpha + \frac{H}{2}p^2 \tag{2}$$

where $tr\,(\alpha) = tr\,(\varepsilon^p) = 0$ and $B,\ H > 0$. The evolution of p will be related to the norm of ε^p.
The state equation concerning the deviatoric stress tensor is easily derived:

$$\sigma_d = \frac{\partial \Psi_d}{\partial \varepsilon_d} = 2\mu\,(\varepsilon_d - \varepsilon^p) \tag{3}$$

and the thermodynamic forces associated to ε^p, α and p are defined by :

$$\begin{cases} \sigma_d = -\frac{\partial \Psi_d}{\partial \varepsilon^p} = 2\mu\,(\varepsilon_d - \varepsilon^p) \\ \mathbf{X} = \frac{\partial \Psi_d}{\partial \alpha} = B\alpha \\ R = \frac{\partial \Psi_d}{\partial p} = Kp \end{cases} \tag{4}$$

One can notice that $tr\,(\mathbf{X}) = 0$. The linearity of the hardening rules $(4)_{2-3}$ follows from the quadratic form assumed for the last two terms in (2). The second principle of thermodynamics can be written as follows [7]:

$$\sigma_d : \dot{\varepsilon}_d - \dot{\Psi}_d \geq 0 \tag{5}$$

By using (4) in (5), the Clausius Duhem inequality is obtained :

$$\sigma_d : \dot{\varepsilon}^p - \mathbf{X} : \dot{\alpha} - R\dot{p} \geq 0 \tag{6}$$

In order to fulfil this inequality, a classical assumption is to impose that $(\dot{\varepsilon}^p, \dot{\alpha}, \dot{p})$ belongs to the subdifferential of a positive convex function ϕ_d^*, equal to zero in zero, called pseudo-potential. In such a case, the evolution of the internal variables is compatible with (6) [3].

The von Mises criterion corresponds to a special choice for the pseudo-potential $\phi_d^* (\sigma_d, \mathbf{X}, R)$ which is equal, in this case, to the indicator function $\mathbb{I}_{f \leq 0}$ of the elastic domain, or, to be more specific, of the set of $(\sigma_d, \mathbf{X}, R)$ such that the so-called yielding function f is non-positive :

$$f = f(\sigma_d, \mathbf{X}, R) = \|\sigma_d - \mathbf{X}\| - \sqrt{\frac{2}{3}}\sigma_y - R \leq 0 \tag{7}$$

where $\|\cdot\|$ is the standard L_2-norm. To impose that $(\dot{\varepsilon}^p, \dot{\alpha}, \dot{p})$ belongs to the subdifferential of $\mathbb{I}_{f \leq 0}$ is equivalent to write:

$$\begin{cases} \dot{\varepsilon}^p = \dot{\lambda}\frac{\partial f}{\partial \sigma_d} = \dot{\lambda}\frac{\sigma_d - \mathbf{X}}{\|\sigma_d - \mathbf{X}\|} \\ \dot{\alpha} = -\dot{\lambda}\frac{\partial f}{\partial \mathbf{X}} = \dot{\lambda}\frac{\sigma_d - \mathbf{X}}{\|\sigma_d - \mathbf{X}\|} \\ \dot{p} = -\dot{\lambda}\frac{\partial f}{\partial R} = \dot{\lambda} \end{cases} \tag{8}$$

with the conditions $\dot{\lambda} \geq 0$, $f \leq 0$ and $\dot{\lambda}f = 0$.
Equations (8) are called generalized associativity conditions or associative flow rules. The relations (8) yield in this case $\dot{\alpha} = \dot{\varepsilon}^p$ and $\dot{\lambda} = \dot{p} = \|\dot{\varepsilon}^p\|$.
¿From $(8)_2$ and $(4)_2$ one obtains the Prager's linear kinematic hardening rule and a linear isotropic hardening rule:

$$\dot{\mathbf{X}} = B\dot{\varepsilon}^p , \qquad \dot{R} = H\dot{p} \tag{9}$$

The coefficient $\dot{\lambda}$ is strictly positive only if $f = 0$. In this case, its value can be derived from the so-called consistency condition $\dot{f} = 0$, i.e.

$$\frac{\partial f}{\partial \sigma_d} : \dot{\sigma}_d + \frac{\partial f}{\partial \mathbf{X}} : \dot{\mathbf{X}} + \frac{\partial f}{\partial R}\dot{R} = 0 \tag{10}$$

The introduction into the previous equation of the state equation (3), as well as the thermodynamic force definitions $(4)_{2-3}$ and the normality (8), yield :

$$\frac{\partial f}{\partial \sigma_d} : \dot{\sigma}_d - \dot{\lambda}B\frac{\partial f}{\partial \mathbf{X}} : \frac{\partial f}{\partial \mathbf{X}} - \dot{\lambda}H\frac{\partial f}{\partial R}\frac{\partial f}{\partial R} = 0 \tag{11}$$

Moreover, in a strain driven approach, Eq. (11) has to be rewritten using again the state equation (3). As a result, by collecting $\dot{\lambda}$, one obtains

$$\begin{aligned} \dot{\lambda} &= \mathcal{H}(f)\frac{2\mu\left\langle \frac{\partial f}{\partial \sigma_d} : \dot{\varepsilon}_d \right\rangle}{2\mu\frac{\partial f}{\partial \sigma_d}:\frac{\partial f}{\partial \sigma_d} + B\frac{\partial f}{\partial \mathbf{X}}:\frac{\partial f}{\partial \mathbf{X}} + H\frac{\partial f}{\partial R}\frac{\partial f}{\partial R}} \\ &= \frac{\mathcal{H}(f)}{1 + \frac{B+K}{2\mu}}\frac{\langle(\sigma_d - \mathbf{X}):\dot{\varepsilon}_d\rangle}{\|\sigma_d - \mathbf{X}\|} \geq 0 \end{aligned} \tag{12}$$

where $\mathcal{H}(f)$ is zero when $f < 0$ and equal to 1 for $f = 0$. The symbol $\langle.\rangle$ represents the MacCauley brackets.

3   A generalization of the four-parameter model

The linear hardening model discussed previously is used here to suggest a generalization of the 4-parameter model cited in the introduction. The tensor α in eq. (2) is replaced by a couple of tensors (α_1, α_2). As a result, the scalar constant B becomes a 2×2 symmetric positive definite matrix, denoted by $\mathbf{B} = [b_{ij}]$. For sake of simplicity, only the thermodynamic potential Ψ_d is considered here and it is defined as:

$$\Psi_d\left(\varepsilon_d, \varepsilon^p, \alpha_1, \alpha_2, p\right) = \frac{2\mu}{2}\left(\varepsilon_d - \varepsilon^p\right) : \left(\varepsilon_d - \varepsilon^p\right) + \frac{1}{2}\alpha^{\mathrm{T}}\, \mathbf{B}\, \alpha + \frac{\mathrm{H}}{2}\mathrm{p}^2 \quad (13)$$

where μ and H have the same meaning as before and α is the column vector defined as $\alpha = [\alpha_1 ; \alpha_2]$. The state equation becomes :

$$\sigma_d = \frac{\partial \Psi_d}{\partial \varepsilon_d} = 2\mu\left(\varepsilon_d - \varepsilon^p\right) \quad (14)$$

and the thermodynamic forces have the following form :

$$\begin{cases} \sigma_d = -\frac{\partial \Psi_d}{\partial \varepsilon^p} = 2\mu\left(\varepsilon_d - \varepsilon^p\right) \\ \mathbf{X}_1 = \frac{\partial \Psi_d}{\partial \alpha_1} = b_{11}\alpha_1 + b_{12}\alpha_2 \\ \mathbf{X}_2 = \frac{\partial \Psi_d}{\partial \alpha_2} = b_{21}\mathrm{r}\alpha_1 + b_{22}\alpha_2 \\ R = \frac{\partial \Psi_d}{\partial p} = H\mathrm{p} \end{cases} \qquad \text{or} \qquad \begin{cases} \sigma_d = 2\mu\left(\varepsilon_d - \varepsilon^p\right) \\ \mathbf{X} = \mathbf{B}\,\alpha \\ R = Hp \end{cases} \quad (15)$$

with $\mathbf{X} = [\mathbf{X}_1 ; \mathbf{X}_2]$. The Clausius -Duhem inequality becomes in this case :

$$\sigma_d : \dot{\varepsilon}^p - \mathbf{X}_1 : \dot{\alpha}_1 - \mathbf{X}_2 : \dot{\alpha}_2 - R\dot{p} \geq 0 \qquad \text{or} \qquad \sigma_d : \dot{\varepsilon}^p - \mathbf{X}^{\mathrm{T}}\,\dot{\alpha} - R\dot{p} \geq 0 \quad (16)$$

Moreover, the loading function f is defined as follows :

$$f = f\left(\sigma_d, \mathbf{X}_1, \mathbf{X}_2, R\right) = \sqrt{\|\sigma_d - \mathbf{X}_1\|^2 + \rho^2\|\mathbf{X}_2\|^2} - \sqrt{\frac{2}{3}}\sigma_y - R \leq 0 \quad (17)$$

with ρ a positive scalar. One can remark that for $H = 0$, the standard von Mises criterion is retrieved for $\rho = 0$ and the 4-parameter model for $\rho = 1$. The flow rules are defined by a normality condition :

$$\left(\dot{\varepsilon}^p, -\dot{\alpha}_1, -\dot{\alpha}_2, -\dot{p}\right) \in \partial\phi_d^* = \partial\mathbb{I}_{f \leq 0} \quad (18)$$

Therefore, the proposed model belongs to the framework of generalized associative plasticity [3] . One can remark that the loading function (17) can be rewritten as

$$f = g\left(\mathbf{Y}_1, \mathbf{Y}_2\right) - \sqrt{\frac{2}{3}}\sigma_y - R \leq 0 \text{ with } \mathbf{Y}_1 = \sigma_d - \mathbf{X}_1 \text{ and } \mathbf{Y}_2 = -\mathbf{X}_2 \text{ . Hence}$$

$$\begin{cases} \dot{\varepsilon}^p = \dot{\lambda}\frac{\partial f}{\partial \sigma_d} = \dot{\lambda}\frac{\sigma_d - \mathbf{X}_1}{\sqrt{\|\sigma_d - \mathbf{X}_1\|^2 + \rho^2\|\mathbf{X}_2\|^2}} \\ \dot{\alpha}_1 = -\dot{\lambda}\frac{\partial f}{\partial \mathbf{X}_1} = \dot{\lambda}\frac{\sigma_d - \mathbf{X}_1}{\sqrt{\|\sigma_d - \mathbf{X}_1\|^2 + \rho^2\|\mathbf{X}_2\|^2}} \\ \dot{\alpha}_2 = -\dot{\lambda}\frac{\partial f}{\partial \mathbf{X}_2} = -\dot{\lambda}\frac{\rho^2\mathbf{X}_2}{\sqrt{\|\sigma_d - \mathbf{X}_1\|^2 + \rho^2\|\mathbf{X}_2\|^2}} \\ \dot{p} = -\dot{\lambda}\frac{\partial f}{\partial R} = \dot{\lambda} \end{cases} \quad \text{or} \quad \begin{cases} \dot{\varepsilon}^p = \dot{\lambda}\frac{\partial f}{\partial \sigma_d} \\ \dot{\alpha} = \dot{\lambda}\nabla g \\ \dot{p} = \dot{\lambda} \end{cases} \quad (19)$$

It can be seen from (19) that $\dot{\alpha}_1 = \dot{\varepsilon}^p$ and $\dot{p} = \dot{\lambda} = \|\dot{\alpha}\| = \sqrt{\|\dot{\alpha}_1\|^2 + \|\dot{\alpha}_2\|^2}$. Moreover, from $(19)_{1-2}$ and $(15)_{2-4}$ one obtains the kinematic and isotropic hardening rules :

$$\dot{\mathbf{X}}_1 = b_{11}\dot{\varepsilon}^p + b_{12}\dot{\alpha}_2, \quad \dot{\mathbf{X}}_2 = b_{21}\dot{\varepsilon}^p + b_{22}\dot{\alpha}_2, \quad \dot{R} = H\dot{p} \tag{20}$$

In [4] it was proved that $\mathbf{B}$ can be written as :

$$\mathbf{B} = \begin{bmatrix} \left(A_\infty + r^2 b\right) & -rb \\ -rb & b \end{bmatrix} \tag{21}$$

where the scalars A_∞ and b are strictly positive and have the dimension of stresses. For $r = 0$, there is no coupling and $\mathbf{B}$ is diagonal, so that the dimensionless scalar r can be seen as a coupling factor in the evolutions of $\mathbf{X}_1$ and $\mathbf{X}_2$:

$$\dot{\mathbf{X}}_1 = A_\infty\ \dot{\varepsilon}^p + rb(r\dot{\varepsilon}^p - \dot{\alpha}_2), \quad \dot{\mathbf{X}}_2 = b(r\dot{\varepsilon}^p - \dot{\alpha}_2), \implies \dot{\mathbf{X}}_1 + r\dot{\mathbf{X}}_2 = A_\infty\ \dot{\varepsilon}^p \tag{22}$$

In the first two flow rules a recalling term appears, as in the non-linear kinematic hardening model of Frederich and Armstrong [8]. As before, the plastic multiplier can be explicitly computed by the consistency condition:

$$\dot{\lambda} = \mathcal{H}\left(f\right)\frac{\left\langle \frac{\partial f}{\partial \sigma_d} : \dot{\varepsilon}_d \right\rangle}{1 + \frac{\nabla g.\mathbf{B}.\nabla g + H}{2\mu}} \geq 0. \tag{23}$$

4   Implementation and some numerical results

In this section, a numerical implementation of the model is proposed. A standard return mapping algorithm is considered (see [6]). The formulation is explicitly described in the univariate case, but the tensorial generalization is straightforward. Let Δt_n be the amplitude of the time step defined by t_n and t_{n+1} and let $\tilde{\alpha}_n = [\alpha_{1,n}, \alpha_{2,n}, p_n]^T$ and $\tilde{\mathbf{X}}_n = [X_{1,n}, X_{2,n}, R_n]^T$ be the vectors collecting the internal variables and the corresponding thermodynamic forces. Moreover, let

$$\mathbf{D} = \begin{bmatrix} \mathbf{B} & 0 \\ 0 & H \end{bmatrix} \tag{24}$$

be the global hardening modulus matrix. In a strain driven approach, knowing the value of all the variables at the time t_n and the *strain increment* $\Delta\varepsilon_n$ occurring during the time step $t_n \rightarrow t_{n+1}$, the numerical scheme computes the variable values at t_{n+1}:

$$\left(\varepsilon_n, \varepsilon_n^p, \tilde{\alpha}_n, \sigma_n, \tilde{\mathbf{X}}_n, f_n\right) + \Delta\varepsilon_n \implies$$
$$\left(\varepsilon_{n+1}, \varepsilon_{n+1}^p, \tilde{\alpha}_{n+1}, \sigma_{n+1}, \tilde{\mathbf{X}}_{n+1}, f_{n+1}\right) \tag{25}$$

The flow equations (19) define a first order differential system, which can be solved by the implicit Euler method. Therefore, the discrete form of the model evolution rules is (the notation $\partial_{\mathbf{w}} f$ is equivalent to $\partial f / \partial \mathbf{w}$) :

$$f_{n+1} := \sqrt{\left(\sigma_{n+1} - X_{1,n+1}\right)^2 + \rho^2 \left(X_{2,n+1}\right)^2} - \left(\sigma_y + R_{n+1}\right) \le 0$$

$$\sigma_{n+1} = E\left(\varepsilon_{n+1} - \varepsilon_{n+1}^p\right) \; ; \qquad \tilde{\mathbf{X}}_{n+1} = \mathbf{D}\tilde{\alpha}_{n+1} \quad \text{with } E = \mu\frac{3\lambda+2\mu}{\lambda+\mu}$$
$$\varepsilon_{n+1}^p - \varepsilon_n^p = \gamma_{n+1}\,\partial_\sigma f_{n+1} \; ; \qquad \tilde{\alpha}_{n+1} - \tilde{\alpha}_n = -\gamma_{n+1}\,\partial_{\tilde{\mathbf{X}}} f_{n+1}$$
$$\gamma_{n+1} \ge 0, \qquad f_{n+1} \le 0, \qquad \gamma_{n+1}\,f_{n+1} = 0.$$

where γ_{n+1} is the discrete counterpart of $\dot{\lambda}$ at t_{n+1}. An elastic-predictor/plastic-corrector algorithm is used to take into account the Kuhn-Tucker conditions [6]. At every time step, in the first predictor phase it holds $f_n < 0$, an elastic behaviour is assumed and a trial value of f_{n+1}, i.e. $f_{n+1}^{(0)}$, is computed. If $f_{n+1}^{(0)} \le 0$, then an elastic behaviour occurs, γ_{n+1} has to be zero and no corrector phase is required. On the other hand, if $f_{n+1}^{(0)} > 0$, then plastic strains occur, the elastic prediction has to be corrected and $\gamma_{n+1} > 0$ has to be computed. This is done by the *return mapping algorithm* described below

i) Initialization
$$k = 0; \; \varepsilon_{n+1}^{p(0)} = \varepsilon_n^p, \tilde{\alpha}_{n+1}^{(0)} = \tilde{\alpha}_n, \gamma_{n+1}^{(0)} = 0$$

ii) Check yield condition and evaluate residuals
$$\sigma_{n+1}^{(k)} := E\left(\varepsilon_{n+1} - \varepsilon_{n+1}^{p(k)}\right) \; ; \; \tilde{\mathbf{X}}_{n+1}^{(k)} := \mathbf{D}\tilde{\alpha}_{n+1}^{(k)} \; ; \; f_{n+1}^{(k)} := f\left(\sigma_{n+1}^{(k)}, \tilde{\mathbf{X}}_{n+1}^{(k)}\right)$$

$$\mathbf{R}_{n+1}^{(k)} := \begin{bmatrix} -\varepsilon_{n+1}^{p(k)} + \varepsilon_n^p \\ \tilde{\alpha}_{n+1}^{(k)} - \tilde{\alpha}_n \end{bmatrix} + \gamma_{n+1}^{(k)}\begin{bmatrix} \partial_\sigma f \\ \partial_{\tilde{\mathbf{X}}} f \end{bmatrix}_{n+1}^{(k)}$$

if : $f_{n+1}^{(k)} < \text{tol}_1$ & $\left\| \mathbf{R}_{n+1}^{(k)} \right\| < \text{tol}_2$ then : EXIT

iii) Elastic moduli and consistent tangent moduli
$$C_{n+1}^{(k)} = E \qquad \mathbf{D}_{n+1}^{(k)} = \mathbf{D}$$

$$\left(\mathbf{A}_{n+1}^{(k)}\right)^{-1} = \begin{bmatrix} \left(C_{n+1}^{-1} + \gamma_{n+1}\partial_{\sigma\sigma}^2 f_{n+1}\right) & \gamma_{n+1}\partial_{\sigma\tilde{\mathbf{X}}}^2 f_{n+1} \\ \gamma_{n+1}\partial_{\tilde{\mathbf{X}}\sigma}^2 f_{n+1} & \left(\mathbf{D}_{n+1}^{-1} + \gamma_{n+1}\partial_{\tilde{\mathbf{X}}\tilde{\mathbf{X}}}^2 f_{n+1}\right) \end{bmatrix}^{(k)}$$

iv) Increment of the consistency parameter
$$\Delta\gamma_{n+1}^{(k)} = \frac{f_{n+1}^{(k)} - \left[\partial_\sigma f_{n+1}^{(k)} \; \partial_{\tilde{\mathbf{X}}} f_{n+1}^{(k)}\right]^T \mathbf{A}_{n+1}^{(k)} \mathbf{R}_{n+1}^{(k)}}{\left[\partial_\sigma f_{n+1}^{(k)} \; \partial_{\tilde{\mathbf{X}}} f_{n+1}^{(k)}\right]^T \mathbf{A}_{n+1}^{(k)}\left[\partial_\sigma f_{n+1}^{(k)} \; \partial_{\tilde{\mathbf{X}}} f_{n+1}^{(k)}\right]^T}$$

v) Increments of plastic strain and internal variables
$$\begin{bmatrix} \Delta\varepsilon_{n+1}^{p(k)} \\ \Delta\tilde{\alpha}_{n+1}^{(k)} \end{bmatrix} = \begin{bmatrix} C_{n+1}^{-1} & 0 \\ 0 & -\mathbf{D}_{n+1}^{-1} \end{bmatrix}^{(k)} \mathbf{A}_{n+1}^{(k)}\left(\mathbf{R}_{n+1}^{(k)} + \Delta\gamma_{n+1}^{(k)}\begin{bmatrix} \partial_\sigma f_{n+1}^{(k)} \\ \partial_{\tilde{\mathbf{X}}} f_{n+1}^{(k)} \end{bmatrix}\right)$$

vi) Update state variables and consistency parameter
$$\varepsilon_{n+1}^{p(k+1)} = \varepsilon_{n+1}^{p(k)} + \Delta\varepsilon_{n+1}^{p(k)} \; ; \; \tilde{\alpha}_{n+1}^{(k+1)} = \tilde{\alpha}_{n+1}^{(k)} + \Delta\tilde{\alpha}_{n+1}^{(k)} \; \gamma_{n+1}^{(k+1)} = \gamma_{n+1}^{(k)} + \Delta\gamma_{n+1}^{(k)}$$

This procedure to determine γ_{n+1} requires the computation, at each iteration, of the gradient and the Hessian matrix of f. Other algorithmic approaches by-pass the need of the Hessian of f, but they are not considered here.

This implementation is used to obtain hysteresis loops in some particular cases. The values of the four parameters E, σ_y, A_∞ and b are the same as those used in [5] and correspond to the identified values of an Inconel alloy ($E = 205580$ MPa, $\sigma_y = 1708,9$ MPa, $A_\infty = 35500$ MPa, $b = 380700$ MPa). The value of the new parameter ρ is $\rho = 1$ and the values of r and H are indicated in the caption of each figure.

Fig. 1 illustrates the hysteresis loops obtained with an increasing amplitude strain history. The effect of the newly introduced isotropic hardening term is highlighted. Fig. 2 refers to a stress input history, with constant amplitude and non-zero mean. The plastic strain accumulation (ratchetting) and the plastic shakedown are modelled by changing only one parameter.

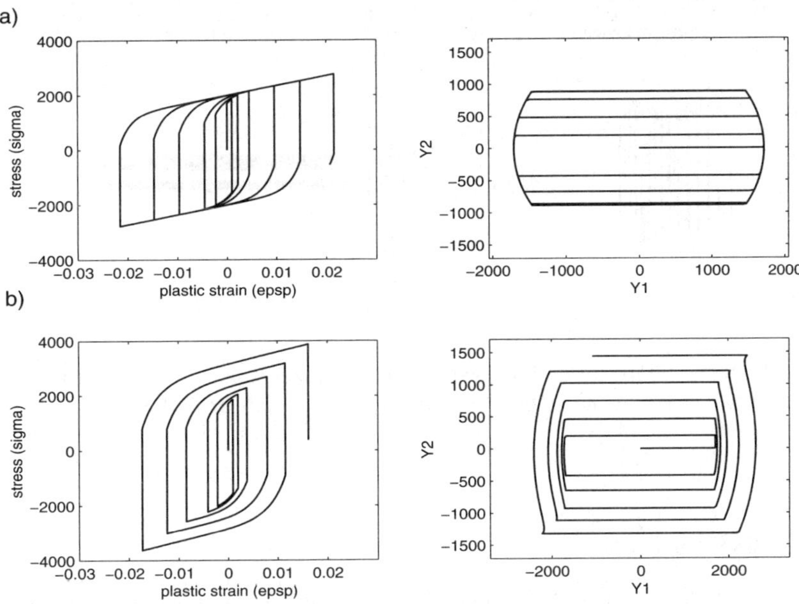

Fig. 1. Hysteresis loops for an imposed history with increasing strain amplitude. a) $r = 0.608$, $H = 0$ MPa b) $r = 0.608$, $H = 6500$ MPa.

5   Conclusions

A model with coupled hardening variables of strain type has been presented. It permits to take into account isotropic hardening and to have an elastic unloading path of varying length depending on the history of the loading. The hysteresis loops are qualitatively similar to the ones of the non-linear kinematic hardening model of Armstrong and Frederick [8]. The simplicity of this model, which depends only on six parameters, seems to be attractive for structural modelling applications with ratchetting effects. To this aim, the proposed return mapping algorithm is a useful numerical tool, which allows numerical simulations to be performed in an effective way.

a)

b)

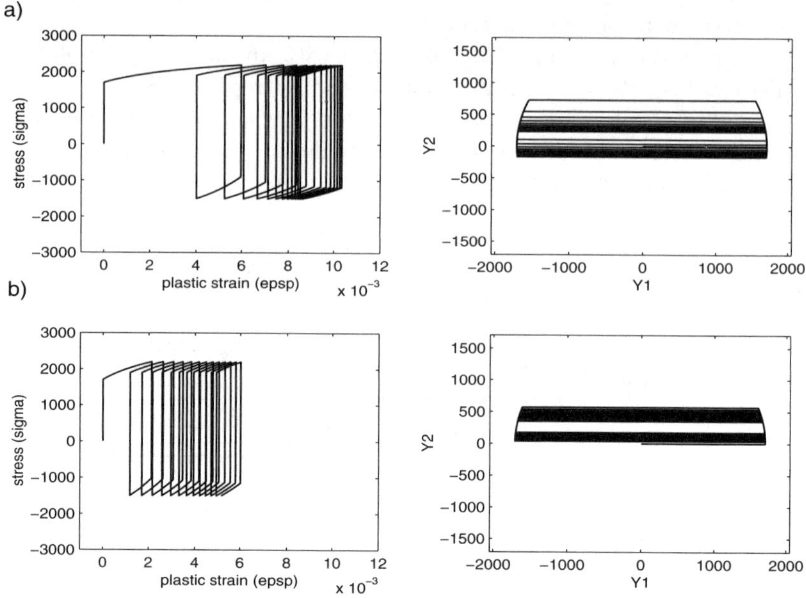

Fig. 2. Hysteresis loops for an imposed history with constant stress amplitude and non-zero mean stress. a) $r = 0.608$, $H = 0$ MPa; b) $r = 0.9$, $H = 0$ MPa.

References

1. Zarka J., Casier J. (1979) Elastic plastic response of a structure to cyclic loadings: practical rules. Mechanics Today, **6**, Ed. Nemat-Nasser, Pergamon Press
2. Khabou M.L., Castex L., Inglebert G. (1990) Effect of material behaviour law on the theoretical shot peening. Eur. J. Mech., A/Solids, **9**, 6 , 537-549.
3. Halphen B., Nguyen Q.S., (1975) Sur les matériaux standards généralisés. J. de Mécanique, **14**, 1 , 39-63
4. Inglebert G., Vial D., Point N. (1999) Modèle micromécanique à quatre paramètres pour le comportement élastoplastique. Groupe pour l'Avancement de la Mécanique Industrielle, **52**, march 1999
5. Vial D., Point N. (2000) A Plasticity Model and Hysteresis Cycles. Colloquium Lagrangianum, 6-9 décembre 2000, Taormina, Italy.
6. Simo J.C., Hughes T.J.R. (1986), Elastoplasticity and viscoplasticity. Computational aspects.
7. J. Lemaitre, J.L. Chaboche (1990), Mechanics of Solid Materials, Cambridge University Press, Cambridge, UK.

8. Armstrong P.J., Frederick C.O. (1966), A mathematical representation of the multiaxial Baushinger effect. CEGB Report, RD/B/N731, Berkeley Nuclear Laboratories.

8. Armstrong P.J., Frederick C.O. (1966), A mathematical representation of the multiaxial Baushinger effect. CEGB Report, RD/B/N731, Berkeley Nuclear Laboratories.

Comparison between Static and Dynamic Criteria of Material Stability

Antonio Grimaldi[1], Raimondo Luciano[2]

[1] Dipartimento di Ingegneria Civile
Università di Roma "Tor Vergata",
via del Politecnico, 1
00133 Roma, Italy
[2] Dipartimento di Meccanica, Strutture, A&T.
Università di Cassino,
via Di Biasio, 43
03043 Cassino, Italia

Abstract. A theoretical investigation on material stability characterisation for incrementally non-linear solids at finite strains is developed. Material stability may be defined by means of static and dynamic approaches leading to different conditions. In the framework of static material stability conditions, the opportunity of defining an infinitesimal stable material behaviour as the positiveness of the scalar product of the Biot strain-rate and the conjugate stress-rate, is taken into consideration. Dynamic conditions of material stability may be formulated in the context of propagation of plane infinitesimal wave (or acceleration waves) in an infinite body, and may lead to conditions similar to those obtained in the context of strain localisation in shear bands. Static and dynamic conditions are thus examined on the basis of the static material stability criterion defined by means of the Biot strain measure. The main objective of the present work is to investigate the interrelations between different material stability criteria. The analysis is illustrated at first by means of a one-dimensional example showing the main features of the problem, subsequently results are obtained for the three-dimensional continuum. Finally, applications are proposed for problems of uniform strain state. Results emphasize the role of the stress state, acting in the examined equilibrium configuration, in the relation between the static and the dynamic material stability conditions.

1 Introduction

In the theory of the constitutive behaviour of solids at finite strains, a suitable material stability definition is of crucial importance. This problem has received a great deal of attention in the literature during the last decades. Accordingly, the material stability concept appears in different contexts. It arises in the context of constitutive inequalities, imposing restrictions on the constitutive response in order to ensure physically admissible behaviour (Hill, 1970; Truesdell & Noll, 1965; Wang & Truesdell, 1973; Hill, 1978). Moreover, for an elastic material a dynamic characterisation of material stability was proposed, corresponding to real speeds for all propagation directions of small-amplitude plane waves superposed on the underlying homogeneous configuration (Hayes & Rivlin, 1961, Sawyers & Rivlin, 1973, Pan & Betty, 1999; Abeyaratne & Knowles, 1999). The material stability problem is also related to the problem of quasi-static bifurcation of a shear band type and of dynamic propagation of acceleration waves (Hill, 1962; Hill & Hutchinson, 1975; Petryk 1992). Strain localisation in the form of shear band zones, in fact, is commonly interpreted as a possible mechanism of material instability, since it may lead to fracture (Rudnicki and Rice, 1975) in an initially continuous medium.

Although the material stability problem has been extensively studied, some aspects have not been sufficiently investigated yet. A main point is: since different criteria, based on static or dynamic approaches, can be adopted to define the intrinsic material stability, which is the most appropriate? In order to reach a better understanding of this problem the present paper analyses the connections between static and dynamic material stability conditions.

The paper is principally devoted to material characterised by an incrementally linear or piecewise-linear constitutive relation, although some results are extended to incrementally non-linear materials characterised by a relation between rates of stress and strain positively homogeneous of degree one.

Static material stability conditions are derived by examining the infinitesimal stability of a homogeneously deformed and uniformly stressed material element subjected to deformation dependent surface tractions. These tractions can be chosen in different way to represent the uniform element stress state and the consequent stability condition depends strictly on this choice. By choosing particular descriptions of surface loading, the stability condition corresponds to the positiveness of the scalar product of a strain-rate and the corresponding conjugate stress-rate. Among this family of material stability conditions, the one defined in terms of the conjugate stress rate-strain rate pair based on the Biot strain measure is considered and proposed as the most appropriate. For an incremental linear material, the adopted material stability condition states that the local tensor of instantaneous moduli associated with the Biot strain measure must be positive definite. Moreover, for an elastic material, this material stability condition coincides with the GCN^+ condition (as called by Truesdell & Noll, 1965), the differential constitutive inequality implying the Generalized Coleman-Noll condition obtained by Truesdell & Toupin (1963), generalising a proposal of Coleman & Noll (1959). In the literature, other criteria which can be derived through the above-mentioned static approach have been proposed, corresponding to different choices of a strain measure. As an example, the logarithmic and the Green-Lagrange strain measures have been utilised. In particular, the first strain measure has been proposed by Hill (1968), whereas the second by Reese & Wriggers (1999). The constitutive restrictions imposed by the material stability condition based on the logarithmic strain measure, have been utilised in several stability and bifurcation analyses (see, for instance, Hill & Hutchinson, 1975; Young, 1976; Needlmann, 1979).

On the other hand, other conditions of material stability are obtained according to a dynamic approach. For example, an incrementally linear material is assumed to be stable if the strong ellipticity (SE) condition holds. This condition is, in fact, associated with infinitesimal plane wave propagation and ensures that for a homogeneously deformed material, for every direction of propagation the superimposed wave speeds corresponding to a real polarisation are real. Likewise, if the SE condition is satisfied the speeds of an acceleration wave related to a real acoustical axis, is real for a given direction of propagation. The SE condition can be opportunely extended to define similar material stability conditions based on acceleration waves propagation for materials characterised by a piecewise-linear incremental constitutive relation (conditions of this kind are derived by Hill, 1962; Petryk, 1992). Moreover, the SE condition arises also in the context of quasi-static bifurcation of shear bands type and acts as an exclusion condition for such kinds of bifurcation mode. Again, similar conditions hold for solids characterised by a non-linearity of the tangent constitutive operator (some conditions for an elastoplastic material can be found, for example, in Bigoni & Zaccaria, 1993). Note that, incidentally, both shear band bifurcation and acceleration or infinitesimal waves propagation lead to the same material stability criterion. This explains the great attention which the SE condition has received as a possible material stability characterisation. As in the case of static material stability conditions, the SE condition has been also utilised to impose restrictions on the constitutive behaviour in stability and bifurcation studies (Sawyers & Rivlin, 1973; Sawyers, 1977; Pan & Betty, 1999).

The present paper examines the connections between static and dynamic material stability conditions. In the static context, the Biot strain measure is considered and comparisons with other strain measures are carried out. After an introductory analysis of a one-dimensional example, results are obtained for the three-dimensional continuum. Necessary and sufficient conditions in order that a statically stable material state could be also dynamically stable, are analysed. From these results, it follows that the static condition of material stability may ensure the dynamic one, provided that suitable restrictions are imposed on the stress state acting in the examined configuration. In conclusion, as an illustration the case of a body under an all-round dead tension stress, is examined. The body has an arbitrary shape and is initially homogeneous and uniformly strained in the considered configuration.

2   A one-dimensional example

In order to introduce the object of this work, at first a one-dimensional example is considered, which, despite its simplicity, shows the basic features of the problem.

Consider a rectangular solid of homogeneous material specified by an incrementally linear constitutive law, whose associated tensor of moduli has orthotropic symmetry with respect to the principal axes of stress. The solid is homogeneously deformed in plane strain in the equilibrium configuration B, by a constant dead surface traction T_{011}, being the boundary value of the corresponding internal stress component. B occupies the region $-L/2 \leq x_1 \leq L/2$, $-h/2 \leq x_2 \leq h/2$, $0 \leq x_3 \leq 1$, where x_i is a rectangular system.

To investigate stability of the equilibrium state under consideration, infinitesimal incremental displacements measured from the current state of deformation, are superimposed upon the finite deformation state B. To avoid algebraic complexities, suppose a beam-like incremental displacement field $\dot{u}(x_1, x_2)$, x_1 and x_2 being the longitudinal and deflection directions, respectively. The components of $\dot{u}(x_1, x_2)$ on the x_i-axes are given by

$$\dot{u}_1 = u(x_1) + x_2 \psi(x_1), \quad \dot{u}_2 = v(x_1), \tag{1}$$

where $u(x_1)$ are the longitudinal displacements at $x_2=0$, $\psi(x_1)$ is the rotation of sections parallel to the x_2-axis and $v(x_1)$ is the deflection function. The considered problem is depicted in Fig. 1. The following notation is used

$$L_{ij} = D_{ij} + W_{ij} = \frac{\partial \dot{u}_i}{\partial x_j}, \tag{2}$$

for the incremental displacement gradient $\boldsymbol{L}$, $\boldsymbol{D}$ and $\boldsymbol{W}$ being the symmetric and antisymmetric part of $\boldsymbol{L}$, respectively and the current configuration is taken as reference. The relevant incremental constitutive equations, relating increments in the Biot strain measure $\boldsymbol{E}^{(1)}$ to increments in the work-conjugate stress measure $\boldsymbol{T}^{(1)}$, therefore reduce to the form:

$$\dot{T}_{11}^{(1)} = E^{(1)} D_{11} \quad \dot{T}_{12}^{(1)} = 2G^{(1)} D_{12}, \tag{3}$$

where $E^{(1)}$ and $G^{(1)}$ are the longitudinal and shear instantaneous moduli, respectively, corresponding to the adopted strain measure.

The incremental condition of stable equilibrium of the considered solid implies that, to the second order, the difference between internal deformation work and surface load work must be positive for every admissible $\dot{u}(x_1, x_2)$, namely

$$\int_B \left[\left(E^{(1)} D_{11}^2 + 4G^{(1)} D_{12}^2 \right) + T_{0\,11} \left(L_{21}^2 - D_{12}^2 \right) \right] dV > 0 \quad \forall \dot{u} \neq 0. \tag{4}$$

The integration is performed over the volume V of the solid in the deformed configuration B. In eqn (4) the internal deformation work is expressed in terms of the first order approximation $\boldsymbol{D}$ of the Biot strain tensor $\boldsymbol{E}^{(1)}$, whereas the work of dead surface tractions is coherently expressed trough the difference, to the second order, between the Biot strain tensor and the incremental deformation gradient, which depends both on the infinitesimal rotation $\boldsymbol{W}$ and on the infinitesimal strain $\boldsymbol{D}$.

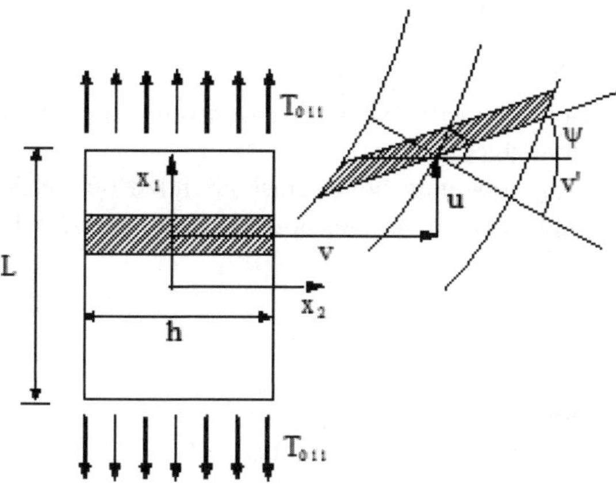

Fig. 1. Uniformly strained rectangular solid subjected to superimposed beam-like incremental displacements

Applying the incremental stability criterion to the assumed displacement field (1), eqn (4) reduces to

$$\int_{-L/2}^{L/2} \left\{ \left[E^{(1)} A u'^2 + E^{(1)} I \psi'^2 + G^{(1)} A \left(\psi + v'\right)^2 \right] + \right.$$
$$\left. + T_{0\,11} A \left[v'^2 - \tfrac{1}{4} \left(\psi + v'\right)^2 \right] \right\} dx_1, \tag{5}$$

where A and I denote area and centroidal moment of inertia of the cross section, respectively, and a prime indicates differentiation with respect to x_1.

In eqn (5) the first term characterises intrinsic material response to incremental superimposed deformations, whereas the second one defines stabilising-instabilising geometrical effects related to the dead surface tractions acting in the equilibrium configuration, the stability of which is tested.

According to a static criterion of material stability, the material is defined stable if the first term in eqn (5) is positive for every admissible displacement field $\dot{u}\,(x_1, x_2)$. For the examined case, this leads to the following requirements:

$$E^{(1)} > 0,\, G^{(1)} > 0. \tag{6}$$

On the other hand, the positiveness of the whole functional (5) corresponds to the stability condition in the examined equilibrium configuration for the solid subjected to dead loading, with reference to the one-dimensional displacement field (1). This requires that

$$E^{(1)} > 0, 0 < T_{011} < 4G^{(1)},\tag{7}$$

where the condition of positiveness of the current stress T_{011} excludes rigid rotations as possible eigenmodes.

Note that such modes may be a-priori excluded by considering an incremental displacement field which does not allow cross-section rotations. In this case, inequalities (7) reduce to:

$$G^{(1)} + (3/4)T_{0\ 11} > 0,\tag{8}$$

which defines the maximum compression that ensures the stability.

For the examined case, the difference between the condition (6) for a stable material response and the conditions (7) or (8), for the stability of the actual equilibrium configuration, is evident. The former is related exclusively to the intrinsic material response, whereas the latter contains a combination of geometrical and material properties.

¿From a dynamic point of view, alternatively, material is defined as stable if the speeds of infinitesimal plane waves superposed to the homogeneously deformed reference configuration, propagating along the x_1 direction and with polarisation contained in the $x_1 - x_2$ plane, are real. In this case the solid is assumed as infinitely extended along the x_1-axis.

The time dependent displacement field representing plane infinitesimal waves, assumes the form

$$u\left(x_1, t\right) = mf\left(x_1 - ct\right),\tag{9}$$

where m is the polarisation vector (taken as a unit vector), whose components are m_1 and m_2 along the x_1 and x_2 axes, respectively, c is the propagation speed and f is an arbitrary infinitesimal function. The propagation condition is

$$Q_0 m = \rho c^2 m,\tag{10}$$

Q_0 being the acoustic tensor associated to the propagation direction x_1, which admits the following matrix representation in terms of the nominal instantaneous moduli C^R_{0ijhk}:

$$Q_0 = \begin{bmatrix} C^R_{01111} & 0 \\ 0 & C^R_{02121} \end{bmatrix}.\tag{11}$$

Consequently, the dynamic material stability corresponds to the positiveness of the acoustic tensor and necessary and sufficient conditions for this are:

$$C^R_{01111} = E^{(1)} > 0, \ C^R_{02121} = G^{(1)} + (3/4)T_{0\ 11} > 0.\tag{12}$$

Conditions (12) are analogous to the strong ellipticity condition for the constitutive law of a three-dimensional continuum. As a consequence, note that these conditions exclude quasi-static shear band bifurcation modes with band normal aligned with the x_1-axis, for the homogeneously deformed solid, with the corresponding incremental displacement field obtained by taking $c = 0$ in eqn (9). At this point, it is clear that the dynamic material stability condition, as the structural stability condition (4), providing restrictions involving both moduli and geometrical effects trough the dead surface traction acting in the examined configuration, does not represent exclusively intrinsic material stability. In fact, dead loading are responsible for geometrical effects which are not related to the intrinsic material behaviour.

In addition it is possible to demonstrate that dynamic and static material stability conditions may be related, if suitable restrictions on the current stress field are imposed. As a matter of fact, if T_{011} is positive, static material stability implies dynamic material stability, namely inequalities (6) imply (12). Vice versa, if T_{011} is negative, then dynamic material stability implies static material stability, namely restrictions (12) leads to (6). Finally, if T_{011} is negative and satisfies the restriction (8), the static material stability implies the dynamic one. Similar results will be determined in the sequel for the general three-dimensional continuum.

3   Static criteria of material stability

In this section static conditions of material stability will be derived. Consider a homogeneous material element of arbitrary shape and unit volume occupying at the time $t = 0$ a region B of the three-dimensional Euclidean point space. The material element in the equilibrium configuration B is uniformly strained and subjected to a uniform stress state specified by the Cauchy stress tensor $\boldsymbol{T}_0$. The material element is therefore loaded only by the traction $\boldsymbol{T}_0\boldsymbol{n}$ on the boundary ∂B, with $\boldsymbol{n}$ denoting the unit outward normal to ∂B. The configuration B is taken as the reference configuration.

Consider now an additional displacement field $\boldsymbol{u}(\boldsymbol{x},\tau)$ superimposed on the reference configuration B, where $\boldsymbol{x}$ is the position occupied by a material particle in the configuration B, τ is a time-like parameter with $0<\tau < \alpha$ and $\boldsymbol{u}(\boldsymbol{x},0)=\boldsymbol{0}$. The additional displacement field deforms the material element quasi-statically from configuration B to configuration $B(\tau)$. The deformation gradient $\boldsymbol{F}(\tau)$ relative to the reference configuration B, defined in component form as

$$F_{ij} = \delta_{ij} + u_{i,j}, \tag{13}$$

where δ_{ij} is the Krönecker delta symbol and indices after a comma denote partial differentiation with respect to x_j coordinates, is assumed to be uniform (i.e. $\boldsymbol{F}(\tau)$ independent of $\boldsymbol{x}$). The material can be defined as stable in configuration B if the difference between the internal deformation work $W(\alpha)$ done in moving from B to $B(t)$ and the work $L(\alpha)$ done by the surface loads for every additional homogeneous deformation $\boldsymbol{u}(\boldsymbol{x},\tau)$, is positive.

Obtaining a proper material stability criterion needs an adequate definition of the loading mechanism representing the uniform stress state acting in B. To this end it is necessary that surface loads do not involve geometrical effects which are not related to the intrinsic material behaviour. Suppose that surface loads may be represented as a function of the deformation gradient and of the initial stress state

$$t_R^* = T_R^* \left(F\left(\tau\right), T_0 \right) n, \tag{14}$$

where $\boldsymbol{t^*}_R$ denotes the surface traction per unit area in the reference configuration B and the uniform tensor $\boldsymbol{T^*}_R$ indicates a fictitious first Piola-Kirchhoff stress tensor in static equilibrium with surface traction $\boldsymbol{t^*}_R$. Note that eqn (14), due to the uniformity of $\boldsymbol{T^*}_R$, represents a class of deformation-dependent surface loads with zero resultant whereas the resultant moment need not to be zero. The surface traction $\boldsymbol{t^*}_R$ in the given equilibrium configuration B at $t = 0$ corresponds to the initial stress state $\boldsymbol{T}_0$, namely

$$T_R^* \left(1, T_0\right) = T_0, \tag{15}$$

where $\boldsymbol{1}$ is the identity tensor. Using the divergence theorem for the rate of work done by surface tractions applied to the element boundary

$$\int_{\partial B} t_R^* \cdot \dot{u}\left(x, \tau\right) ds = \int_{\partial B} T_R^* n \cdot \dot{u}\left(x, \tau\right) ds = T_R^* \cdot \dot{F}, \tag{16}$$

where ds indicates the reference area elements, a superposed dot denotes partial differentiation with respect to τ at fixed $\boldsymbol{x}$ (i.e. a quasi-static rate) and a dot between vectors denotes the scalar product, leading to the following expression for the difference between internal deformation work and surface load work

$$W(\alpha) - L(\alpha) = \int_0^\alpha \left\{ \left[T_R\left(\tau\right) - T_R^*\left(\tau\right)\right] \cdot \dot{F}(\tau) \right\} d\tau. \tag{17}$$

It is evident that a surface loading remaining dead during the additional deformation

$$t_R^* = T_0 n, \tag{18}$$

should induce geometrical effects related to material rotation and thus must be rejected as a proper loading mechanism representing the uniform stress state $\boldsymbol{T}_0$. As a matter of fact, it is well known that in this case eqn (17), which specialises in

$$W(\alpha) - L(\alpha) = \int_0^\alpha \left\{ \left[T_R\left(\tau\right) - T_0\right] \cdot \dot{F}(\tau) \right\} d\tau, \tag{19}$$

represents the well-known structural stability criterion for dead loading (Hill, 1978) and does not characterise the intrinsic behaviour of the material since dead loads can do work even in a pure rotation.

3.1   Adoption of a material stability criterion

Clearly, defining a material stability condition needs a specification of the surface loading mechanism (14). A proper surface loading mechanism may be defined by choosing the surface traction distribution $T_R^* n$ which does work only through pure deformation, namely by choosing T_R^* in such a way to satisfy the following equation:

$$T_R^* \cdot \dot{F} = T_0 \cdot \dot{E}^{(1)} \ \forall \dot{u}\,(x,\tau)\,, \tag{20}$$

where $E^{(1)} = U\text{-}1$ denotes the Biot strain tensor and U the right stretch tensor. In other words T_R^* represents the initial uniform stress state T_0 if the stress power is measured in terms of the Biot strain rate $\dot{E}^{(1)}$. In addition to eqn (20), to avoid geometrical effects which may arise in rigid rotations, T_R^* must satisfy the rotational balance

$$T_R^* F^T = F T_R^{*T}. \tag{21}$$

A particular form of deformation dependence for T_R^* is now derived. By eqn (21) it follows that

$$T_R \dot{F} = R^T T_R \dot{U}, \tag{22}$$

where R is rotation tensor, and eqn (20) furnishes

$$T_0 = \frac{1}{2}(R^T T_R^* + T_R^{*T} R). \tag{23}$$

When T_0 commutes with U, the solution for T_R^* of eqs (23) and (21) is

$$T_R^* = R T_0. \tag{24}$$

The surface loading defined by eqn (24) follows the material rotation (therefore the tractions can be denoted as "follower tractions") and are represented in Fig. 2.

For a generic deformation process $F(\tau)$ according to eqn (20), the following potential can be defined for T_R^*

$$L(t) = \int_0^t T_0 \cdot \dot{E}^{(1)} d\tau = T_0 \cdot E^{(1)}, \ T_R^* = \frac{\partial L}{\partial F}, \tag{25}$$

where $\partial L/\partial F$ denotes a second order tensor whose components are $\partial L/\partial F_{ij}$. Note that if $F(\tau)$ does not ensure that T_0 and U commute, then T_R^* does not assume the special form (24). A general expression for T_R^*, in this case, will be derived in the sequel to the first order in τ.

Using eqn (20), eqn (17) specialises in

$$W(\alpha) - L(\alpha) = \int_0^\alpha \left\{ \left[T^{(1)}(\tau) - T_0 \right] \cdot \dot{E}^{(1)}(\tau) \right\} d\tau, \tag{26}$$

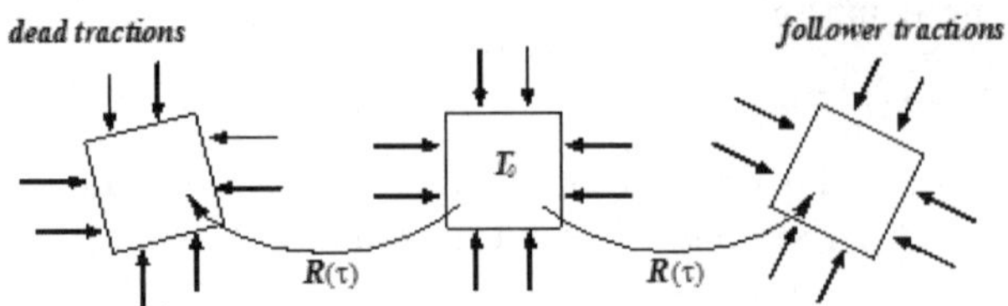

Fig. 2. Representation of follower tractions related to the loading mechanism in eqn (24) and of dead tractions

where $\boldsymbol{T}^{(1)}$ denotes the stress work-conjugate (Hill, 1968; 1978) to $\boldsymbol{E}^{(1)}$ (called Biot stress tensor). Eqn (26) represents a global sufficient material stability condition but in this work, according to the incremental character of the constitutive description of the material, attention is restricted to incremental material stability conditions obtained when $B(\alpha)$ is confined in some neighbourhood of B. To this end, suppose that the additional deformation $\boldsymbol{u}(\boldsymbol{x},\tau)$ is infinitesimal when α is sufficiently small and that the material is characterised by an incremental constitutive law positively homogeneous of degree one:

$$\dot{T}_R = C_0^R(L), \tag{27}$$

with

$$C_0^R(\tau L) = \tau C_0^R(L) \ \forall \tau > 0, \tag{28}$$

where L is the gradient of the deformation rate $\dot{u}(x,\tau)$ at $\tau{=}0$ (denoted in the sequel as $\dot{u}(x,\tau)_0$), the subscript zero denotes differentiation at $\tau{=}0$. Materials governed by the constitutive law (27) will be referred to as incrementally non-linear material. The constitutive law (27) is able to model incrementally linear material (elastic or hypoelastic), for which the response $\dot{T}_R = C_0^R[L]$ is characterised by the fourth-order tensor of instantaneous moduli C_0^R, as well as piecewise incrementally linear material (elastoplastic material with associative and nonassociative flow rules). In addition, (27) includes the class of material admitting a homogeneous of degree two potential of the velocity gradient (Hill, 1978):

$$\dot{T}_R = \frac{\partial U}{\partial L} = C_0^R \left(L/\sqrt{L \cdot L} \right) [L], \quad U = \frac{1}{2}\dot{T}_R \cdot L, \tag{29}$$

where the tensor of instantaneous moduli C_0^R depends non-linearly only on the direction of the velocity gradient L and satisfies the symmetry conditions $C_{0ijkl}^R = C_{0klij}^R$. Consequently, this class of material may model elastoplastic response with associative flow rules. For an incrementally linear material the symmetry condition of C_0^R derives from eqn $(29)_2$ since it follows that

$$C_{0\ ijhk}^R = \frac{\partial^2 U}{\partial L_{ij}\partial L_{hk}}.$$

Using the following expansions

$$T^{(1)}(\tau) = T_0 + C_0^{(1)}(L)\tau + o(\tau), \quad \dot{E}_0^{(1)} = D + O(\tau),$$

where $o(\tau)$ denotes terms approaching zero faster than τ and $O(\tau)$ denotes terms of first order in τ, leads to the following second order approximation of eqn (26):

$$W(\alpha) - L(\alpha) = \frac{1}{2}\left(\dot{T}_0^{(1)} \cdot D \right)\alpha^2 + o\left(\alpha^2\right), \tag{30}$$

where D is the symmetric part of L. This, in turn, leads to the following incremental material stability condition

$$C_0^{(1)}(D) \cdot D > 0 \forall D \neq 0, \tag{31}$$

where $C_0^{(1)}$ denotes the incremental response function relative to the conjugate stress-strain pair $(T^{(1)}, E^{(1)})$, namely

$$\dot{T}_0^{(1)} = C_0^{(1)}(D).$$

For an incrementally linear material eqn (31) states that the tensor of instantaneous moduli $C_0^{(1)}$ is positive definite and for elastic materials coincides with the GCN$^+$ condition (Truesdell & Noll, 1965). In this work eqn (31) is proposed as a possible incremental material stability condition and other material stability definitions will be compared to this one in order to analyse the effectiveness of the above choice.

3.2 Other static material stability criteria

Different conditions which have been taken as incremental material stability criteria in the literature can be obtained similarly from eqn (17). A particular class of conditions may be obtained by assuming that surface tractions correspond to T_0 if they are work-conjugate to a general measure of strain:

$$T_R^* \cdot \dot{F} = T_0 \cdot \dot{F}(U) \ \forall \dot{u}\,(x,\tau), \tag{32}$$

where the strain tensor $F(U)$ is coaxial with U and has principal values $f(\lambda_i)$ with f a monotonic increasing function of the principal values λ_i of U such that $f(1)=0$ e $df/d\lambda_i(1)=1$ (Hill, 1968). In this case eqn (17) leads to the positivity of the following expression

$$\frac{1}{2}C_0^f(D)\cdot D\alpha^2, \tag{33}$$

where C_0^f is the incremental response function related to the stress-strain measure pair $(T_f,\, F(U))$. Among these stress-strain measure pairs, the class $(T^{(m)},\, E^{(m)})$ corresponding to $f(\lambda_i)=(\lambda_i^m\text{-}1)/m$, where m is an integer, has been largely used in the literature (Ogden, 1984). Eqn (33) leads to the following family of material stability conditions:

$$C_0^f(D)\cdot D > 0\ \forall D \neq 0, \tag{34}$$

which correspond to the positiveness of the incremental material response measured in terms of a generic strain tensor $F(U)$. Taking the first order approximation of eqn (32) leads to the following expression for the derivative at $\tau=0$ of $T_R^*(\tau)$:

$$\dot{T}_R^* = \frac{1}{2}[f''(\lambda) - 1](DT_0 + T_0D) + LT_0. \tag{35}$$

By taking the Biot strain measure in eqn (35), it is possible to derive a first order approximation of the stress tensor T_R^* acting as a loading mechanism in the proposed material stability criterion eqn (31) (namely, defined by eqn (20)). This results in

$$T_R^* = T_0 + \left[\frac{1}{2}(DT_0 - T_0D) + WT_0\right]\tau + o(\tau), \tag{36}$$

showing that if T_0 does not commute with D, a shear stress rates $\frac{1}{2}(DT_0 - T_0D)$ arise in order to restore incremental rotational balance destroyed by the additional incremental deformation superposed to B. To see this, consider the derivative at $\tau=0$ of the fictitious Cauchy stress tensor, denoted by T^*, corresponding to T_R^*

$$\dot{T}^* = \dot{T}_R^* - T_0 tr L + T_0 L^T = \dot{T}_1^* + \dot{T}_2^*,$$

where

$$\dot{T}_1^* = WT_0 - T_0 tr L + T_0 L^T,\ \ \dot{T}_2^* = \frac{1}{2}(DT_0 - T_0D),$$

together with eqn (36). The symmetry of $\dot{T}^*$ is restored by the shear stress rates $\dot{T}_2^*$, since $\dot{T}_1^*$, which represents the Cauchy stress rate when T_0 commutes with D, should lead to an unsymmetrical $\dot{T}^*$. On the other hand, when T_0 commutes with D, eqn (36) leads to the first order version of eqn (24):

$$T_R^* = T_0 + WT_0\tau + o(\tau). \tag{37}$$

Eqn (36) can be also derived by taking the derivative of the potential (25) with respect to the incremental deformation gradient $\boldsymbol{L}_T$. The family of static material stability criteria represented by eqn (34) shows the connection with constitutive inequalities which have been analysed in the literature as a-priori restrictions upon material response function consistent with the physical material behaviour (Truesdell & Noll, 1965; Wang & Truesdell, 1973).

Using the logarithmic strain measure in the condition (34) leads to the incremental inequality proposed by Hill (1968) by means of considerations based on the behaviour of incompressible elastic materials, as the most suitable to impose restrictions on the incremental material response function:

$$C_0^{(0)}(D) \cdot D > 0 \ \forall D \neq 0. \tag{38}$$

The stress rate conjugate to this strain measure coincides with the Jaumann rate of the Kirchhoff stress tensor, the current configuration coinciding with the reference one. The material stability condition (38) has been largely utilised to model the incremental material behaviour of incrementally linear and non-linear solids in the literature. On the other hand, the choice of the Green-Lagrange strain measure leads to the material stability condition proposed by Reese & Wriggers (1997). Obviously, these choices do not correspond to surface tractions working on pure material deformation as in the case of the Biot strain measure. Moreover, from eqn (35) note that in the class $\boldsymbol{E}^{(m)}$, the Biot strain measure is the only one leading to the first order approximation of tractions following material rotation (37) when $\boldsymbol{T}_0$ commutes with $\boldsymbol{D}$.

4   Dynamic criteria of material stability

Dynamic criterions of material stability have been proposed in the literature in the context of the propagation of plane infinitesimal waves or of acceleration waves (Hayes & Rivlin, 1961; Hill, 1962; Petryk 1992). Here some aspects of these conditions are briefly recalled for the sake of clarity in the subsequent analysis.

Consider a homogeneous material of infinite extension, homogeneously deformed and with vanishing body forces occupying the equilibrium configuration B, assumed as reference configuration. The material admits an incrementally linear constitutive law. An additional infinitesimal time dependent displacement field $\boldsymbol{u}(\boldsymbol{x},t)$ is superimposed on the reference configuration. The propagation condition for plane infinitesimal waves

$$u(x,t) = mg(n \cdot x - ct), \tag{39}$$

with propagation direction $\boldsymbol{n}$, polarization $\boldsymbol{m}$ (taken as a unit vector) and speed of propagation c, is

$$Q_0(n)m = \rho c^2 m, \tag{40}$$

where $Q_{0ik} = C^R_{0ijkl} n_j n_l$ is the acoustic tensor associated to the propagation direction $\boldsymbol{n}$, g is a twice differentiable function and ρg is the mass density of the material in B(Truesdell & Noll, 1965). Eqn (40) together with the associated characteristic equation, determines possible polarisations and wave speeds for any given direction $\boldsymbol{n}$. If for a given $\boldsymbol{n}$, $\boldsymbol{m}$ is a real proper vector of $\boldsymbol{Q}_0(\boldsymbol{n})$, the wave speed is given in the following form

$$\rho c^2 = [Q_0(n)\, m] \cdot m, \tag{41}$$

obtained from eqn (40). In general, eigenvalues and corresponding eigenvectors of eqn (40) may not be real and, consequently, the wave speeds are not necessarily real, except when the acoustic tensor is symmetric and positive-definite.

The dynamic material stability conditions proposed in the literature are based on the nature of the eigenvalues ρc^2. For example, in the case of an hyperelastic material the existence of a negative eigenvalues ρc^2 for any direction of propagation of a plane wave, has been indicated as a condition of material instability (Hayes & Rivlin, 1961), since in this case the infinitesimal wave tends to grow indefinitely with time. The propagation condition (40) is the same as that for plane acceleration waves in a deformed body and a non-real acceleration wave speed is usually taken as a condition denoting material instability for incrementally non-linear constitutive laws (the connection between acceleration waves with non-real speeds and material instability has been considered, for instance, by Sawyers & Rivlin, 1973 for an hyperelastic material; by Hill, 1962 for a piece-wise incrementally linear material; by Petryk, 1992 for an incremental constitutive law positively homogeneous of degree one which derives from a potential).

On the basis of the previous considerations, an incrementally linear material can be defined stable in a dynamic sense, if for any propagation direction $\boldsymbol{n}$ the speeds of plane infinitesimal (acceleration waves) related to each real polarisation $\boldsymbol{m}$ (acoustical axis), are real and nonzero. If the acoustic tensor is symmetric, the strong ellipticity condition (SE), which corresponds to the positiveness of the right hand side of eqn (41), is necessary and sufficient for a material to be stable. Analogously, for an incrementally non-linear material characterised by eqn (28), the following condition, similar to the SE, for the tangent constitutive operator,

$$C^R_0 (m \otimes n) \cdot m \otimes n > 0 \forall m, n \neq 0, \tag{42}$$

where $\otimes$ denotes the dyadic product, ensures real and nonzero acceleration wave speeds associated with each real acoustical axis m, when the velocity gradient vanishes at one side of acceleration wave surface (Hill, 1962; Petryk ,1992). Moreover, for an elasto-plastic material adminitting an incrementally linear comparison material (see Hill, 1958, for associative flow rules or Raniecki and Bruhns, 1981, for non associative flow rules), the SE condition for the tensor of instantaneous moduli of the comparison material $^*C_0^R$

$$^*C_0^R[m \otimes n] \cdot m \otimes n > 0 \forall m, n \neq 0, \tag{43}$$

ensures real and nonzero wave speeds for each real acoustical axis also when the velocity gradient is nonzero at one side of the wave. This condition does not seem to be in the literature.

The SE condition has received considerable attention in the context of material stability characterisation not only due to its connection with the dynamic instability phenomena related to wave propagation, but also because of its relationship with strain localisation phenomena. Strain localisation of shear band types, in fact, has been often considered in the literature as a form of material instability and the SE condition represents an exclusion condition for shear band bifurcation modes. The SE condition excludes bifurcation modes of shear band type for an incrementally linear material, whereas its analogous eqn (42), excludes shear band bifurcation modes for homogeneous all-round dead load boundary conditions along a uniformly strained deformation path. On the other hand, the SE condition of the tensor of instantaneous moduli for the linear comparison material excludes localisation in the form of shear bands in the elastoplastic solid.

5   Some connections among material stability criteria

In this section some interrelations among the proposed static material stability condition eqn (31) (referred to as the MS condition in the sequel) and dynamic matcrial stability criteria (referred to as the DMS condition in the sequel) based on the SE condition, will be derived. In addition, connections between the family of material stability conditions eqn (34) (referred to as the MS^f conditions in the sequel) will be analysed. By examining these connections, the role which geometrical effects play in the difference among different criteria will be investigated. Results will show that the MS condition may lead to the DMS condition only if opportune restrictions are imposed upon the stress state acting in the equilibrium configuration B, whereas the DMS condition is not so strong, in general, to ensure the MS condition. On the other hand, analogous restrictions can be imposed on the stress state in order that the MS^f conditions ensure the DMS condition. This confirms that dynamic material stability criteria are affected by some geometrical effects.

5.1   Relations between MS and DMs conditions

At first, note that, generally, the MS condition implies nonzero real speeds only for eventual longitudinal waves ($m=n$). This can be shown by appropriate specialisation of the scalar product between the Piola-Kirchhoff stress rate and the deformation rate gradient:

$$C_0^R(L) \cdot L = C_0^{(1)}(D) \cdot D + T_0 \cdot \left(L^T L - DD\right), \tag{44}$$

to symmetric deformation rate gradient $\boldsymbol{L}=\boldsymbol{D}$. Eqn (44) can be derived by considering a second order expansion analogous to eqn (36) of the work-conjugacy requirement $T_R \cdot \dot{F} = T^1 \cdot \dot{E}^1$. In eqn (44) the first term is related to the material stability condition, whereas the second to geometrical effects associated with dead loading in B. Eqn (44) can be extended also to incompressible materials by including the arbitrary stress rate arising from the incompressibility constraint in the response functions. Evaluating eqn (44) along deformation rate gradient of rank one $\boldsymbol{m} \otimes \boldsymbol{n}$, $\boldsymbol{m}$ and $\boldsymbol{n}$ being arbitrary unit vectors, leads to

$$
C_0^R (m \otimes n) \cdot m \otimes n = C_0^{(1)} \left(\widehat{m \otimes n} \right) \cdot \widehat{m \otimes n} +
$$
$$
+ T_0 \cdot \left[(n \otimes m)(m \otimes n) - \left(\widehat{m \otimes n} \right) \left(\widehat{m \otimes n} \right) \right], \tag{45}
$$

where a head over a tensor denotes its symmetric part. Note that if $\boldsymbol{m} \otimes \boldsymbol{n}$ is symmetrical (i.e. $\boldsymbol{m}=\boldsymbol{n}$), the geometrical effects arising from dead loading vanish.

Suppose that the scalar functions in the right hand side of eqn (44) are restricted to rank one tensors. If the minimum of the scalar function representing geometrical effects in eqn (44) is greater than the minimum of the opposite of the scalar function representing material stability, then the MS condition implies the DMS condition:

$$
\left. \begin{array}{l} MS \\ min \left\{ T_0 \cdot \left[(n \otimes m)(m \otimes n) - \left(\widehat{m \otimes n} \right) \left(\widehat{m \otimes n} \right) \right] \right\} > \\ > -min \left\{ C_0^{(1)} \left(\widehat{m \otimes n} \right) \cdot \widehat{m \otimes n} \right\} \end{array} \right\} \Rightarrow DMS \tag{46}
$$

Let us derive an estimate of the restrictions imposed by the above sufficient condition for dynamic material stability, based on the lower bound of the scalar function representing geometrical effects. By using the invariance of scalar product and expressing this scalar function in terms of components on the axes of principal stresses (t_1, t_2, t_3), the following expression can be obtained

$$
t_i \delta_{ij} \left[m_k m_k n_i n_j - \tfrac{1}{4} (m_k n_i + n_k m_i)(m_k n_j + n_j m_k) \right] =
$$
$$
= t_i \left[n_i^2 - \tfrac{1}{4} \left(n_i^2 + m_i^2 + 2 m_k n_k m_i n_i \right) \right], \tag{47}
$$

where summation over repeated indices is intended ($i=1,2,3$). The scalar function (47) admits the following inferior limit

$$
t_i \left[n_i^2 - \tfrac{1}{4} \left(n_i^2 + m_i^2 + 2 m_k n_k m_i n_i \right) \right] \geq
$$
$$
\geq min \{t_i\} \sum_i n_i^2 - max \{t_i\} \sum_i \alpha_i^2, \tag{48}
$$

where $\alpha_i^2 = (n_i^2 + m_i^2 + 2 m_k m_k m_i n_i)/4$ with no summation over i. Since

$$
\frac{1}{2} \leq \sum_i \alpha_i^2 \leq 1 ;
$$

$$-max\{t_i\}\sum_i \alpha_i^2 \geq \begin{cases} -max\{t_i\}\, max\left\{\sum_i \alpha_i^2\right\} \text{ if } max\{t_i\} \geq 0 \\ -max\{t_i\}\, min\left\{\sum_i \alpha_i^2\right\} \text{ if } max\{t_i\} \leq 0 \end{cases},$$

consequently

$$T_0 \cdot \left[(n\otimes m)(m\otimes n) - \left(\widehat{m\otimes n}\right)\left(\widehat{m\otimes n}\right)\right] \geq min\{t_i\} - \gamma max\{t_i\} \quad (49)$$

where

$$\gamma = \begin{cases} 1 \text{ if } max\{t_i\} \geq 0 \\ \frac{1}{2} \text{ if } max\{t_i\} \leq 0 \end{cases}.$$

If the inferior limit in eqn (49) is greater than the opposite of the scalar function representing material stability restricted to rank one tensors, then

$$\left.\begin{array}{l} MS \\ min\{t_i\} - \gamma max\{t_i\} > -min\left\{C_0^{(1)}\left(\widehat{m\otimes n}\right)\cdot\widehat{m\otimes n}\right\} \end{array}\right\} \Rightarrow DMS, (50)$$

namely the sufficient condition (46) is satisfied. For incompressible material, since $m\bullet n$ vanishes, the summation of α_i^2 is equal to $1/2$ and, therefore, γ is equal to $1/2$ also. Moreover, by considering that

$$min\{t_i\} - \gamma\ max\{t_i\} \geq -2\ max\{|t_i|\},$$

the following restrictions on the maximum principal stress can be obtained:

$$max\{|t_i|\} < \frac{1}{2}min\left\{C_0^{(1)}\left(\widehat{m\otimes n}\right)\cdot\widehat{m\otimes n}\right\}, \tag{51}$$

which, by (46), ensures that the MS condition implies the DMS condition. For piecewise incrementally linear material admitting a linear comparison solid, the sufficient condition (50) may be replaced by

$$\left.\begin{array}{l} MS^* \\ \quad min\{t_i\} - \gamma max\{t_i\} > \\ \\ > -min\left\{{}^*C_0^{(1)}\left[\widehat{m\otimes n}\right]\cdot\widehat{m\otimes n}\right\} \end{array}\right\} \Rightarrow SE^* \Rightarrow DMS, \tag{52}$$

where MS* denotes the material stability condition for the comparison solid in terms of the tensor instantaneous moduli ${}^*C_0^{(1)}$

$${}^*C_0^{(1)}[D]\cdot D > 0\ \forall D \neq 0, \tag{53}$$

SE* indicates the condition (43) and DMS the condition (42).

Note that due to the property of relative convexity (Hill, 1978) between the incremental response of the real solid and that of the comparison one, the MS* condition implies MS

$$MS^* \Rightarrow MS.$$

Consider now the restrictions imposed by the DMS condition on the stress state $\boldsymbol{T}_0$. By looking at the condition (42) and eqn (44), it follows that a necessary condition in order that a stable material in static sense is stable also in a dynamic one, is that the minimum (maximum) of the scalar function representing geometrical effects is greater than the opposite of the maximum (minimum) of the scalar function representing the MS condition, both being evaluated along rank one tensors:

$$\left.\begin{array}{c} MS \\ DMS \end{array}\right\} \Rightarrow \begin{array}{c} min(max)\left\{T_0 \cdot \left[(n \otimes m)(m \otimes n) - \left(\widehat{m \otimes n}\right)\left(\widehat{m \otimes n}\right)\right]\right\} > \\ \\ > -max(min)\left\{C_0^{(1)}\left(\widehat{m \otimes n}\right) \cdot \widehat{m \otimes n}\right\} \end{array} \tag{54}$$

An estimate of the superior limit of the scalar function representing geometrical effects, is now needed in order to make more explicit the condition (54). This scalar function admits the following upper bound

$$\begin{array}{c} t_i\left[n_i^2 - \frac{1}{4}\left(n_i^2 + m_i^2 + 2m_k n_k m_i n_i\right)\right] \leq \\ \\ \leq max\{t_i\}\sum_i n_i^2 - min\{t_i\}\sum_i \alpha_i^2, \end{array} \tag{55}$$

Since

$$-min\{t_i\}\sum_i \alpha_i^2 \leq \begin{cases} -min\{t_i\}\,min\left\{\sum_i \alpha_i^2\right\} & \text{if } min\{t_i\} \geq 0 \\ \\ -min\{t_i\}\,max\left\{\sum_i \alpha_i^2\right\} & \text{if } min\{t_i\} \leq 0 \end{cases},$$

then the following inferior limit can be obtained

$$T_0 \cdot \left[(n \otimes m)(m \otimes n) - \left(\widehat{m \otimes n}\right)\left(\widehat{m \otimes n}\right)\right] \leq max\{t_i\} - \delta min\{t_i\}, \tag{56}$$

where

$$\delta = \begin{cases} \frac{1}{2} & \text{if } min\{t_i\} \geq 0 \\ 1 & \text{if } min\{t_i\} \leq 0 \end{cases}. \tag{57}$$

Hence, we obtain an explicit version of the necessary condition (54)

$$\left.\begin{array}{c} MS \\ DMS \end{array}\right\} \Rightarrow \begin{array}{c} max\{t_i\} - \delta min\{t_i\} > \\ \\ -max(min)\left\{C_0^{(1)}\left(\widehat{m \otimes n}\right) \cdot \widehat{m \otimes n}\right\}. \end{array} \tag{58}$$

In general, the superior limit estimated in the inequality (56) is non negative and the condition (58) does not provide useful information. On the contrary, for incompressible material the value of δ is equal to $1/2$ and condition (58) can furnish some information. Analogously to the case of the sufficient condition, considering that $max\{t_i\}-\delta min\{t_i\}\leq 2max\{|t_i|\}$ a

rougher estimate for the inferior limit can be obtained involving only the maximum principal stress.

For piecewise incrementally linear material admitting a linear comparison solid, the necessary condition (58) may be replaced by

$$\left.\begin{array}{l} MS^* \\ DMS^* \end{array}\right\} \Rightarrow \begin{array}{l} max\{t_i\} - \delta min\{t_i\} > \\ \\ > -max(min)\left\{{}^*C_0^{(1)}\left(\widehat{m\otimes n}\right)\cdot\widehat{m\otimes n}\right\}, \end{array} \tag{59}$$

where DMS* denotes the dynamic material stability condition for the comparison solid (43).

On the other hand, a condition on the stress state sufficient for the MS condition, once the DMS condition is satisfied, cannot be generally obtained. To this end, note that the positive definiteness of eqn (44) implies the MS condition, while the DMS condition, being a structural stability condition restricted to rank one tensors, is not so strong to imply the MS condition.

The above results, obtained in terms of the MS condition, show that the dynamic material stability condition is not a proper characterisation of the intrinsic behaviour of the material, involving some geometrical effects arising from dead loads acting in the examined equilibrium configuration.

5.2 Relations between Ms^f and DMs conditions

Conditions similar to (50) and (58), can be obtained for the family of material stability conditions expressed by inequalities (34). To this end, the following equation must be considered

$$C_0^R\left(L\right)\cdot L = C_0^{(1)}\left(D\right)\cdot D + T_0\left(L^T L + \beta DD\right), \tag{60}$$

where β is equal to $f''(1)-1$, which can be derived by a second order expansion of the work-conjugacy requirement $T_R^{..}\dot{F} = T^f\cdot\dot{F}(U)$.

The sufficient condition for DMS is now

$$\left.\begin{array}{l} MS^f \\ min\left\{T_0\cdot\left[(n\otimes m)(m\otimes n) + \beta\left(\widehat{m\otimes n}\right)\left(\widehat{m\otimes n}\right)\right]\right\} > \\ \\ > -min\left\{C_0^f\left(\widehat{m\otimes n}\right)\cdot\widehat{m\otimes n}\right\} \end{array}\right\} \Rightarrow DMS \tag{61}$$

where MS^f denotes inequality (34); whereas the inferior limit for the scalar function containing T_0 is

$$T_0\cdot\left[(n\otimes m)(m\otimes n) + \beta\left(\widehat{m\otimes n}\right)\left(\widehat{m\otimes n}\right)\right] \geq min\{t_i\}+\gamma min\{\beta t_i\} \tag{62}$$

with

$$\gamma = \begin{cases} \frac{1}{2} & \text{if } min\left\{\beta t_i\right\} \geq 0 \\ 1 & \text{if } min\left\{\beta t_i\right\} \leq 0 \end{cases}.$$

Consequently, the sufficient condition transforms to

$$\left. \begin{array}{l} MS^f \\ min\left\{t_i\right\} + \gamma min\left\{\beta t_i\right\} > -min\left\{C_0^f \left(\widehat{m \otimes n}\right) \cdot \widehat{m \otimes n}\right\} \end{array} \right\} \Rightarrow DS. \quad (63)$$

In the case of the conjugate stress-strain pairs $(\boldsymbol{T}^{(m)}, \boldsymbol{E}^{(m)})$, βg is equal to $m - 2$. For instance, restrictions imposed by (63) to ensure the DS condition, are expressed by

$$min\left\{t_i\right\} - 2\gamma max\left\{t_i\right\} > -min\left\{C_0^{(0)} \left(\widehat{m \otimes n}\right) \cdot \widehat{m \otimes n}\right\},$$

in the case of the logarithmic strain measure, and by

$$min\left\{t_i\right\} > -min\left\{C_0^{(2)} \left(\widehat{m \otimes n}\right) \cdot \widehat{m \otimes n}\right\},$$

in the case of the Green-Lagrange strain measure.

An analogous condition holds for piecewise incrementally linear material admitting a linear comparison solid:

$$\left. \begin{array}{l} MS^{*f} \\ \quad min\left\{t_i\right\} + \gamma max\left\{\beta t_i\right\} > \\ \\ > -min\left\{{}^*C_0^f \left[\widehat{m \otimes n}\right] \cdot \widehat{m \otimes n}\right\} \end{array} \right\} \Rightarrow SE^* \Rightarrow DMS, \quad (64)$$

where MS^{*f} denotes the material stability condition for the comparison solid in terms of the tensor instantaneous moduli ${}^*C_0^f$.

On the other hand, the necessary condition (54) now reads

$$\left. \begin{array}{l} MS^f \\ DMS \end{array} \right\} \Rightarrow \begin{array}{l} min(max)\left\{T_0 \cdot \left[(n \otimes m)(m \otimes n) + \beta \left(\widehat{m \otimes n}\right)\left(\widehat{m \otimes n}\right)\right]\right\} > \\ \\ > -max(min)\left\{C_0^f \left(\widehat{m \otimes n}\right) \cdot \widehat{m \otimes n}\right\} \end{array} \quad (65)$$

The estimate of the superior limit of the scalar function representing geometrical effects in the generalised strain measure sense, is

$$T_0 \cdot \left[(n \otimes m)(m \otimes n) + \beta \left(\widehat{m \otimes n}\right)\left(\widehat{m \otimes n}\right)\right] \leq max\left\{t_i\right\} + \delta max\left\{\beta t_i\right\} \quad (66)$$

where

$$\delta = \begin{cases} \frac{1}{2} & \text{if } max\left\{\beta t_i\right\} \leq 0 \\ 1 & \text{if } max\left\{\beta t_i\right\} \geq 0 \end{cases}. \quad (67)$$

Hence, we obtain an explicit version of the necessary condition (65)

$$\left.\begin{array}{l} MS^f \\ DMS \end{array}\right\} \Rightarrow \begin{array}{l} max\left\{t_i\right\} + \delta max\left\{\beta t_i\right\} > \\ \\ -max(min)\left\{C_0^f\left(\widehat{m \otimes n}\right) \cdot \widehat{m \otimes n}\right\} \ . \end{array} \tag{68}$$

The restrictions imposed by this condition specialises to

$$max\left\{t_i\right\} - 2\delta min\left\{t_i\right\} > -max(min)\left\{C_0^{(0)}\left(\widehat{m \otimes n}\right) \cdot \widehat{m \otimes n}\right\} \ ,$$

in the case of the logarithmic strain measure, and to

$$max\left\{t_i\right\} > -max(min)\left\{C_0^{(2)}\left(\widehat{m \otimes n}\right) \cdot \widehat{m \otimes n}\right\} \ ,$$

in the case of the Green-Lagrange strain measure. Note that the former restrictions are a-priori satisfied.

For piecewise incrementally linear material admitting a linear comparison solid, the necessary condition (68) may be replaced by

$$\left.\begin{array}{l} MS^{*f} \\ DMS^* \end{array}\right\} \Rightarrow \begin{array}{l} max\left\{t_i\right\} + \delta max\left\{\beta t_i\right\} > \\ \\ > -max(min)\left\{{}^*C_0^f\left(\widehat{m \otimes n}\right) \cdot \widehat{m \otimes n}\right\} \ , \end{array} \tag{69}$$

A condition on the stress state sufficient for the MS^f condition, once the DMS condition is satisfied, cannot be generally obtained.

5.3   The uniformly stressed body

A useful application of the above-obtained conditions is now presented for a body of an arbitrary shape homogeneously deformed and subjected to a hydrostatic dead load (namely $t_1 = t_2 = t_3 = \sigma$). In this circumstance we have

$$C_0^R\left(L\right) \cdot L = C_0^{(1)}\left(D\right) \cdot D + \sigma W \cdot W \tag{70}$$

and the geometrical term assumes the sign of the uniform stress σ. From eqn (70) it is evident that the sufficient condition (46) is satisfied when σ is non negative, in fact

$$min\left\{t_i\right\} - \gamma max\left\{t_i\right\} = 0 > -min\left\{C_0^{(1)}\left(\widehat{m \otimes n}\right) \cdot \widehat{m \otimes n}\right\} \ ,$$

whereas in the case of a negative σ , we obtain

$$\sigma > -2min\left\{C_0^{(1)}\left(\widehat{m \otimes n}\right) \cdot \widehat{m \otimes n}\right\} . \tag{71}$$

Thus, the material is stable both from the dynamic and static point of view in tension, whereas in compression if the material is statically stable then the material is dynamically stable only when the condition (71) is satisfied. The necessary condition (54), furnishes banal information. On the other hand, for the logarithmic strain measure the sufficient condition (63) is satisfied when σ is non negative if

$$\sigma < min \left\{ C_0^{(0)} \left(\widehat{m \otimes n} \right) \cdot \widehat{m \otimes n} \right\},$$

whereas in the case of a negative σ it is always satisfied, since it furnishes

$$0 > -min \left\{ C_0^{(0)} \left(\widehat{m \otimes n} \right) \cdot \widehat{m \otimes n} \right\}.$$

For the Green-Lagrange strain measure we have

$$\sigma > -min \left\{ C_0^{(2)} \left(\widehat{m \otimes n} \right) \cdot \widehat{m \otimes n} \right\},$$

for both tension and compression loading.

The necessary condition (68) identically satisfied for the logarithmic strain measure when σ is non negative and results in

$$\sigma < max(min) \left\{ C_0^{(0)} \left(\widehat{m \otimes n} \right) \cdot \widehat{m \otimes n} \right\},$$

for compression loading. On the other hand, in the case of the Green-Lagrange strain measure the necessary condition (68) furnishes:

$$\sigma > -max(min) \left\{ C_0^{(2)} \left(\widehat{m \otimes n} \right) \cdot \widehat{m \otimes n} \right\}.$$

6   Conclusions

Some connections between static and dynamic conditions for a stable material response, are investigated. In the static context, surface traction representing the uniform stress state in the examined equilibrium configuration, are defined in such a way to work exclusively on pure deformations and, consequently, to exclude geometrical effects. This leads to defining a stable material behaviour as the positiveness of the scalar product between the Biot strain-rate and the conjugate stress-rate. The proposed static criterion is analysed through comparisons with different material stability conditions, which can be formulated by using static or dynamic approaches. Within the former approach, conditions expressed in terms of different strain measures are examined. According to the latter, conditions based on wave propagation or on strain localisation in shear bands are considered.

Connections between different material stability criteria emphasize the role of geometrical effects in the difference between static and dynamic material stability conditions. As a matter of fact, the proposed static condition of material stability may ensure the dynamic one, only provided that suitable restrictions are imposed on the stress state acting in the examined configuration, whereas the converse is, in general, not possible. These connections show that some geometrical effects are included in the dynamic material stability criterion.

References

1. Abeyaratne R., Knowles J.K., On the stability of thermoelastic materials, Journal of Elasticity, Vol. 53, pp. 199-213, 1999.
2. Bigoni D., Zaccaria D., On strain localization analysis of elastoplastic materials at finite strains, Int. J. of Plasticity, vol. 9, pp. 21-33, 1993.
3. Biot M. A., Mechanics of incremental deformation; Wiley, New York, 1965.
4. Coleman B.D., Noll W., On the thermostatics of continuous media . Arch. Rational Mech. Anal., 4, pp. 97-128, 1959.
5. Hayes M., Rivlin R.S., Propagation of a plane wave in an isotropic elastic material subjected to pure homogeneous deformation, Arch. Rat. Mech. Anal., Vol. 16, pp. 238-242, 1961.
6. Hill R., Aspects of invariance in solid mechanics; in *Advances in applied mechanics*, Vol. 18, Academic Press, pp.1-72, 1978.
7. Hill R., Hutchinson J. W., Bifurcation phenomena in the plane tension test, J. Mech. Phys. Solids, Vol. 23, pp. 239-264, 1975.
8. Hill R., Acceleration waves in solids, J. Mech. Phys. Solids, Vol. 10, pp. 1-16, 1962.
9. Hill R., Constitutive inequalities for isotropic elastic solids under finite strain, Proc. R. Soc. Lond. A., 314, pp. 457-472, 1970.
10. Hill R., A general theory of uniqueness and stability in elastic/plastic solids, J. Mech. Phys. Solids, Vol. 6, pp. 236-249, 1958.
11. Li G.-C., Zhu C., Formation of shear bands in plane sheet, Int. J. of Plasticity, Vol. 11, pp. 605-622, 1995.
12. Miehe C., Schröder J., Becker M., Computational homogenization analysis in finite elasticity: material and structual instabilities on the micro- and macro-scales of periodic composites and their interaction, Comput. Methods Appl. Mech. Engrg., Vol. 191, pp. 4971-5005, 2002.
13. Needleman A., Non-normality and bifurcation in plane strain tension and compression, J. Mech. Phys. Solids, 1979, Vol. 27, pp.231-254.
14. Ogden R. W., Non-linear elastic deformations; Ellis Horwood L.T.D., Chichester, U.K., and John Wiley and Sons, Chichester, U.K., 1984.
15. Pan F., Beatty M.F., Instability of an internally constrained hyperelastic material, Int. J. Non-Linear Mech., 1999, vol. 34, pp. 168-177.
16. Petryk H., Material instability and strain-rate discontinuities in incrementally nonlinear continua, J. Mech. Phys. Solids, 1992, Vol. 40 No. 6, pp. 1227-1250.
17. Raniecki B., Bruhns O. T., Bounds to bifurcation stresses in solids with non-associated plastic flow law at finite strain, J. Mech. Phys. Solids, Vol. 29 no. 2, pp. 153-172, 1981.
18. Rice J. R., Rudnicki J. W., A note on some features of the theory of localization of deformation, Int. J. Solids Structures, Vol. 16, pp. 597-605, 1980.
19. Reese S., Wriggers P., Material instabilities of an incompressible elastic cube under triaxial tension, Int. J. Solids Structures, Vol. 34, pp. 3433-3454, 1997.
20. Sawyers K. N., Rivlin R.S.: Instability of an elastic material, Int. J. Solids Structures, pp. 607-613, 1973.
21. Sawyers K.N.: Material Stability and bifurcation in finite elasticity. In: Finite Elasticity Applied Mechanics Symposia Series Vol. 27 (Ed. R.S. Rivlin) American Society of Mechanical Engineers. Pp. 103-123, 1977.

22. Truesdell C., Toupin R.A., Static grounds for inequalities in finite elastic strain. Arch. Rational Mech. Anal, 12, pp. 1-33, 1963.
23. Truesdell C., Noll W., The non-linear field theories of mechanics, Encyclopedia of Physics, Vol. III/3, Springer Verlag, Berlin, 1965.
24. Wang C.-C., Truesdell C., Introduction to rational elasticity; Noordhoff International Publishing, Leyden, The Netherlands, 1973.
25. Young N. J. B., Bifurcation phenomena in the plane compression test, J. Mech. Phys. Solids, Vol. 2, pp. 77-91, 1975.

Material Damage Description via Structured Deformations

Marc François[1], Gianni Royer-Carfagni[2]

[1] Laboratoire de Mécanique et Technologie
61, Avenue du President Wilson, 94235, Cachan CEDEX

[2] Dipartimento di Ingegneria Civile, dell'Ambiente, del Territorio e Architettura
Università degli Studi di Parma,
Parco Area delle Scienze 181/A, I-43100 Parma, Italy
and
Istituto per le Applicazioni del Calcolo "M. Picone"
Consiglio Nazionale delle Ricerche
Viale del Policlinico 137, I-00161 Rome, Italy

Abstract. The proposed micro-macro model describes the behavior of damaged materials with rough fractures, whose cohesive-frictional sliding is considered at the continuum level by using the recent "Structured deformation theory", now set in the classical thermodynamic formulation for generalized standard materials. The shear-induced dilatation due to crack surmounting is here considered by assuming an equivalent shape function for crack-lip roughness, and the material internal variables are reduced to one scalar parameter, associated with the smeared-crack slip. Remarkably, the structured-deformation approach renders the model fully-associated in type, even in the presence of friction. Even if no evolution of crack density and orientation is supposed, a good description of the response of cracked masonry or concrete walls under seismic-like shear loading are provided by the model. Interesting instabilities in the shearing path are exhibited, suggesting a possible generalization of the Mohr-Coulomb criterion.

1 Introduction

The mechanical response of a wide class of materials (concrete, geomaterials, ceramics, masonry) presents distinguishing peculiarities, that may be summarized by the name "quasi-brittle" with which they are usually referred to [1]. After a first pseudo-elastic stage under moderate loads, an increase in the stress level causes the progressive appearance of micro-fractures, almost uniformly distributed throughout the body. The reason why material does not break at once into neat pieces (hence the name quasi-brittle) is due to an underlying microstructural effect. Various mechanisms (grain bridging, fiber bridging, aggregate frictional interlock, crack overlap) cause that the lips of developing fractures continue to transmit cohesive forces, even after that matrix separation occurs. If the micro-cracked (but not yet broken) body is first released and then moderately reloaded, its response is quite different from that of a virgin body since sliding of the oriented microcrack lips gives an important contribution to the deformation.

Crack sliding reaction forces, in general of cohesive and frictional nature, induce a stress field which, for assigned boundary data, depends upon the number, the length, the orientation and, in particular, the roughness of the fractures. If the magnitude of the incremental stress is moderate, crack length and orientation remain practically unaltered and the damaged materials appears macroscopically homogeneous but strongly anisotropic, due to the presence of the fractures. Eventually, failure occurs when the material is significantly further stressed. The purpose of this paper is not to model the complete non-linear behavior up to failure. Indeed, the first phase (undamaged material) and the last phase (prior to failure) are not considered at this point, but rather the attention is focused upon the response of materials uniformly damaged by diffuse cracking and moderately loaded.

The serviceability state of structural elements made of quasi brittle materials, e.g.concrete and masonry, occurs in this condition, in which opening and sliding of micro-fractures affects the structural performance [2]. According to a well established approach [3], it may be supposed that the crack pattern is fixed and coincides with that pattern originally generated once the tensile-strength has been overcome for the first time. For the evaluation of the orientation of the initial cracks a few criteria may be suggested. For example, according to some authors [2], crack orientation is orthogonal to the lines of principal tensile stress in the undamaged material element subjected to the design loads (in this analysis the material is usually assumed to be linear elastic, homogeneous and isotropic). Another possibility is offered by no-tension elasticity theory [4], according to which external actions are equilibrated by compressed material struts delimited by cracks. However, this aspect will not be examined at this stage, but rather the crack pattern will be supposed to be assigned *a priori*.

In general, the disarrangements due to crack gliding render the deformation field discontinuous, but in recent years there have been several attempts to describe the fracture process with a continuum approach, in view of numerical implementation. In general, most of the models in the literature are "smeared", that is, crack openings are distributed (smeared) over the representative volume element to generate an equivalent continuum, whose constitutive relationship should take into account crack sliding. However, this procedure leads to some theoretical difficulties which, yet, appear to be still partially unappreciated. First of all, the continuous displacement fields describing averaged responses cannot sort out, in general, the physical sources of the deformation, e.g. crack sliding or elastic strain. This is instead very importance to keep track of the individual mechanical effects. If, on the one hand, consideration of a full micromechanical model is complicated, on the other hand, it might be sufficient and effective to have a "mesoscopic" view of the phenomena, where the influence of cracks is interpreted through the definition of proper damage parameters.

But the smearing process should also be questioned on theoretical grounds. Because of crack sliding the equivalent continua often exhibit a strain-softening response, but implementation of the resulting continuum model in FEM codes regularly produces strain localization in regions of zero volume, disappearance of fracture energy, lack of mesh objectivity [1]. As discussed at length in [5], the assumption that the "average-stress vs. average-strain" response of a characteristic RVE of finite size is a good approximation of the "local-stress vs. local-strain" constitutive relationship for a continuous body, as usually given for granted, may lead to the aforementioned inconsistencies. Moreover, what is still uncertain is a systematic settling of the aforementioned theories in a consistent continuum-damage-mechanics framework, accounting for the thermodynamical characterization of irreversible processes.

The theory of *structured deformations*, recently proposed by Del Piero and Owen [6], represents a rigorous mathematical construct that allows to treat infinitesimal discontinuous displacements as continuous fields, though preserving a neat distinction with other "regular" contribution, i.e. the elastic strain of integer material lamellae comprised between consecutive cracks. Roughly speaking, a structured deformation is the combination of a classical "regular" deformation with a "singular" deformation, produced by micro-disarrangements (such as in a sliding deck of cards). This distinction is important because crack sliding, and the consequent induced crack-opening, lead to a plastic-like deformation splitting, while preserving their precise significance. Here, the structured deformation approach is set in a broader continuum damage mechanics theory, founded upon the thermodynamics of irreversible processes [7]. The thermodynamical forces associated with each part of the deformation are consistent with the physics of the problem, characterized by the dichotomy between elastic strain and crack sliding. In particular, the dissipation and yield criteria are conveniently stated as functions of the thermodynamical forces associated with the disarrangements caused by crack sliding.

A mesoscopic view of the complex micromechanical phenomena is obtained through the introduction of an equivalent crack roughness. The underlying idea is that the leading parameter governing the gross response of micro-cracked bodies is the crack-shape profile and, now, we conceive a mesoscopic model where all non-linear effects, traditionally settled in a plasticity or damage framework, are caught by the sliding of rough cracks. Material properties required for this theory, apart from elasticity, are crack shape and Coulomb-like friction. The contribution of other microstructural effects (e.g. grain or fiber bridging) can also be taken into account from this physical point of view through the introduction of equivalent cohesive forces. In general, the crack profile is described by a periodic "wave function", so that lip sliding induces a dilatation due to the geometrical surmounting of displaced rough surfaces. Sliding-induced dilatation will be proved to be a key-effect to reproduce experimental evidences under shear and confinement, such as the non-linear and hysteretic response of masonry in-filled r.c. or steel frames.

Despite a generalization is still possible, some convenient hypotheses have been introduced in the model. One of them is that crack surfaces remain in contact during the deformation, so that the model is particularly suitable for confined brittle-solids or quasi-brittle materials. In the former case it is the confining pressure, in the latter the cohesive stresses, that maintain structural integrity. Besides, this study is limited so far to the analysis of elements in generalized plane stress or plain strain, for which it is assumed that crack spacing and orientation are uniform. This particular case is the first step for the implementation a FEM code for general purposes, since it indicates the response of a representative volume element under various boundary conditions. Nevertheless, despite the model in its actual form is minimal and can only represent a first attempt to catch the essential aspects of the phenomena, it is to reproduce, with good agreement, the experimental observations of relevant tests.

2   Micromechanically-motivated kinematical description via structured deformation theory

Consider a panel completely damaged by a diffuse distribution of micro-cracks. We suppose that in any region $\mathcal{R}$, whose diameter is much larger than the panel thickness and crack spacing, but much smaller than the panel size, the orientation, spacing and characteristic length of the micro-cracks is uniform. We surmise that loading does not affect this crack pattern, i.e. neither new cracks are nucleated nor crack length varies, although gliding is allowed. Now, it is clear that, due to the rough geometry (i.e. micro-crack overlapping and bridging effects), the punctual stress and strain consequent to crack-lip gliding will be very complicated. Nevertheless, an average view of such complex phenomena may be taken by considering an equivalent crack layout for the panel, where cracks, of regular shape, parallel and equidistant, go through the element from one side to the other.

The geometric characterization of the translation-induced shear response of fractured panels has been recently considered by Del Piero & Owen as an example of structured deformation. Following their idea ([8], part. II Section 5.1), such a deformation may be described by assigning the shape of the equivalent through-crack. For this, it is sufficient to assign a shape function $\omega : \mathbb{R} \to \mathbb{R}$ for the crack lips, supposed to be a continuously differentiable, symmetric, even, periodic function, with period p. With respect to a (ξ, η) reference system such that ξ coincides with the crack average plane, the fracture shape is identified by the graph of $\eta = a\omega(\xi/a)$, represented with a bold face line in Figure 1a. Here a is a reference length, so that crack wave-length results equal to pa.

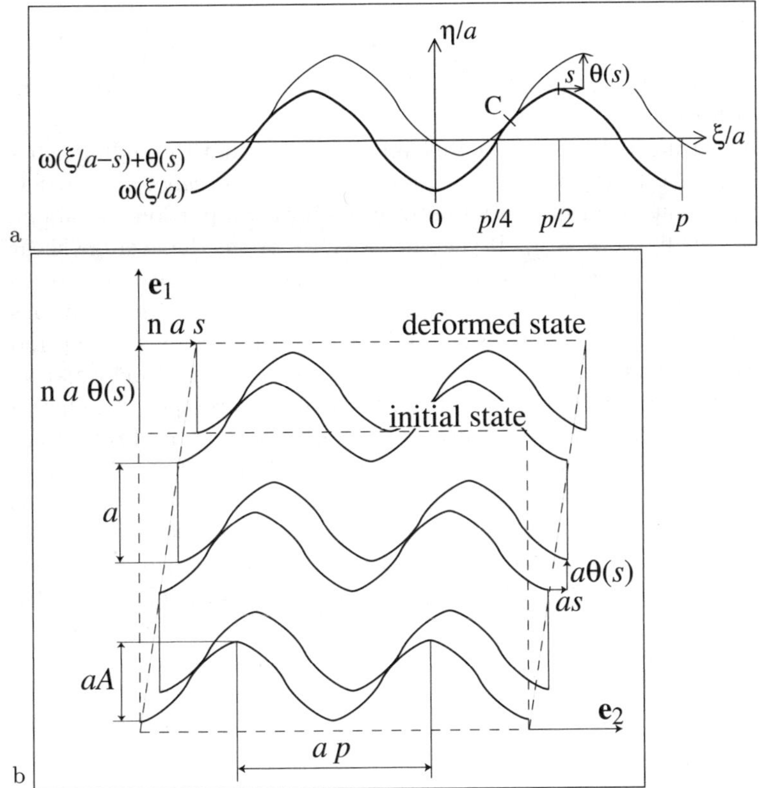

Fig. 1. Structured deformation of the damaged material.
a) The crack wave function and the translation-induced shear dilatation. b) The deformation of the representative volume element.

In the case at hand, the crack wave can be made to coincide with itself after the combination of a reflection in its average line and a translation of half the wave-length. This means that an observer cannot distinguish, at the scale of the RVE, the two crack surfaces after material separation. This symmetry implies that $\omega(p/4 + \xi/a) = -\omega(p/4 - \xi/a)$. Let us suppose that the upper lip of the crack is sliding with respect to the lower (Figure 1a), while remaining in contact. If sa is the relative horizontal translation, with s a non-dimensional parameter, the relative vertical translation will be denoted by $a\theta(s)$, where $\theta(\cdot)$ will be referred to as the separation function. Using the symmetry of $\omega(\cdot)$, one finds

$$\theta(s) = \quad 2\,\omega(p/4 + s/2), \quad \text{if } s \in\,]\,0,\ p/2] + kp,$$
$$\theta(s) = -2\,\omega(p/4 + s/2), \quad \text{if } s \in\,]-p/2, 0] + kp. \tag{1}$$

The function θ is periodic of period p and its derivative is, in general, discontinuous at $s = kp$. For example, when $\omega(\xi/a) = -A/2\cos(2\pi\xi/ap)$, one obtains $\theta(s) = A|\sin(\pi s/p)|$. For a triangle function $\omega(\xi/a) = -A/2 + A\mathrm{Tr}(2\xi/ap)$ one finds $\theta(s) = A\mathrm{Tr}(2s/p)$. Recall that $\mathrm{Tr}(\cdot)$ is the periodic extension of $\mathrm{T}(\cdot) : [0,2] \to [0,1]$ defined as $\mathrm{T}(t) = t$ for $t \in [0,1)$ and $\mathrm{T}(t) = 2 - t$ for $t[1,2]$.

Assume that in the RVE cracks are equidistant and that the sliding is homogeneous, i.e., s is the same for each crack. Figure 1b represents a microview of the discontinuous deformation induced by sliding for a portion containing n cracks (in this figure $n = 3$). If each crack slides of sa, the average shear in the RVE is $(nsa)/(na) = s$ and the average shear-induced dilatation is $na\theta(s)/(na) = \theta(s)$. Observe that whereas the discontinuous deformations depend upon the length scale a (the crack spacing), the averaged shear and shear-induced dilatation do not depended upon a. Structured deformation theory furnishes the mathematical setting to treat the average deformation as the limit of discontinuous deformations as $a \to 0$. To be noticed is that the shear s of the elementary cell can be theoretically infinite (we shall however assume to remain in infinitesimal deformation), but the translation $\theta(s)$ has a maximum value A, i.e., the amplitude of the separation function θ.

Referring the whole panel to an orthogonal reference system (x, y), in a neighborhood of the point (x, y) fracture orientation is indicated by the vector $\mathbf{e}_1(x, y)$, orthogonal to the crack plane as in Figure 1b. Let $\mathbf{e}_2(x, y)$ the unit vector orthogonal to $\mathbf{e}_1(x, y)$. Following [8], in a neighborhood of the point (x, y) the average deformation can be considered as the deformation of an equivalent continuum with infinite fractures for which the disarrangements, due to fracture sliding, induce the deformation gradient

$$\mathbf{M}(x, y) = \theta\left(s(x, y)\right)\mathbf{e}_1(x, y) \otimes \mathbf{e}_1(x, y) + s(x, y)\mathbf{e}_2(x, y) \otimes \mathbf{e}_1(x, y) \ . \quad (2)$$

This represent not only a simplified notation, but could have been as well the starting point of the study since, at least in principle, it is possible to experimentally identify the average separation function $\theta(s)$ with respect to the average shear s (this will be mentioned in the concluding Section). This "macro" point of view allows to forget about the length-scale a, despite here the equivalent crack has been introduced to furnish a micro-mechanical motivation to the model. An experimental study of the fracture profile by *post mortem* point-by-point measurement of the fracture profile has been considered by [9] in order to determine experimentally $\theta(s)$.

Let $u(x, y)$ and $v(x, y)$ represent the components of the macroscopic displacement $\mathbf{u}(x, y)$ in the x and y directions respectively. The pairing $(\mathbf{g}, \mathbf{G})$ with

$$\mathbf{g}(x, y) = (x + u(x, y), y + v(x, y)) \ , \quad \mathbf{G}(x, y) = \nabla\mathbf{g}(x, y) - \mathbf{M}(x, y) \ , \quad (3)$$

completely describes the deformation of the micro-cracked body. In particular, $\mathbf{g}$ represents the macroscopic deformation, $\mathbf{M}$ the deformation gradient due to the disarrangements, and $\mathbf{G}$ that part of deformation gradient due to material elasticity. Del Piero and Owen [6] call the pairing $(\mathbf{g}, \mathbf{G})$ a structured deformation and constructed the mathematical framework for the relevant theory.

Formula (3) implies that the displacement field varies smoothly throughout the panel, i.e., in any region of dimensions comparable with those of the RVE the macroscopic strain is homogeneous and the rotation uniform. Assuming that the gradients of $u(x,y)$ and $v(x,y)$, as well as $s(x,y)$ and $\theta(s(x,y))$ are infinitesimal quantities of the first order, the Green's strain $(\mathbf{G}^T\mathbf{G} - \mathbf{I})/2$, due to the elasticity of the material, may be approximated by the elastic infinitesimal strain

$$\boldsymbol{\epsilon}^e(x,y) = \frac{1}{2}(\nabla\mathbf{g} + \nabla\mathbf{g}^T) - \mathbf{I} - \frac{1}{2}(\mathbf{M} + \mathbf{M}^T)$$
$$= \boldsymbol{\epsilon}(x,y) - \theta\left(s(x,y)\right)\mathbf{E}_{11}(x,y) - s(x,y)\mathbf{E}_{12}(x,y), \tag{4}$$

where $\boldsymbol{\epsilon}(x,y) = (\nabla\mathbf{u}(x,y) + \nabla^T\mathbf{u}(x,y))/2$ represents the *macroscopic infinitesimal strain*, while $\mathbf{E}_{11}(x,y) = \mathbf{e}_1(x,y)\otimes\mathbf{e}_1(x,y)$ and $\mathbf{E}_{12}(x,y) = (\mathbf{e}_1(x,y)\otimes\mathbf{e}_2(x,y) + \mathbf{e}_2(x,y)\otimes\mathbf{e}_1(x,y))/2$ represent the crack local orientation. The quantity

$$\boldsymbol{\epsilon}^s(x,y) = \theta\left(s(x,y)\right)\mathbf{E}_{11}(x,y) + s(x,y)\mathbf{E}_{12}(x,y), \tag{5}$$

represents the (infinitesimal) *singular* part of the strain, i.e. that part due to the microscopic disarrangements. The macroscopic strain $\boldsymbol{\epsilon}$ is thus obtained with respect to a classical hypothesis of its partition in the elastic part $\boldsymbol{\epsilon}^e$ (due to the elasticity of the undamaged *lamellae* separated by cracks) and the structured deformation part $\boldsymbol{\epsilon}^s$ due to the disarrangements, leading to the plastic-like deformation splitting

$$\boldsymbol{\epsilon} = \boldsymbol{\epsilon}^e + \boldsymbol{\epsilon}^s. \tag{6}$$

Of course, this theory can account neither for the local stress fields consequent to Hertzian contact, nor for crack profile modifications due to surface wear. Consequently, the present model can only apply to moderately-loaded damaged materials.

Compatibility conditions must be added in order that the complete strain field $\boldsymbol{\epsilon}$ derives from a displacement field (u,v) and from a field of disarrangements which do not interfere with one another. These give rise to mathematical conditions for the fields $\theta(x,y)$, $s(x,y)$, $\mathbf{E}_{11}(x,y)$ and $\mathbf{E}_{12}(x,y)$. However, here we shall remain at the RVE scale, supposing that fractures are straight and parallel and that boundary conditions are homogeneous. Thus, there will be no need to deal with the problem of the compatibility of deformation, which will become indeed of crucial importance in the general case.

3 Structured deformations in a thermodynamical framework

The deformation splitting of (6) allows to use a classical formulation [10] to set the damage mechanics approach into a thermodynamic framework. In this approach, the state variables are the total strain ϵ and the slip s, which defines the structured part of the deformation as *per* (5) and is the *internal* state variable. At constant temperature, the model is defined by assigning the Helmholtz's free energy per unit mass Ψ and the yield function f, which determines the thermodynamic force S associated with s. The tensors $\mathbf{E}_{11}(x,y)$ and $\mathbf{E}_{12}(x,y)$ of (5), as well as the function $\theta(s)$ of (1), are determined by the initial microcrack configuration for the damaged material and, since the crack pattern is supposed to be invariable during the loading process, such quantities are retained as problem data.

Considering isothermal evolution, the Helmholtz free energy is defined as the reversible elastic energy. Let $\mathbb{C}$ represent the fourth order elasticity tensor for the undamaged (uncracked) material, supposed linear elastic. Since elastic energy is stored in the integer material *lamellae* comprised between adjacent microcracks, using the expression (4) for the elastic component of the strain ϵ^e the free energy per unit mass reads

$$\Psi[\epsilon,s] = \frac{1}{2\rho}(\epsilon - \theta(s)\mathbf{E}_{11} - s\mathbf{E}_{12}) \cdot \mathbb{C}\,[\epsilon - \theta(s)\mathbf{E}_{11} - s\mathbf{E}_{12}]\ , \tag{7}$$

where ρ is the material density. When the material matrix is isotropic, it is convenient to split each tensor into hydrostatic and deviatoric (traceless) parts, denoted with the apexes "H" and "D", respectively. Observing that $\mathbf{E}_{11} \cdot \mathbf{E}_{12} = 0$, and that $\mathbf{E}_{12}$ is deviatoric, i.e., $\mathbf{E}_{12}^D \equiv \mathbf{E}_{12}$ and $\mathbf{E}_{12}^H \equiv 0$, the previous equation can be re-written in a simpler form with respect to the shear modulus μ and the compression modulus κ as

$$2\rho\Psi[\varepsilon,s] = 2\mu[\varepsilon^D - \theta(s)\mathbf{E}_{11}^D - s\mathbf{E}_{12}]^2 + 3\kappa[\varepsilon^H - \theta(s)\mathbf{E}_{11}^H]^2\ , \tag{8}$$

where $\mathbf{E}_{11}^D$ and $\mathbf{E}_{11}^H$ are the deviatoric and hydrostatic part of $\mathbf{E}_{11}$, respectively.

Similarly, the stress $\boldsymbol{\sigma}$ is split into the deviatoric and hydrostatic parts $\boldsymbol{\sigma}^D$ and $\boldsymbol{\sigma}^H$. For an isotropic material, these correspond to the thermodynamic forces associated with the deviatoric ϵ^D and hydrostatic ϵ^H parts of the *macroscopic* strain, obtained by differentiation of the free energy with respect to ϵ^D and ϵ^H, i.e.,

$$\boldsymbol{\sigma}^D = 2\mu[\epsilon^D - \theta(s)\mathbf{E}_{11}^D - s\mathbf{E}_{12}]\ , \quad \boldsymbol{\sigma}^H = 3\kappa[\epsilon^H - \theta(s)\mathbf{E}_{11}^H]\ . \tag{9}$$

When $s = 0$, from (1) also $\theta(s) = 0$ and isotropic elasticity is recovered. Observe, in passing, that an alternative point of view may consist in regarding $\epsilon^s = \theta(s)\mathbf{E}_{11} + s\mathbf{E}_{12}$ to be equivalent to a plastic strain since, as in classical plasticity theory, from (7) the stress is given by $\mathbb{C}(\epsilon - \epsilon^s)$. However, the similarity stops here as the evolution law presented in the forthcoming part will be different.

Following the standard generalized material framework [10], the thermodynamic force S associated with s is $S = -\partial(\rho\Psi)/\partial s$, where the minus sign is due to a thermodynamical convention. The quantity S is the decrease in elastic energy per unitary variation of s and thus represents the driving force for crack slip. In case of isotropic materials, from (8) one obtains

$$S = 2\mu \left[(\theta' \mathbf{E}_{11}^D + \mathbf{E}_{12}) \cdot \boldsymbol{\epsilon}^D - \mathbf{E}_{11}^D \cdot \mathbf{E}_{11}^D \theta \, \theta' - \mathbf{E}_{12} \cdot \mathbf{E}_{12} \, s \right]$$
$$+ \, 3\kappa\theta' \left[\mathbf{E}_{11}^H \cdot \boldsymbol{\epsilon}^H - \mathbf{E}_{11}^H \cdot \mathbf{E}_{11}^H \theta \right] \, , \tag{10}$$

where θ' denotes the derivative of θ. Using (9), such an expression can be rewritten in the form

$$S = \boldsymbol{\sigma} \cdot (\theta' \mathbf{E}_{11} + \mathbf{E}_{12}) \, , \tag{11}$$

where $\boldsymbol{\sigma} \cdot \mathbf{E}_{12}$ is the shear stress at the crack level while $\boldsymbol{\sigma} \cdot \mathbf{E}_{11} \equiv \mathbf{e}_1 \cdot \boldsymbol{\sigma} \mathbf{e}_1$ represents the normal component of stress to the crack plane. Thus, we deduce from (11) that not only is the driving force S produced by the shear $\boldsymbol{\sigma} \cdot \mathbf{E}_{12}$, but a significant role is also played by the term $\theta' \boldsymbol{\sigma} \cdot \mathbf{E}_{11}$, which represents the "dilation-induced-shear".

The convex of elasticity is defined by the function $f(\boldsymbol{\sigma}, S)$, which describes the elasticity domain with respect to the thermodynamic forces $\boldsymbol{\sigma}$ and S through condition $f(\boldsymbol{\sigma}, S) < 0$. Here, only the particular function

$$f(\boldsymbol{\sigma}, S) = |S| - (B - \varphi \, \boldsymbol{\sigma} \cdot \mathbf{E}_{11}) \, , \tag{12}$$

is considered, which corresponds to a cohesive-frictional response with frictional coefficient φ and cohesive constant B. Notice that $\boldsymbol{\sigma} \cdot \mathbf{E}_{11} < 0$ implies that the crack is "in the average" compressed ($\boldsymbol{\sigma}$ is a macroscopic stress defined at a material scale much larger than the crack roughness), so that the borderline case when $B = 0$ corresponds to unilateral Coulomb's frictional law. In fact, when $B = 0$ condition $\boldsymbol{\sigma} \cdot \mathbf{E}_{11} > 0$ is not admissible: the body would split into pieces when pulled across the crack plane. Besides, the compressive normal stress $\boldsymbol{\sigma} \cdot \mathbf{E}_{11} < 0$, weighted by φ, increases the size of the elastic domain. More precisely, this formulation is not the exact Coulomb's law, which should act according the tangent plane at the contact point C (Figure 1b), but an "average" frictional law, effective at the level of average crack plane. The second borderline case, when $\varphi = 0$, corresponds to cohesive sliding: $|S|$ remains constants at yielding, independently of the compressive stress. The combinations of these two cases furnishes the admissibility condition

$$\boldsymbol{\sigma} \cdot \mathbf{E}_{11} < B/\varphi \, . \tag{13}$$

In words, the simultaneous presence of cohesion and friction allows to withstand also moderate tensile stress at right angle to the crack plane.

At yielding ($f = 0$), the evolution for the state variable s is determined from the consistency equation $df = 0$ and generalized normality rule [10]

$$\dot{s} = \frac{\partial f}{\partial S} \dot{\lambda} = \mathrm{sgn}(S) \dot{\lambda} \, , \tag{14}$$

where $\dot{\lambda}$ is analogous to the classic plastic multiplier with $\dot{\lambda} > 0$. To this respect, the model can be considered *associated*, i.e., the dissipation pseudo-potential coincides with the yield function. This is a remarkable property of the structured-deformation approach, which renders the model fully-associated in type even in the presence of friction.

In particular, since from (14) $\dot{s}$ has the same sign of S, the yield condition deriving from (12) can be re-written in the form,

$$\begin{cases} S = B - \varphi\,\boldsymbol{\sigma}\cdot\mathbf{E}_{11} & \text{if } \dot{s} > 0\,, \\ S = -(B - \varphi\,\boldsymbol{\sigma}\cdot\mathbf{E}_{11}) & \text{if } \dot{s} < 0\,, \\ |S| \leqslant B - \varphi\,\boldsymbol{\sigma}\cdot\mathbf{E}_{11} & \text{if } \dot{s} = 0\,, \end{cases} \tag{15}$$

where $B - \varphi\boldsymbol{\sigma}\cdot\mathbf{E}_{11}$ is positive. This is a further indication that S is correlated with the effective shearing force in the average crack plane.

The second principle of thermodynamics states that the intrinsic dissipation $\dot{D}$ must be positive. From the general theory [7], $\dot{D}$ equals the sum of the products $A_k V_k$, where A_k denotes the thermodynamic force associated with the internal state variable V_k. For the case at hand $A_k \equiv S$ and $\dot{V}_k \equiv \dot{s}$ and, from (14), we notice that the second principle is naturally fulfilled, i.e., $\dot{D} \equiv S\dot{s} > 0$. Moreover, from (15), the intrinsic dissipation takes the simple form

$$\dot{D} = (B - \varphi\boldsymbol{\sigma}\cdot\mathbf{E}_{11})\,|\dot{s}|\,, \tag{16}$$

from which the admissibility condition (13) is again found. In other words, condition $\boldsymbol{\sigma}\cdot\mathbf{E}_{11} < B/\varphi$ is strictly correlated to the positiveness of the dissipation.

4   Response of compressive/tensile panels under shear

The model is now used in a paradigmatic example: the cyclic response of shear-wall-type panels, contemporarily subjected to axial load. Assume that the panel is rectangular, in generalized plane stress, and uniformly damaged by uniformly distributed *horizontal* microcracks. Let $\mathbf{e}_1$ and $\mathbf{e}_2$ be unit vectors orthogonal and parallel to the crack average planes as in Figure 1b so that, in this case, $\mathbf{e}_1$ and $\mathbf{e}_2$ are parallel to the height and width of the panel, respectively. The state of stress is supposed to be homogeneous and of the type $\boldsymbol{\sigma} = \sigma_0\mathbf{e}_1 \otimes \mathbf{e}_1 \otimes +\tau(\mathbf{e}_1 \otimes \mathbf{e}_2 + \mathbf{e}_2 \otimes \mathbf{e}_1)$. Moreover, from (11) the thermodynamical variable S associated with the slip s reads

$$S = \sigma_0\theta'(s) + \tau\,. \tag{17}$$

Using (12) and (17), the elastic domain at a given state s is defined by

$$|\sigma_0\theta'(s) + \tau| \leqslant B - \varphi\,\sigma_0\,, \tag{18}$$

where $B - \varphi\sigma_0 > 0$ from the admissibility condition (13).

Assume that the strain in the RVE is homogeneous and the crack slip uniform. If γ denotes the average shear strain in the panel, from (9) we have

$$\gamma = \frac{\tau}{\mu} + s \ , \tag{19}$$

and, for fixed s and σ_0, the yield conditions read

$$\tau = (B - \sigma_0\varphi) - \sigma_0\theta'(s), \quad \text{or} \quad \tau = -(B - \sigma_0\varphi) - \sigma_0\theta'(s), \ . \tag{20}$$

Consideration of the dissipation equation and of (17) lead to retain the first (second) of (20) if $\dot{s}$ is positive (negative).

For the sake of simplicity, only the case when the crack shape function ω is a saw-tooth function will be considered, i.e., $\omega(\xi/a) = -A/2 + A\mathrm{Tr}(2\xi/ap)$. From (1), the associated separation function $\theta(s)$ is the triangle function $A\mathrm{Tr}(2s/p)$ and its derivative is piecewise constant, i.e. $\theta'(s) = \pm 2A/p =: \pm M$, with jumps at $s = kp$, $k \in \mathbb{Z}$. It is then convenient to distinguish the compression from the tensile cases.

4.1 Shear response of compressed panels ($\sigma_0 < 0$)

When the average crack plane is compressed, the point representative of shear stress and strain in the RVE will follow the path O-A-B-C-D-E-F-G-H-A... in the (γ, τ) plane of Figure 2a. The materials starts to yield at point A for $\tau_A = B - \sigma_0(\varphi + M)$ and, afterwards, it follows a pseudo-plastic *plateau*. During unloading, starting at point B, a path parallel to the initial elastic branch is followed until the materials yields again at C under the shear $\tau_C = -B + \sigma_0(\varphi - M)$. The distance from the γ-axis of AB is larger than the distance of the reverse yielding path CD because in the first case, due to the slope of the crack teeth, the compression load acts against yielding, while the contrary is true for path CD. This is clear from the schematic representation of the crack profile and opening in Figure 2a. At D, which is collinear with O and A, the crack has been closed ($s = 0$), and now the material can again gain stiffness. The material yields again at point E, whose abscissa equals a value opposite to that of point A. Discussion for points G and H is similar.

When the stress level at D is greater than zero, the hysteresis loop is as depicted in Figure 2b. This case is attained when $-B + \sigma_0(\varphi - M) > 0$, i.e., when $M > \varphi$ and $\sigma_0 < -B/(M-\varphi)$ (recall that $\sigma_0 < 0$). Recalling that $M := 2A/p$ is the slope of the teeth in the saw-tooth cracks, this situation is for high compression combined with either low friction or sharp crack teeth. In fact, in the sliding process there is competition between the driving shear stress τ and the effect of the compressive load. When φ is small when compared to the crack-tooth slope as in the case at hand ($\varphi < M$), sliding may result *reversible*, i.e. for high compression the crack closes by itself without the need of a shear force τ. This is why on the equilibrium paths CD and GH of Figure 2b the shearing stress *tau* must act in a direction opposite to that of crack sliding, to refrain the crack closure. The result is a pseudo-elastic response.

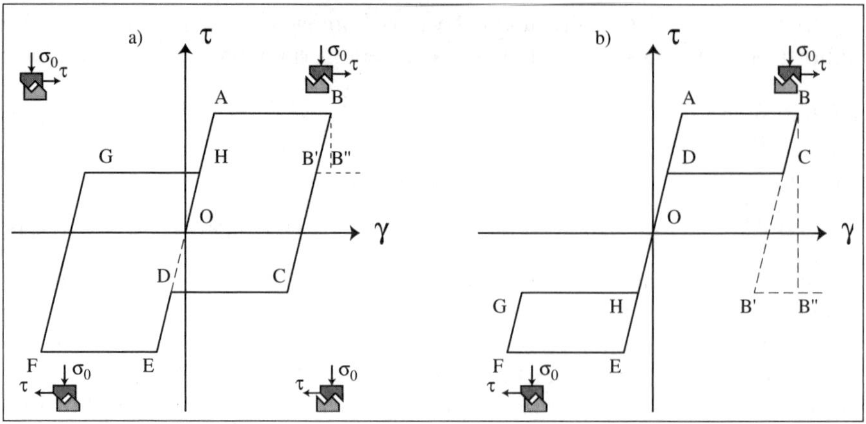

Fig. 2. Hysteresis loop in a shear test on compressed panels.
a) Case of moderate compression. b) Pseudoelastic response for reversible sliding
(high compression and low friction).

When the sliding s exceeds the value p/2, i.e. for $\gamma > \tau_A/\mu + p/2$, the teeth
of the crack lips have overcome the point when their vertexes are in contact,
so that now the compressive stress tends to move the crack profiles towards a
configuration shifted of one tooth-length with respect to the original one. For
what this response is concerned, the material acts similarly to the case $s < 0$,
but an instability is now involved. To illustrate, suppose that in Figures
2a (or 2b) point B corresponds to $s = p/2$ and $\gamma = \tau_A/\mu + p/2$. If the
test was performed with a closed loop testing machine by controlling the
crack slip s, the representative point would follow the equilibrium path B-B′,
parallel to H-A of the re-loading path. But if the test is strain-driven, point
B would directly jump to point B″, thus exhibiting a material instability.
In conclusion, increasing γ the representative points would follow the path
indicated with O-A-B-B″ in Figures 2a-b. The cases $s = p/2 + kp$, $k \in \mathbb{Z}$, is
analogous.

4.2   Shear response of tensile panels ($\sigma_0 > 0$)

Cohesive forces prevent complete material separation under moderate tensile
loads, so that crack lips are supposed to remain always in contact during
sliding. Observe from Figures 1 that if the contact points cannot separate,
a tensile load at right angle to the crack plane acts in favor of crack glide
("dilation-induced-shear"). It is convenient to treat separately the situations
$0 < \sigma_0 < B/(\varphi + M)$ and $B/(\varphi + M) < \sigma_0 < B/\varphi$. Referring to [11] for a
detailed treatment of the various possibilities, the main peculiarities of this
case are now briefly reported.

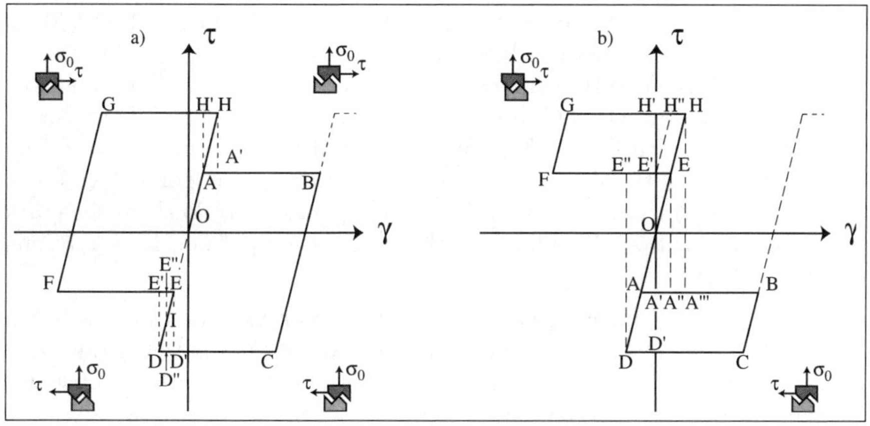

Fig. 3. Hysteresis loop in the (γ, τ) plane for tensile panels.
a) Case $0 < \sigma_0 < B/(\varphi + M)$. b) Case $B/(\varphi + M) < \sigma_0 < B/\varphi$.

When $0 < \sigma_0 < B/(\varphi + M)$, the (γ, τ) graph associated with the panel response is represented in Figure 3a. The material starts to yield at point A at $\tau_A = B - \sigma_0(\varphi + M)$, with $\tau_A > 0$, and reverse yielding occurs at point C for $\tau_C = -B + \sigma_0(\varphi - M) < 0$. This is because tensile load favors sliding on path AB but acts against sliding on path CD, so that $|\tau_A| < |\tau_C|$. This explains as well why path AB is lower than GH and CD is lower than EF. The loop O-A-B-C-D-E-F-G-H-A in Figure 3a is made of equilibrium paths, obtainable with a closed loop testing machine while controlling the crack slip in the specimen. On the base of the control signal, the machine should reverse the stroke movement at branch DE during path CDEF and the same at branch HA for path GHAB. If, on the other hand, the test is strain-driven, at D the point (γ, τ) representative of stress and strain in the specimen, would jump on branch EF at E′. In fact, on DC the tensile load acts against shearing, but at D the crack closes and immediately afterwards the tensile loads favors shearing. Consequently, the material will exhibit a snap-through-like instability. Similar considerations show that there is a jump from H to A′ during the re-loading path. It might be questioned if, for example, the jump should occur at a point different from D on the segment DD′.Indeed, it can be demonstrated [11] that the highest dissipation is obtained when jumping from D to E′, while the lowest dissipation is from D′to E . Analogously, the highest (lowest) dissipation is when the jump is from H (H′) to A′ (A). Since materials tend to dissipate as much energy as possible, the path followed in a strain driven test would be O-A-B-C-D-E′-F-G-H-A′-B.

Observe that during the first loading branch OA, there is an equilibrium bifurcation right at point A, since both paths AH and AB are attainable. No dissipation is done along AH since $\dot{s} = 0$, whereas the power expended following the branch AA′, recalling (18), equals $(\sigma_0 M + \tau_A)\dot{s} \equiv (\sigma_0 M + \tau_A)\dot{\gamma}$.

If materials tend to dissipate as much as possible, branch AH should be considered a locus of meta-stable equilibrium states. It can also be verified that a jump from A to H$'$ can never occur, because it would correspond to a negative dissipation [11]. This is clear also from a physical point of view since, in order to jump from A to H$'$, the crack would have to slip in a direction opposite to that of the shear stress (Figure 3a). In any case, point H$'$ could be attained through the action of an external "operator", picking up the crack lips against the action of τ. Of course the opposite jump, from H$'$ to A, occurs during the hysteresis loop.

The instability phenomena are even more complex in the case when $B/(\varphi + M) < \sigma_0 < B/\varphi$. The corresponding equilibrium paths in the (γ, τ) plane are represented in Figure 3b. Notice that, here, the paths AB and EF change the sign of their level with respect to Figure 3a, i.e., when $s > 0$ $(s < 0)$ equilibrium can exist only when $\tau < 0$ $(\tau > 0)$. This is because the tensile load is so high that friction alone cannot arrest the sliding of the crack, so that the action of a shearing force acting in the opposite direction is necessary in order to re-establish equilibrium. This may be thought of as a confinement effect, since it is necessary to apply a load τ opposite to γ in order to keep the material integrity.

In particular, the origin $(s = \gamma = \tau = 0)$ does not correspond to a point of stable equilibrium. In fact, there exist other four points, i.e., H$'$, E$'$, A$'$, D$'$ which are in equilibrium at $\gamma = 0$. It can be demonstrated [11] that the same energy is dissipated when jumping to either H$'$, or E$'$, or A$'$, or D$'$. Due to symmetry, the equivalence of points E$'$ and A$'$, as well as H$'$ and D$'$, is not surprising. The equivalence between E$'$ and H$'$ is due to the antagonistic effect of σ_0 and τ: the slip s at H$'$ is greater than at E$'$, but while σ_0 produces a positive dissipation, τ is responsible of a negative dissipation and the two contributions compensate one another. Indeed, the correlation between points E$'$ and H$'$ (or between A$'$ and D$'$) may be explained paraphrasing the behavior of soil, i.e., the difference between the confining effect of τ at points, say, E$'$ and H$'$ recalls the same difference between *active* and *passive* confining pressure for a soil. In fact, at E$'$ the shear restrains the crack from sliding towards the configuration of point F, whereas at H$'$ the shear τ produces sliding towards the configuration of point H. Clearly $\tau_H > \tau_E$ because at H$'$ the frictional forces have to be overcome.

Point O is not stable because the effect of the tensile stress σ_0 is sufficient to provoke the sliding of the inclined crack teeth and, consequently, if no shearing stress τ was applied, the material would irremediably break (branch DAOEH is unstable for a perturbation of the internal variable s). A suitable confining constraint should be applied such to maintain $\gamma = 0$ and the question arises whether, in this conditions, the material would jump towards E$'$ or H$'$ (the discussion for points A$'$ or D$'$ is symmetrical). Paraphrasing again the difference between active and passive soil pressure, observe that at E$'$ it is the material that pushes the constraint ("active pressure"), whereas at H$'$

it is the constraint that pushes the material ("passive" pressure). If the constraint is not an actuator, its reaction would be the lowest to withstand the material pressure, similarly to the way in which a retaining wall withstands the action of a soil. Thus, the system would naturally jump towards E$'$ (or its symmetric A$'$). This behavior is also confirmed by observing that elastic energy at H$'$ (or D$'$) is greater than at E$'$ (or A$'$): the system should naturally tend towards the configuration which, for the same dissipation, corresponds to the lowest energy level.

Suppose now that, after applying σ_0, the material jumps to the configuration A'. Increasing γ, the path A$'$-B is followed. Reversing γ, branches BC and CD are attained, but at point D a snap-through instability phenomenon analogous to that discussed previously occurs, so that the point representative of the material state directly jumps from D to branch EF at E$''$ (a similar-in-type instability occurs from H to A$'''$). Therefore, in a strain-driven cycle the material would follow the path O-A$'$-B-C-D-E$''$-F-G-H-A$'''$-B. On the other hand, suppose that at the beginning the material had jumped to the configuration E$'$. It can be demonstrated [11] that the path followed in Figure 3b would be O E$'$-H$''$-H-A$'''$-B-C-D-E$''$-F-G-H-A$'''$-B. Again, this behavior is pseudo-elastic in kind, but characterized by complex instability phenomena.

The case where s exceeds $p/2$, is represented in dashed lines on Figure 3. In general, it leads to pure elastic paths at $s = p/2 + kp$, $k \in \mathbb{Z}$.

5   Proposal for an experimental calibration of the model

The cases just discussed can only illustrate some basic features of the model, whose potentiality is yet to be fully exploited. In our opinion, the strength of the present approach consists in the description of the *shear-induced dilatation*, a crucial property to interpret the shear response of quasi-brittle materials. Such a phenomenon is essentially due to the congenital *roughness* of developing cracks. In concrete, for example, cracking occurs through the cement matrix along the circumference of the aggregates since the contact area between the two materials,i.e., the "bond zone", is the weakest link of the system. The present model proposes a mesoscopic interpretation of the complex interaction between fracture surfaces through the definition of an *"equivalent" crack roughness*, whose shape interprets and defines the shear-induced dilatation.

There are a few ways to *indirectly* assess such an "equivalent" roughness by means of experimental measurements. Figures 4 represent, for example, the device conceived of by Walraven and Reinhardt [12] for their experiments on concrete under shear. When the specimens are loaded as represented by the arrow in Figure 4a, shear without moment is produced in the crack plane. Crack sliding induces the specimen dilatation that is impeded by the restraint bars (Figure 4b), so that crack plane is compressed.

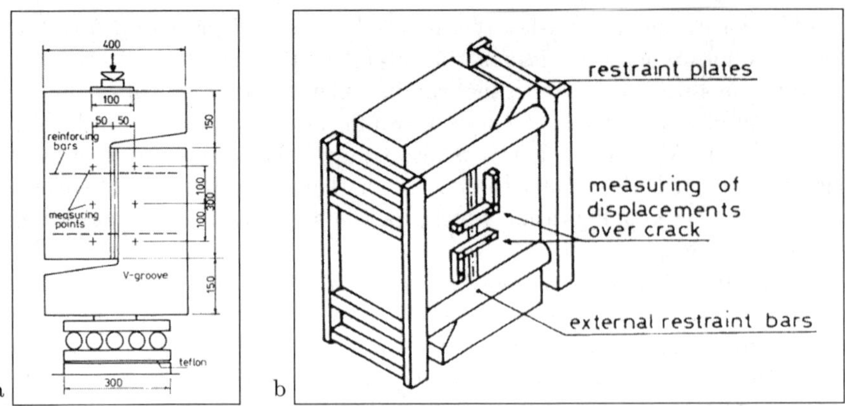

Fig. 4. Experimental apparatus proposed by Walraven and Reinhardt [12].
a) Geometry of the test specimen. b) Measuring gages and arrangement of restraint bars and plates.

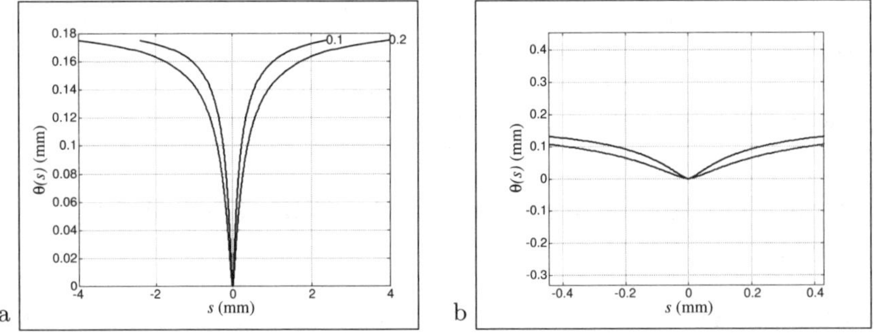

Fig. 5. Separation function $\theta(s)$ obtained from the experimental data by Walraven and Reinhardt.
a) Compressed panels by either $\sigma_0 = 0.1 f_{cc}$ or $\sigma_0 = 0.2 f_{cc}$. b) Magnification in a neighborhood of the origin.

The normal stress acting on this plane can be estimated by measuring the elastic elongation of the restraint bars using strain gauges. The relative displacement over the crack plane, and in particular its tangential and normal components Δ and w, can be directly measured by a set of gages placed as in Figure 4b. If δ denotes the thickness of the material layer amenable to cracking (comprised between the two massive L-shaped blocks of Figures 4), the sliding s and the separation function θ, defined as per Section 2 (Figure 1b), may be estimated by $s = \Delta/\delta$ and $\theta = w/\delta$. Experimentally, one finds a definite, albeit complicated, relationship among shear stress τ, normal compression stress σ_0, slip s and separation θ, from which the equivalent crack-roughness can be consequently calculated. Figure 5a represents the separation function $\theta(s)$ as derived from measured data reported in [12]. The two

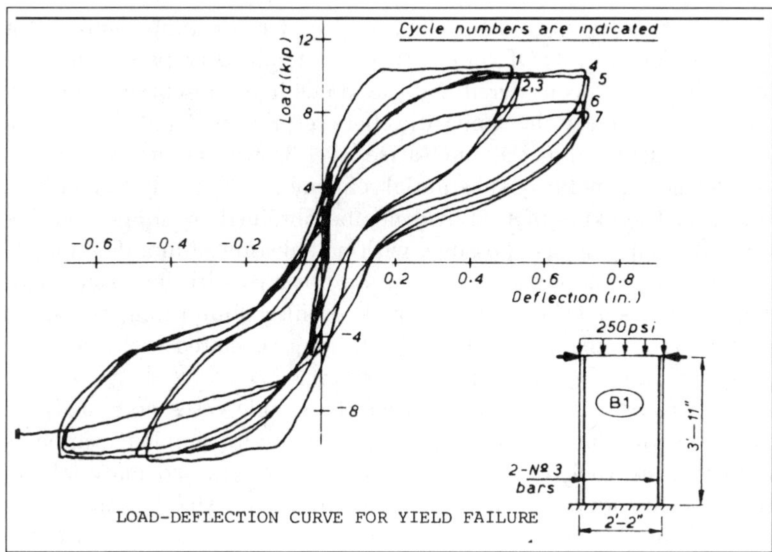

Fig. 6. Typical cyclic response of masonry walls under compression and shear (Mayes & Clough, 1975).

graphs here represented, labelled as "0.1" and "0.2", correspond to compressive stress equal to $0.1f_{cc}$ and $0.2f_{cc}$, where f_{cc} denotes the concrete cubic strength. The experimental evidence indicates that the separation function depends upon the normal stress level. By insight this is not surprising, since the crack profile is surely modified by the presence the normal stress because of Hertzian contact and/or material wearing off. The similarity between the two curves in Figure 5a is quite evident and, since Hertzian contact blunts the crack asperities, it is logical that the lower curve corresponds to the higher compression. Notice as well that, although the different scale used for the axes in Figure 5a might suggest the contrary, in a neighborhood of the origin both profiles are smooth as illustrated in Figure 5b, which represents a magnification in a neighborhood of the origin.

The present model could be complicated to account for the dependence of θ upon σ_0, but when the compressive stress is far lower than f_{cc}, such dependence is not decisive. In any case, if σ_0 is kept constant during the load-history, one may select the separation function associated with such a stress level and the analysis is identical to that of Section 4.

A micromechanically-motivated analysis of the possible dependence of the separation function upon the normal stress will be considered in further work, but even without such a refinement the present model can interpret a wide range of structural responses. In particular, the response of masonry or concrete panels under shear and confinement shows significant similarities with the cases considered in Section 4. Figure 6 shows, for example, the load-deflection curve for a reinforced masonry wall under constant compression and cyclic shear forces [13]. Of course, in this introductory paper only the simplest cases have been considered and the model here presented is nothing but minimal. Consequently the similarities between Figure 6 and, say, Figure 2a, can only be considered at the qualitative level. However, the present theory is versatile and amenable of substantial refinements. Not only can a closer representation of the experimental evidence be obtained by simply varying the crack profile, but it is also possible, with no substantial modification, to allow that the convex of elasticity varies its properties with increasing sliding. In fact, sliding smoothes out the crack profiles diminishing the shear resistance, thus producing a material degradation, correlated with the crack sliding s, that cumulates during the load history.

Albeit tentatively, it is our opinion that the simple cases here analyzed are sufficient to show that the present model, despite its simplicity, appears to be an appropriate tool to describe the response of cracked materials, in particular the shear-induced dilatation. Structured crack sliding s appears a natural internal variable to describe the material state, to which a consistent thermodynamic force is associated. Such force allows to describe, in a consistent way, the evolution laws (flow rules) to cover a wide spectrum of material responses.

References

1. Bažant, Z.P., Planas, J. (1998) Fracture and Size Effect in Concrete and Other Quasi-Brittle Materials. CRC press, New York.
2. Belletti, B., Cerioni, R., Iori I. (2001) Physical approach for reinforced-concrete (PARC) membrane elements. J. Struct. Engrg. ASCE **127**(12), 1412-1426.
3. Pang, X. B., Hsu T. T. (1996) Fixed angle softened truss model for reinforced concrete. ACI struct. J. **93**(2), 197-207.
4. Steigman D. (1991) Analysis of a theory of elasticity for masonry solids. Journ. Mech. Phys. Solids **39**, 534-553.
5. Gelli M.S., Royer-Carfagni, G. (2004) Separation of scales in fracture mechanics. From molecular to Continuum theory via Γ-convergence. ASCE J. Eng. Mech. **130**, 204-215.
6. Del Piero, G., Owen D. (1993) Structured deformations of continua. Archive for Rat. Mech. & Anal. **124**, 99-155.
7. Lemaitre, J. (1992) A Course on Damage Mechanics, Springer-Verlag, Berlin.
8. Del Piero, G., Owen, D. (2000) Structured Deformations. Quaderni dell'Instituto Nazionale di Alta matematica, **58**, Firenze.
9. Schmittbuhl, J., Roux, S., Berthaud, Y. (1994) Development of roughness in crack propagation. Europhysic Letters **28**(8), 585-590.
10. Halphen, B., Nguyen, Q. S. (1975) Sur les matriaux standards gnraliss. J. de Mcanique **14**(1), 38-63.
11. François, M., Royer Carfagni G. (2005) Structured deformation of damaged continua with cohesive-frictional slidiig rough fractures. Eur. J. Mech. A-Solids, in press.

12. Walraven, J. C., Reinhardt, H. W. (1981). Concrete Mechanics, part A - Theory and experiments on the mechanical behavior of cracks in plain and reinforced concrete subjected to shear loading. Heron **26**(1a), 1-68.

13. Mayes, R. L., Clough, R. W. (1975) State-of-the-art in seismic shear strength of masonry - an evaluation and review. NSF report n. EERC 75-21, College of Engineering, University of California at Berkeley.

Singular equilibrated stress fields for no-tension panels

Massimiliano Lucchesi[1], Miroslav Silhavy[2], Nicola Zani[1]

[1] Dipartimento di Costruzioni
   Università di Firenze
   50121 Firenze, Italy
[2] Dipartimento di Matematica
   Università di Pisa
   56100 Pisa, Italy

Abstract. In this work we study the equilibrium problem for rectangular panels made of a no-tension material, clamped at the bottom, subjected to distributed vertical loads on the top, and to different types of lateral loads. Admissible and equilibrated stress fields are interpreted as tensor-valued measures with zero divergence. Such stress fields are explicitly determined under the assumption that the measure is absolutely continuous outside a smooth curve which supports a δ type singularity of the stress.

1 Introduction

In this paper, which follows [7–9], we study the equilibrium problem of a panel made of a no tension material [4]. The panel is free from body forces, clamped at its bottom and subjected to loads prescribed on the boundary. The stress field is plane and negative-semidefinite. If horizontal and vertical loads are distributed on the panel's top and the stress determinant is null [5], the equilibrium equations constitute a system of conservation laws, formally identical to the nonlinear system ruling the dynamics of the one-dimensional isentropic flow of a pressureless compressible gas [2].

For sufficiently regular distributions of loads with sufficiently small ratio between the tangential and normal component, the stress field can be explicitly determined by means of a representation formula [8]. As this ratio increases, it can happen that the representation formula loses its validity despite the fact that no collapse of the panel occurs. Under these circumstances the solution of the equilibrium equations cannot be unique and it displays some singularities. In [7,9], some cases are examined in which the solution is regular except on a finite number of curves where the stress field is unbounded. The solution is based on the fact that under appropriate hypotheses, the system of equilibrium equations is equivalent to a single scalar conservation law [1,2]. Then the singularity curves can be determined by means of the Rankine-Hugoniot jump condition corresponding to this scalare quation. This method is not directly applicable if distributed loads are present on the lateral sides of the panel.

In this paper we suppose that the stress field is a tensor measure with divergence measure in the interior Ω of the panel [3,6] to account for the singularities in the stress field. We limit ourselves to the cases in which only a single singularity curve γ is present. Thus the stress field $\mathbf{T}$ is the sum of a measure absolutely continuous with respect to Lebesgue's measure with a smooth density $\mathbf{T}_r$ on $\Omega\backslash\gamma$, and a measure concentrated on γ, whose density is a smooth superficial tensor field $\mathbf{T}_s$. The equilibrium requires that $\mathbf{T}_r$ has null divergence outside γ, and that the surface divergence of $\mathbf{T}_s$ be balanced by the jump of the normal component of $\mathbf{T}_r$ across γ. In the examples presented in this paper, the form of the curve γ and superficial stress field $\mathbf{T}_s$ are obtained by means of this relation, once $\mathbf{T}_r$ has been determined.

2   Preliminaries

Let us designate $\mathcal{E}$ as the two-dimensional Euclidean point space, $\mathcal{V}$ as the associated real vector space with standard basis $\mathbf{e}_1$, $\mathbf{e}_2$, Lin, the space of all linear transformations of $\mathcal{V}$ and $\Omega \subset \mathcal{E}$, an open set. Let $\gamma : [0, \bar{\tau}] \to \Omega$ be a smooth curve in Ω, with unit normal vector $\mathbf{n}$, and $\mathbf{u} : \gamma \to \mathcal{V}$ a smooth vector field on γ. The surface gradient of $\mathbf{u}$ is the map $\nabla_\gamma \mathbf{u} : \gamma \to$Lin, such that

$$\nabla_\gamma \mathbf{u}(\mathbf{p})\mathbf{n}(\mathbf{p}) = \mathbf{0}, \frac{d}{d\tau}\mathbf{u}(\gamma(\tau)) = \nabla_\gamma \mathbf{u}\left(\gamma(\tau)\right)\left(\frac{d}{d\tau}\gamma(\tau)\right), \tag{1}$$

for all $\mathbf{p} \in \gamma$ and for all $\tau \in [0, \bar{\tau}]$, respectively. Moreover, the surface divergence of $\mathbf{u}$ is the trace of its surface gradient,

$$\mathrm{div}_\gamma \mathbf{u} = \mathrm{tr}(\nabla_\gamma \mathbf{u}). \tag{2}$$

Similarly, if $\mathbf{T} : \gamma \to$Lin is a smooth tensor field on γ, its surface divergence is defined by

$$\mathbf{a} \cdot \mathrm{div}_\gamma \mathbf{T} = \mathrm{div}_\gamma(\mathbf{T}^T \mathbf{a}), \tag{3}$$

for every $\mathbf{a} \in \mathcal{V}$.

We define the vector field $\mathbf{u}$ as tangential if $\mathbf{u} \cdot \mathbf{n} = 0$, in γ, and tensor field $\mathbf{T}$ to be superficial if $\mathbf{Tn} = \mathbf{0}$, in γ.

Now, let s be the natural (arc) parameter of γ, $\mathbf{t}(s)$ its unit tangent vector and $\mathbf{u}(s) = u(s)\mathbf{t}(s)$ a tangential vector field on γ. In view of $(1)_2$ we have

$$(\nabla_\gamma \mathbf{u})\mathbf{t} = \frac{d}{ds}(u\mathbf{t}). \tag{4}$$

Then, with the help of $(1)_1$, from (2) we get

$$\mathrm{div}_\gamma \mathbf{u} = \mathrm{tr}(\nabla_\gamma \mathbf{u}) = \mathbf{t} \cdot (\nabla_\gamma \mathbf{u})\mathbf{t} = \mathbf{t} \cdot \frac{d}{ds}(u\mathbf{t}) = \frac{du}{ds}, \tag{5}$$

because $\mathbf{t} \cdot \mathbf{t} = 1$ and $\mathbf{t} \cdot \frac{d\mathbf{t}}{ds} = 0$.

If $\mathbf{T} : \gamma \to \mathrm{Lin}$ is a symmetric tensor field on γ it is a simple matter to verify that $\mathbf{T}$ is superficial if and only if we have

$$\mathbf{T}(s) = \sigma(s)\,\mathbf{t}(s) \otimes \mathbf{t}(s), \tag{6}$$

with σ a scalar field on γ. Then, in view of (3) and (5), for every $\mathbf{a} \in \mathcal{V}$, we deduce

$$\mathbf{a} \cdot \mathrm{div}_\gamma \mathbf{T} = \mathbf{a} \cdot \mathrm{div}_\gamma\,(\sigma\,\mathbf{t} \otimes \mathbf{t}) = \mathrm{div}_\gamma\,(\sigma\,(\mathbf{a} \cdot \mathbf{t})\mathbf{t}) = \frac{d}{ds}(\sigma\,(\mathbf{a} \cdot \mathbf{t})) =$$

$$= \mathbf{a} \cdot \frac{d}{ds}(\sigma\mathbf{t}) \tag{7}$$

and thus

$$\mathrm{div}_\gamma \mathbf{T} = \frac{d}{ds}(\sigma\mathbf{t}). \tag{8}$$

Let $\mathcal{B}(\Omega)$ be the σ algebra of all Borel subsets of Ω and X be a finite dimension real vector space with a dot product. We say that $M\colon \mathcal{B}(\Omega) \to X$, is a X valued (Borel) measure on Ω if M is a σ additive function on $\mathcal{B}(\Omega)$.

In particular, we call M either a (finite) signed measure, a vector measure or a tensor measure on Ω, if $X = \mathbb{R}$, $X = \mathcal{V}$ or $X = \mathrm{Lin}$, respectively. Thus, if $\mathbf{v} = v_1\mathbf{e}_1 + v_2\mathbf{e}_2$ belongs to $C_0(\Omega, \mathcal{V})$, the space of all continuous vector fields with compact support in Ω, and $\boldsymbol{\mu}$ is a vector measure on Ω, there there are two unique signed measures, μ_1 and μ_2, on Ω such that

$$\int_\Omega \mathbf{v} \cdot d\mu = \int_\Omega v_1 d\mu_1 + \int_\Omega v_2 d\mu_2. \tag{9}$$

Similarly, we interpret $\int_\Omega \mathbf{S} \cdot d\mathbf{T}$, with $\mathbf{S} \in C_0(\Omega,\mathrm{Lin})$ and $\mathbf{T}$ a tensor measure on Ω.

If $\varphi \in C_0(\Omega, \mathbb{R})$ is a real valued function with compact support in Ω and $\mathbf{M}$ is a vector (resp. tensor) measure, then

$$\int_\Omega \varphi\, d\mathbf{M} \tag{10}$$

is the element of $\mathcal{V}$ (resp. Lin) defined by

$$\mathbf{a} \cdot \int_\Omega \varphi\, d\mathbf{M} = \int_\Omega (\varphi\mathbf{a}) \cdot d\mathbf{M}, \tag{11}$$

for every $\mathbf{a} \in \mathcal{V}$ (resp. Lin). Moreover, if $\mathbf{T}$ is a tensor measure and $\mathbf{v} \in C_0(\Omega, \mathcal{V})$, then

$$\int_\Omega \mathbf{v}\, d\mathbf{T} \tag{12}$$

is the element of $\mathcal{V}$ such that

$$\mathbf{a} \cdot \int_\Omega \mathbf{v}\, d\mathbf{T} = \int_\Omega \mathbf{a} \otimes \mathbf{v} \cdot d\mathbf{T} \tag{13}$$

for every $\mathbf{a} \in \mathcal{V}$.

Now let $\mathbf{T}: \mathcal{B}(\Omega) \to \mathrm{Lin}$ be a tensor measure on Ω. We say that $\mathbf{T}$ is a tensor measure with divergence measure if there exists a vector measure $\boldsymbol{\mu}$ on Ω such that, for every $\varphi \in C_0^\infty(\Omega, \mathbb{R})$,

$$\int_\Omega \nabla\varphi \, d\mathbf{T} = -\int_\Omega \varphi \, d\boldsymbol{\mu}, \tag{14}$$

or, in view of (13) and (11), if

$$\int_\Omega \mathbf{a} \otimes \nabla\varphi \cdot d\mathbf{T} = -\int_\Omega (\varphi\mathbf{a}) \cdot d\boldsymbol{\mu}, \tag{15}$$

for every $\varphi \in C_0^\infty(\Omega, \mathbb{R})$ and every $\mathbf{a} \in \mathcal{V}$. If this is the case, we write $\mathrm{div}\,\mathbf{T} = \boldsymbol{\mu}$, and say that $\mathbf{T}$ is a balanced tensor measure if $\mathrm{div}\,\mathbf{T} = \mathbf{0}$.

Let $\Omega \subset \mathcal{E}$ be an open set and γ be a smooth curve in Ω, with unit normal $\mathbf{n}$. If $\mathbf{T} : \Omega\backslash\gamma \to \mathrm{Lin}$ is a tensor field, we say that $\mathbf{T}$ is piecewise differentiable if $\mathbf{T}$ is continuously differentiable on $\Omega\backslash\gamma$ and if for every point $\mathbf{p} \in \gamma$ the limits

$$\mathbf{T}^\pm(\mathbf{p}) = \lim_{t\to 0, t>0} \mathbf{T}(\mathbf{p}\pm t\mathbf{n}(\mathbf{x})) \tag{16}$$

exist and the functions $\mathbf{T}^\pm$ are continuously differentiable on γ.

For our current aims, we need the following result [6]. Let $\Omega \subset \mathcal{E}$ be an open set, $\partial\Omega$ its boundary and $\gamma : [0, \bar{\tau}] \to \mathcal{E}$ a smooth curve such that $\gamma(]0, \bar{\tau}[) \subset \Omega$, $\gamma(0) \in \partial\Omega$ and $\gamma(\bar{\tau}) \in \partial\Omega$. Let $\mathbf{T}$ be a tensor measure on Ω defined by

$$\int_\Omega \mathbf{S} \cdot d\mathbf{T} = \int_\Omega \mathbf{S} \cdot \mathbf{T}_r dA + \int_\gamma \mathbf{S} \cdot \mathbf{T}_s ds, \tag{17}$$

for each $\mathbf{S} \in C_0(\Omega, \mathrm{Lin})$, where $\mathbf{T}_r$ is a piecewise differentiable tensor field on $\Omega\backslash\gamma$ and $\mathbf{T}_s$ is a continuously differentiable tensor field on γ. Then, $\mathbf{T}$ is a tensor measure with divergence measure on Ω, if and only if $\mathbf{T}_s$ is superficial. Moreover, if this is the case, then $\mathbf{T}$ is balanced, if and only if

$$\mathrm{div}\,\mathbf{T}_r = 0 \quad \text{on} \quad \Omega\backslash\gamma \tag{18}$$

and

$$[\mathbf{T}_r]\mathbf{n} + \mathrm{div}_\gamma\mathbf{T}_s = 0 \quad \text{on} \quad \gamma, \tag{19}$$

with $[\mathbf{T}_r] = \mathbf{T}_r^+ - \mathbf{T}_r^-$ the jump of $\mathbf{T}_r$ across γ.

Now let us suppose that curve γ is the graph of a function $y = \omega(x)$, with $x \in [x_0, x_1]$, i.e. $\gamma = \{(x, \omega(x)) \in \mathcal{E} : x \in [x_0, x_1]\}$. Then, for $\mathbf{e}_1$ and $\mathbf{e}_2$ the unit normal vectors of the standard basis in $\mathcal{V}$, the unit tangent $\mathbf{t}$ and the unit normal $\mathbf{n}$ of γ are

$$\mathbf{t} = J^{-1}(\mathbf{e}_1 + \omega'\mathbf{e}_2), \mathbf{n} = J^{-1}(-\omega'\mathbf{e}_1 + \mathbf{e}_2), \tag{20}$$

where a prime denotes differentiation with respect to x and

$$J = \frac{ds}{dx} = \sqrt{1 + (\omega')^2}. \tag{21}$$

Moreover, from (6) we get

$$\mathbf{T}_s = \sigma J^{-2} \left\{ \mathbf{e}_1 \otimes \mathbf{e}_1 + \omega'(\mathbf{e}_1 \otimes \mathbf{e}_2 + \mathbf{e}_2 \otimes \mathbf{e}_1) + (\omega')^2 \mathbf{e}_2 \otimes \mathbf{e}_2 \right\}. \tag{22}$$

Now, denoting

$$[\mathbf{T}_r] = \delta_{11} \mathbf{e}_1 \otimes \mathbf{e}_1 + \delta_{12}(\mathbf{e}_1 \otimes \mathbf{e}_2 + \mathbf{e}_2 \otimes \mathbf{e}_1) + \delta_{22} \mathbf{e}_2 \otimes \mathbf{e}_2, \tag{23}$$

from $(20)_2$ we obtain

$$[\mathbf{T}_r]\mathbf{n} = J^{-1}\{(\delta_{12} - \omega'\delta_{11})\mathbf{e}_1 + (\delta_{22} - \omega'\delta_{12})\mathbf{e}_2\} \tag{24}$$

and from (8) and (21)

$$\mathrm{div}_\gamma \mathbf{T}_s = \frac{d}{ds}\left(\frac{\sigma}{J}\mathbf{e}_1 + \frac{\sigma}{J}\omega'\mathbf{e}_2\right) = J^{-1}\frac{d}{dx}\left(\frac{\sigma}{J}\mathbf{e}_1 + \frac{\sigma}{J}\omega'\mathbf{e}_2\right). \tag{25}$$

Then, for

$$\frac{\sigma}{J} = \beta, \tag{26}$$

from (19), (24) and (25) we deduce the system of ordinary differential equations

$$\beta' - \omega'\delta_{11} + \delta_{12} = 0, \tag{27}$$

$$(\beta\omega')' - \omega'\delta_{12} + \delta_{22} = 0, \tag{28}$$

some applications of which are illustrated in the following.

3   Rectangular panels

Let us now consider a rectangular panel of width b and height h, clamped at its base $y = h$ and subjected to a vertical load, p, distributed on its top, $y = 0$, a horizontal load, q, distributed along its right side, $x = 0$, and a force, $\mathbf{f} = f_1\mathbf{e}_1 + f_2\mathbf{e}_2$, concentrated at the point o with coordinates $x = 0$, $y = 0$, (Fig. 1).

Denoting by Ω the inner part of the panel,

$$\Omega = \{(x, y) \in \mathcal{E} : 0 < x < b,\, 0 < y < h\}, \tag{29}$$

we aim to determine a curve γ

$$y = \omega(x), \quad \text{with} \quad \omega(0) = 0, \tag{30}$$

and a continuously differentiable, negative-semidefinite superficial stress field $\mathbf{T}_s$ on γ, such that, denoted by $\Omega^+ = \{(x, y) \in \Omega : 0 < x < \omega^{-1}(y)\}$ and

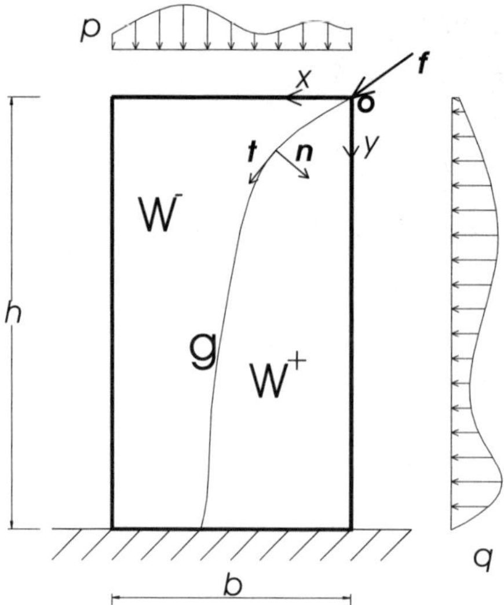

Fig. 1. The panel under general load conditions

$\Omega^- = \{(x, y) \in \Omega : \omega^{-1}(y) < x < b\}$ the two regions in which γ divides Ω, and assumed for $\mathbf{T}_r$ the expression

$$\mathbf{T}_r = \begin{cases} -p(x)\,\mathbf{e}_2 \otimes \mathbf{e}_2 & \text{in } \Omega^- \\ -q(y)\,\mathbf{e}_1 \otimes \mathbf{e}_1 & \text{in } \Omega^+, \end{cases} \tag{31}$$

the stress tensor measure $\mathbf{T}$ defined by (17) are balanced and in equilibrium with the external loads. Since, according to (31), $\mathbf{T}_r$ is equilibrated with distributed loads p and q and satisfies (18), it is sufficient to determine γ and $\mathbf{T}_s$ to satisfy (19) and the equilibrium boundary condition

$$\mathbf{T}_s(\mathbf{o})\mathbf{t}(\mathbf{o}) = -\mathbf{f}. \tag{32}$$

To this end, note that in this case (23), (31) and $(30)_1$ give us

$$\delta_{11} = -q(\omega(x)), \quad \delta_{22} = p(x), \quad \delta_{12} = 0 \tag{33}$$

and therefore from (27) and (28) we deduce

$$\beta' + q(\omega(x))\omega' = 0, \tag{34}$$

$$(\beta\omega')' + p(x) = 0, \tag{35}$$

from which, designating P and Q as the respective primitives of p and q, such that $P(0) = 0$ and $Q(0) = 0$, we get

$$\beta(x) = \beta(0) - Q(\omega(x)), \tag{36}$$

$$\beta(x)\omega'(x) = \beta(0)\omega'(0) - P(x). \tag{37}$$

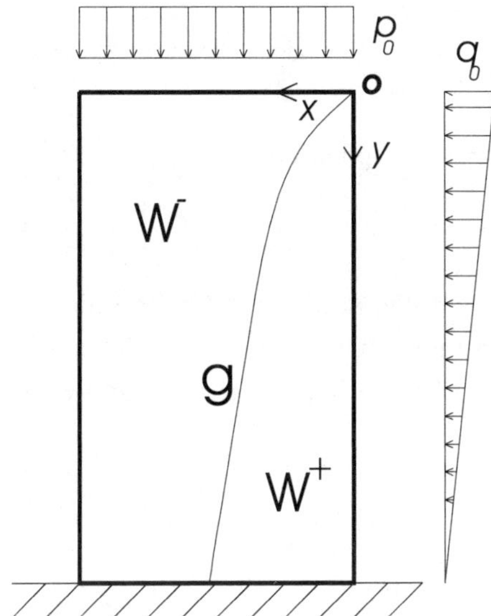

Fig. 2. Load-distribution laws on the boundary of the panel. Case 1

Now, with the help of (36), (37) becomes

$$(Q(\omega(x)) - \beta(0))\,\omega'(x) = P(x) - \beta(0)\omega'(0). \tag{38}$$

And, moreover, in view of (6), $(20)_1$ and (26), the equilibrium boundary condition (32) becomes

$$\beta(0)(\mathbf{e}_1 + \omega'(0)\mathbf{e}_2) = -(f_1\mathbf{e}_1 + f_2\mathbf{e}_2), \tag{39}$$

from which we deduce

$$\beta(0) = -f_1, \qquad \beta(0)\omega'(0) = -f_2. \tag{40}$$

Then, (38) implies

$$(Q(\omega(x)) + f_1)\,\omega'(x) = P(x) + f_2 \tag{41}$$

that can be integrated with the help of initial condition $(30)_2$. We can see that, since $\mathbf{T}_s$ is negative-semidefinite, then in view of (6), it holds that $\sigma(0) \leq 0$, and from (21) and (26) we obtain $\beta \leq 0$. Moreover, since curve γ (except for its ends) is wholly contained within Ω, it must hold that $\omega'(0) \geq 0$. Therefore from (40) it follows that both f_1 and f_2 must be non-negative, that is to say, that the force $\mathbf{f}$ must be directed towards the inside of the panel [5].

3.1   Example 1

In this example we suppose that the vertical distributed load is uniform, the horizontal one is linear and the concentrated force is null (Fig. 2),

$$p(x) = p_0, \quad q(x) = q_0\left(1 - \tfrac{y}{h}\right), \quad \mathbf{f} = \mathbf{0}. \tag{42}$$

Under such conditions

$$Q(\omega) = q_0\omega\left(1 - \frac{\omega}{2h}\right), \quad P(x) = p_0 x \tag{43}$$

and from (41) and (30)$_2$ we obtain for γ the implicit equation

$$q_0\omega^2\left(1 - \frac{\omega}{3h}\right) = p_0 x^2 \tag{44}$$

It can be seen that γ intersects the panel base at $x = h\sqrt{\frac{2q_0}{3p_0}}$. In order for such a solution to be valid, the intersection point must be within the panel's base, that is to say, $b \geq h\sqrt{\frac{2q_0}{3p_0}}$ and this, fixed p_0, requires that q_0 not exceed the value

$$q_m = \frac{3}{2}p_0\left(\frac{b}{h}\right)^2, \tag{45}$$

whose attainment would cause the panel to overturn around the corner at coordinates $x = b$, $y = h$. Now, from (44) we deduce $x = \omega\sqrt{\frac{q_0}{p_0}\left(1 - \frac{\omega}{3h}\right)}$, and then, from (41) and (43) we obtain $\omega' = \left(\frac{2h}{2h-\omega}\right)\sqrt{\frac{p_0}{q_0}\left(1 - \frac{\omega}{3h}\right)}$.

In particular, at the panel bottom we have $\omega = h$ and therefore, taking (21) and (36) into account, it follows

$$\omega' = 2\sqrt{\frac{2p_0}{3q_0}}, \quad J = \sqrt{1 + \frac{8p_0}{3q_0}}, \quad \beta = -\tfrac{1}{2}q_0 h \tag{46}$$

from which, with the help of (20) and (26), we obtain the reaction $\mathbf{T}_s\mathbf{t} = \sigma\mathbf{t}$ at the end of γ whose intensity is $\frac{1}{2}q_0 h\sqrt{1 + \frac{8p_0}{3q_0}}$.

3.2   Example 2

In this second example, we once again suppose that the vertical load is uniform, while the horizontal one in this case is null. Moreover, we assume a concentrated horizontal force to be acting (Fig. 3),

$$p = p_0, \quad q = 0, \quad \mathbf{f} = f\mathbf{e}_1, \tag{47}$$

so that $P(x) = p_0 x$ and $Q(\omega) = 0$. Therefore, from (41), for $f_1 = f$, and also taking (30)$_2$ into account , we deduce

$$\omega(x) = \frac{p_0 x^2}{2f} \tag{48}$$

and therefore, in view of (21), (36), (40)$_1$ and (26) we obtain

$$J = \sqrt{1 + \left(\frac{p_0 x}{f}\right)^2}, \quad \beta = -f, \quad \sigma = -f\sqrt{1 + \left(\frac{p_0 x}{f}\right)^2}, \tag{49}$$

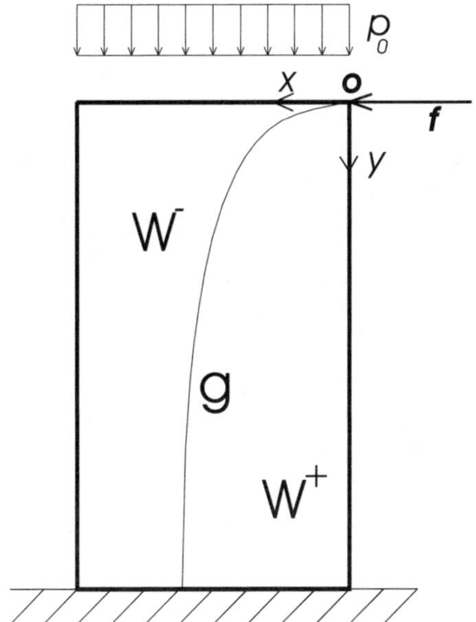

Fig. 3. Load-distribution laws on the boundary of the panel. Case 2

from which, with the help of (22), we can determine $\mathbf{T}_s$. It should be noted that the horizontal component of $\mathbf{T}_s\mathbf{t}$ is $-f$ and then constant along γ. Moreover, from (48), for $x = b$ and $\omega = h$, we get the maximum intensity of force $\mathbf{f}$ compatible with the equilibrium $f_m = \frac{p_0 b^2}{2h}$, while in view of (48) and (49)$_3$ the intensity of the concentrated reaction at the panel's base is $f\sqrt{1 + \frac{2p_0 h}{f}}$.

3.3   Example 3

As a last example, let us consider the case in which the panel is subjected only to the sole action of the uniform distributed vertical load $p = p_0$. Using the results of the previous example, we wish to verify that, beyond the regular stress state, $\mathbf{T} = -p_0 \mathbf{e}_2 \otimes \mathbf{e}_2$, defined throughout Ω, it is possible to determine infinite equilibrated and compatible stress fields in the class of measures with divergence measure, each of which characterized by

- a superficial stress $\mathbf{T}_s$ defined on a curve γ $(y = \omega(x))$ that is symmetric with respect to y axis (Fig. 4), which it intersects for $y = \lambda$, $(0 \leq \lambda < h)$ and that also intersects the panel bottom for $|x| = \mu$ $(0 < \mu \leq \frac{1}{2}b)$,

$$\omega(0) = \lambda, \quad \omega(|\mu|) = h, \quad \omega'(0) = 0, \tag{50}$$

- a stress field

$$\mathbf{T}_r = \begin{cases} -p_0\, \mathbf{e}_2 \otimes \mathbf{e}_2 & \text{in } \Omega^- \\ 0 & \text{in } \Omega^+, \end{cases} \tag{51}$$

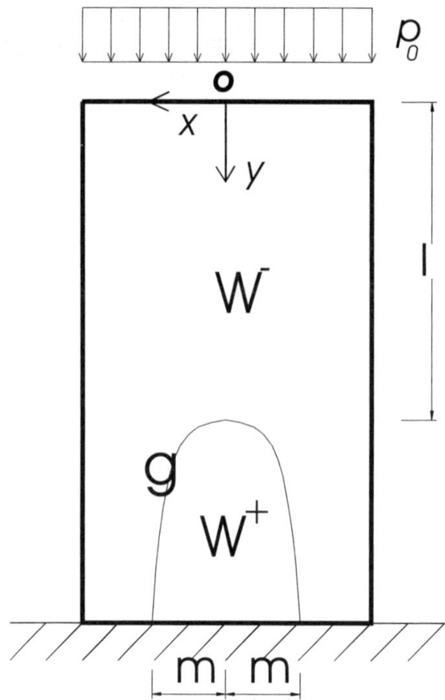

Fig. 4. Load-distribution laws on the boundary of the panel. Case 3

where $\Omega^+ = \{(x, y) \in \Omega : |x| < \omega^{-1}(y), \lambda < y < h\}$ is the region below γ and Ω^- is the interior of its complementary in Ω.

In fact, from (34) and (35), with $p = p_0$ and $q = 0$, with the help of (50), we obtain

$$\omega(x) = \lambda + \frac{h - \lambda}{\mu^2}x^2, \quad \beta = -\frac{p_0\mu^2}{2(h - \lambda)}, \tag{52}$$

from which, in view of (21), (26) and (22), it is a simple matter to calculate J, σ and $\mathbf{T}_s$. It can be seen that the interaction between the two parts of the panel, separated by the symmetry axis, $x = 0$, is made up solely of a horizontal force concentrated at the γ apex, whose intensity β is an increasing function of λ, and which becomes unbounded when λ tends to h.

Similar solutions can be used to study the equilibrium problem of panels with openings [9].

References

1. Bouchut, F., James, F. (1999) Duality solutions for pressureless gases, monotone scalar conservation laws, and uniquess. Comm. P.D.E., 24, 2173-2190.

2. Brenier Y., Grenier E. (1998) Stickly particles and scalar conservation laws. SIAM J. Numer. Anal., 35, 2317-2328.

3. Chen, G.Q., Frid, H. (1999) Divergence-Measure Fields and Hyperbolic Conservation Laws, Arch. Rational Mech. Anal., 147, 89-118.

4. Del Piero G. (1989) Constitutive equations and compatibility of the external loads for linear elastic masonry-like materials. Meccanica, 24, 150-162.

5. Di Pasquale S. (1984) Statica dei solidi murari teorie ed esperienze. Dipartimento di Costruzioni, Pubblicazione 27.

6. Lucchesi M., Silhavy M., Zani N. (2004) In preparation.

7. Lucchesi M., Zani N. (2002) On the collapse of masonry panel. VII International Seminar on Structural Masonry for Developing Countries, Belo Horizonte, Brazil, 315-323.

8. Lucchesi M., Zani N. (2003) Some explicit solutions to equilibrium problem for masonry like bodies. Structural Engineering and Mechanics 16, 295-316.

9. Lucchesi, M., Zani, N., Stati di sforzo per pannelli costituiti da materiale non resistente a trazione. XVI Congresso AIMETA di Meccanica Teorica e Applicata, Ferrara, Italia.

Damage of Materials: Damaging Effects of Macroscopic Vanishing Motions

Elena Bonetti[1], Michel Frémond[2]

[1] Dipartimento di Matematica "F.Casorati"
  Universit degli Studi di Pavia
  via Ferrata 1, 27100 Pavia, Italy
[2] Laboratoire Central des Ponts et Chaussées
  58, boulevard Lefbvre
  75732 Paris Cedex, France

Abstract. We investigate a mechanical model describing the evolution of damage in elastic and viscoelastic materials. The state variables are macroscopic deformations and the volume fraction of sound material. The equilibrium equations are recovered by refining the principle of virtual power including also microscopic forces. After proving an existence and uniqueness result for a regularized problem, we investigate the behaviour of solutions, in the case when a vanishing sequence of external forces is applied. By use of a rigorous asymptotics analysis, we show that macroscopic deformations can disappear at the limit, but their damaging effect remains in the equation describing the evolution of damage at a microscopic level. Moreover, it is proved that the balance of the energy is satisfied at the limit.

1   Introduction

We investigate a macroscopic mechanical model describing the evolution of damage in elastic and viscoelastic bodies. This model is written in terms of the basic laws of continuum mechanics and accounts for macroscopic effects of microscopic motions, which are responsible for the process of damage [9]. Indeed, it is known that damage is caused by microfractures and microcavities resulting in the decreasing of the material stiffness. As a consequence, towards the aim to provide a good and predictive macroscopic theory, the model accounts for a generalized version of the power of interior forces, including the power of microscopic forces. It is assumed that the power of interior forces depends on the damage rate, which is clearly related to microscopic motions, on the gradient of damage rate, accounting for local microscopic interactions. The equations are recovered by continuum mechanics laws and they correspond to equations for macroscopic and microscopic motions, the latter accounting for the evolution of the damage. It has been show that this model is coherent with civil engineering application, as it can predict correctly the behaviour of concrete structures, when compared with experience [7]. In particular, let us point out that it predicts the structural size effect. This is mainly due to the presence of the gradient of the damage parameter in the state variables.

Concerning the analytical investigation of the above model, only some partial results have been obtained. However, they turn out meaningful both from a mathematical and mechanical point of view. The difficulty is mainly due to the degeneracy of elasticity modulus in the macroscopic equilibrium equation and the presence of a quadratic nonlinearity involving macroscopic deformations in the double nonlinear equation describing the evolution of damage. Although numerical results show that the above system describes a behaviour of the solutions in accordance with experimental results in any space dimension, analytical existence results have been obtained only for some simplified versions. In the one dimensional case, in [6] the authors investigate the evolution of the damage to elastic materials caused by compression or tension that is governed by a slightly modified version of the original model we are going to discuss in this paper. The existence and uniqueness of local weak solutions are proved, in a quasi-static situation. We point out that it seems quite natural in this case to get a local-in-time existence result, since once the damage reaches the value zero, the system snaps (which corresponds to a blow up of the quadratic source of damage related to the strain becoming unbounded due to the degeneracy of the material stiffness). An analogous result has been obtained in the case when macroscopic accellerations are considered. Nonetheless, the arguments exploited in [6] do not extend to higher dimension situations, as they are based on the one-dimensional structure of the equations and regularity theorems holding only in the one-dimensional case. In [2] the authors restrict themselves to look at the evolution of damage till the material is not completely damaged, which corresponds to the fact that in the model the damage parameter, say χ, remains strictly greater than a positive constant δ. By use of a truncation technique, combined with suitable a priori estimates, they prove a local in time existence result and uniqueness of the solution. Concerning the separation between macroscopic and microscopic motions, it turns out to be interesting to investigate what occurs when a macroscopic motion becomes microscopic in the evolution. In a recent paper [4], the author provides an example of a situation in which a sequence of vanishing macroscopic motions retains at the limit its damaging effect as a source of damage in the equation governing at a microscopic level the evolution of the damage parameter. In this note, we show that this behaviour can be proved in more general situations. Towards this aim, we approximate the model and get existence and uniqueness of the solution to the regularized version. Hence, we apply a sequence of exterior forces vanishing at the limit, so that the corresponding motions become smaller and smaller. By use of a rigorous asymptotics analysis, we show that at the limit the macroscopic motions disappear, while their damaging effect remains as a weak limit of some deformation energy.

2   The model

In this section, we briefly recall the derivation of the mechanical model and write the corresponding analytical formulation in terms of a PDE's system. We assume that the temperature is constant and write the model in terms of the state variables (for the sake of simplicity small perturbations are assumed)

$$\mathcal{E} := \{\varepsilon(\mathbf{u}), \chi, \nabla\chi\}, \tag{1}$$

where $\mathbf{u}$ is the vector of small displacements, $\varepsilon(\mathbf{u})$ is the linearized symmetric strain tensor, and χ is the damage parameter such that

$$\chi \in [0, 1]. \tag{2}$$

More precisely, we let $\chi = 1$ correspond to the undamaged structure and $\chi = 0$ to the completely damaged material. In damage theory χ is often thought as the volume fraction of microvoids and microcracks or the quotient of the modulus of the damaged material divided by the modulus of the undamaged material. Hence, dissipation (related to microscopic velocities) is described by the following dissipative variables

$$\mathcal{E}_D := \{\chi_t, \nabla\chi_t\}. \tag{3}$$

We introduce the free energy functional

$$\Psi(\mathcal{E}) := \frac{1}{2}\varepsilon(\mathbf{u})\chi K\varepsilon(\mathbf{u}) + w(1 - \chi) + \frac{k}{2}|\nabla\chi|^2 + \frac{b}{4}|\varepsilon(\mathbf{u})|^4 + I_{[0,1]}(\chi), \tag{4}$$

$b > 0$ if we consider some higher order nonlinear elasticity contribution [9], K being the elasticity matrix, $k > 0$, $w > 0$ accounting for the cohesion energy of the material, and $I_{[0,1]}$ is the indicator function of the interval $[0, 1]$ and accounts for the constraint on the admissible values of χ, as it is $I_{[0,1]}(\chi) = 0$ if $\chi \in [0, 1]$ and $I_{[0,1]}(\chi) = +\infty$ otherwise. Note that by the presence of $I_{[0,1]}$ Ψ is a generalized free energy functional, as it is defined for any value of χ, also for those that are physically impossible, but it is equal to $+\infty$ if χ does not take values in the mechanically admissible range. In this sense the internal constraint on χ is considered as a physical property of the system. Hence, we specify the pseudo-potential of dissipation

$$\Phi(\mathcal{E}_D) := \frac{c}{2}|\chi_t|^2 + \frac{a}{2}|\varepsilon(\mathbf{u_t})|^2 + I_-(\chi_t), \tag{5}$$

where $c > 0$, $a > 0$ for viscoelastic materials, and $I_-(\chi_t)$ is the indicator function of $(-\infty, 0]$ and it is defined as $I_-(\chi_t) = 0$ if $\chi_t \leq 0$, $I_-(\chi_t) = +\infty$ if $\chi_t > 0$. Note that this implies that the damage process is irreversible, i.e. the material can damage (χ decreases towards 0) but it cannot recover its orignal state (χ cannot increase towards 1). This is the typical behaviour, e.g., of concrete. Obviously, this term is not present if we describe a reversible damage evolution, as it occurs for instance in polymers. Concerning the choice of defining dissipative forces, assuming that there exists a pseudo-potential of dissipation in the sense by Moreau, we refer to [12]. Here, we recall that a pseudo-potential of dissipation is a positive convex function, which is zero in the case when dissipative variables are zero.

In this paper, we restrict ourselves to describe the damage phenomenon in a bitumineous material. Thus, we consider a reversible damage process and omit I_- in the expression of Φ. Hence, we let $b = 1$ and $a = 1$ in Ψ and Φ, respectively. The body is located in a smooth bounded domain $\Omega \subset \mathbf{R}^3$, with boundary $\Gamma := \partial\Omega$, and we look at the damage evolution during a finite time interval $[0, T]$; let $Q := \Omega \times (0, T)$. The principle of virtual power written for microscopic forces provides the following variational inclusion governing the evolution of χ (we assume for the sake of simplicity K is the identity tensor)

$$c\chi_t - k\Delta\chi + \partial I_{[0,1]}(\chi) \ni w - \frac{1}{2}|\varepsilon(\mathbf{u})|^2, \tag{6}$$

where $\partial I_{[0,1]}(\chi)$ is a reaction of the system to avoid χ to assume values not physically admissible. Note that the reaction is zero if χ is in the interior of admissible values and it appears whenever χ is on the boundary of its mechanically admissible range. In the case of viscous materials, we add some viscosity contribution in the macroscopic equilibrium equation. Moreover, we have introduced an higher order elasticity term, which turns out to be usefull to investigate the asymptotics behaviour of microscopic forces as macroscopic ones vanish. We finally get

$$\mathbf{u}_t - \operatorname{div}\,\varepsilon(\mathbf{u}_t) - \operatorname{div}\left(|\varepsilon(\mathbf{u})|^2\varepsilon(\mathbf{u}) + \chi\varepsilon(\mathbf{u})\right) = \mathbf{f}, \tag{7}$$

$\mathbf{f}$ representing an exterior volumic applied force. The above equations are complemented by physically meaningful initial and boundary conditions. We assume

$$\chi(0) = 1 \quad \text{a.e. in } \Omega, \tag{8}$$

i.e. the material is undamaged before the evolution, and

$$\mathbf{u}(0) = \mathbf{0}. \tag{9}$$

Then, we assume that no energy is provided from the outside

$$\partial_n\chi = 0 \quad \text{a.e. in } \Gamma \times (0, T), \tag{10}$$

and zero traction acts on the boundary

$$(\varepsilon(\mathbf{u}_t) + |\varepsilon(\mathbf{u})|^2\varepsilon(\mathbf{u}) + \chi\varepsilon(\mathbf{u}))\mathbf{n} = 0 \quad \text{on } \Gamma. \tag{11}$$

Now, we are interested in considering the behaviour of the solutions $(\chi, \mathbf{u})$ to the above initial and boundary values problem, in the case when a sequence of vanishing external volume forces are applied, i.e. $\mathbf{f}_\tau \searrow 0$. We proceed as follows. We first write the abstract version of the problem we are investigating in a suitable variational framework. From a mechanical point of view this corresponds to specify the equilibrium equations in a duality pairing, i.e., solving the problem in the weak sense corresponding to the formulation of the principle of virtual power. Hence, we apply the vanishing sequence $\mathbf{f}_\tau$ to the system and investigate the asymptotic behavior of the corresponding solutions $(\chi_\tau, \mathbf{u})$ as $\tau \searrow 0$. This result is obtained by use of an a priori estimate technique and weak compactness results. Finally, we discuss an energy balance relation holding at the limit.

3   The variational problem and the weak existence result

Let us first introduce a suitable abstract setting. In regards of simplicity, but without loss of generality, in the following we take $\mathbf{u}$ as a scalar quantity u. In addition, fix $c = k = 1$. Hence, we start by considering the Hilbert triplet (cf. [10] for a definitions and properties of the following Sobolev spaces)

$$V \subset H \subset V',$$

where

$$V := H^1(\Omega) \quad \text{and} \quad H := L^2(\Omega),$$

H being identified as usual with its dual space. Notice that we use the same notation $\|\cdot\|_X$ both for the norm in a functional space X and in X^3. Hence, let

$$W := W^{1,4}(\Omega).$$

Finally, we introduce the following abstract operators

$$A : V \to V' \quad {}_{V'}\langle Au, v\rangle_V = \sum_{i=1}^{3} \int_\Omega D_i u D_i v = \int_\Omega \nabla u \cdot \nabla v \quad u, v \in V, \ (12)$$

$$\mathcal{A} : W \to W' \quad {}_{W'}\langle \mathcal{A}u, v\rangle_W = \sum_{i=1}^{3} \int_\Omega |D_i u|^2 D_i u D_i v \quad u, v \in W, \ (13)$$

$$B : (L^{4/3}(\Omega))^3 \to W' \quad {}_{W'}\langle B\mathbf{v}, v\rangle_W = \sum_{i=1}^{3} \int_\Omega u_i D_i v = \int_\Omega \mathbf{v} \cdot \nabla v,$$

$$\mathbf{v} \in (L^{4/3}(\Omega))^3, v \in W, \ (14)$$

where D_i stands for the usual partial derivative operator with respect to the variable x_i, $i = 1, 2, 3$. In particular, let us observe that A corresponds to the laplacian operator with associated Neumann homogeneous boundary condition in the duality pairing between V' and V. Analogously, $\mathcal{A}$ corresponds to the abstract formulation of the bi-laplacian operator with associated homogeneous boundary assumption. Now, we can make precise the abstract problem we are dealing with and state a corresponding existence and uniqueness result.

Problem P_a   Find (u, χ) such that

$$u(0) = 0, \quad \chi(0) = 1 \quad \text{a.e. in } \Omega,$$

and fulfilling for a.a. t in $(0, T)$

$$u_t + Au_t + \mathcal{A}u + B(\chi \nabla u) = f \quad \text{in } W', \tag{15}$$

$$\chi_t + A\chi + \xi = w - \frac{1}{2}|\nabla u|^2 \quad \text{in } H, \tag{16}$$

where $\xi \in L^2(0, T; H)$ is such that $\xi \in \partial I_{[0,1]}(\chi)$ a.e. in Q. The following proposition holds.

Theorem 31 *Let*
$$f \in L^2(0, T; V') \cap L^1(0, T; H).$$

Then, Problem P_a admits a unique solution (u, χ) with regularity

$$u \in H^1(0, T; V) \cap L^\infty(0, T; H^2(\Omega)), \tag{17}$$
$$\chi \in H^1(0, T; H) \cap L^\infty(0, T; V) \cap L^2(0, T; H^2(\Omega)). \tag{18}$$

The existence result stated by the above theorem is proved exploiting a fixed point argument, i.e. applying the Schauder fixed point theorem. We do not detail the proof, for which the reader can refer to [1]. The uniqueness result follows by contradiction and a suitable contracting argument.

4   Passage to the limit for vanishing external forces

We denote by (u_τ, χ_τ) the solution of Problem P corresponding to the applied force f_τ. Then, we take a sequence f_τ such that the sequence converges weakly to 0 in V' ($f_\tau \rightharpoonup 0$), i.e. for any test function $v \in V$ we have

$$_{V'}\langle f_\tau, v \rangle_V \to 0 \quad \text{as } \tau \searrow 0. \tag{19}$$

Then, the following theorem characterizes the asymptotics behaviour of (u_τ, χ_τ).

Theorem 41 *Let (19) holds. Then, the following convergences are verified*

$$u_\tau \overset{*}{\rightharpoonup} 0 \quad in \ H^1(0, T; V) \cap L^\infty(0, T; W), \tag{20}$$

$$|\nabla u_\tau|^2 \overset{*}{\rightharpoonup} d \quad in \ L^\infty(0, T; H), \tag{21}$$

$$\chi_\tau \overset{*}{\rightharpoonup} \chi \quad in \ H^1(0, T; H) \cap L^\infty(0, T; V) \cap L^2(0, T; H^2(\Omega)). \tag{22}$$

Moreover, d and χ fulfill, a.e. in $(0, T)$,

$$\chi_t + A\chi + \xi = w - \frac{1}{2}d \quad in \ H, \tag{23}$$

for some ξ belonging to $\partial I_{[0,1]}(\chi)$ a.e. in Q.

The proof of Theorem 41 is perfomermed by use of an a priori estimates-passage to the limit technique. In particular, we prove some uniform estimates on the solutions (u_τ, χ_τ) independent of the parameter τ, and then pass to the limit by compactness arguments. We do not detail the estimating and passage to the limit procedures. However, let us briefly discuss the mechanical meaning of the above convergences. At the limit $\tau = 0$ we get that there are not macroscopic displacements or deformations ($u = 0$ a.e.). On the contrary, the function d in (23) is obtained as the weak limit of the deformation energy associated to the vanishing sequence of macroscopic motions and, in general, d may be different from 0. Thus, this function, which represents a source of damage in (23), can be interpreted as the remaining damage effect of macroscopic motions, actually acting at a microscopic level. As a consequence, we are allowed to deduce that a sequence of vanishing macroscopic motions can retain its damaging effect at the limit, as a source of damage in the equation governing the equilibrium of microscopic forces. In order to support this argument, in the next section we discuss the balance of the energy of the problem.

Remark 42 *An example of such a vanishing macroscopic motion is a highly oscillating sequence of space motion becoming smaller and smaller (see [4] for a precise example in 1-D, where the author considers the sequence of macroscopic motions*

$$u_n(x) = \frac{1}{n} \sin nx.$$

5   Balance of the energy

Let us introduce the work provided by the exterior forces during a time interval $(0, t)$, i.e.

$$T_\tau(t) = \int_0^t {}_{V'}\langle f_\tau, u_{\tau_t} \rangle, \tag{24}$$

which can be rewritten, due to the equation, as follows

$$T_\tau(t) = \int_0^t \int_\Omega |u_{\tau_t}|^2 + \int_0^t \int_\Omega |\nabla u_{\tau_t}|^2 + \sum_{i=1}^3 \frac{1}{4} \int_\Omega |D_i u_\tau(t)|^4$$
$$+ \frac{1}{2} \int_0^t \int_\Omega \chi_\tau \frac{d}{dt} |\nabla u_\tau|^2 . \tag{25}$$

We let

$$A_\tau(t) = \int_0^t \int_\Omega |u_{\tau_t}|^2 + \int_0^t \int_\Omega |\nabla u_{\tau_t}|^2 + \sum_{i=1}^3 \frac{1}{4} \int_\Omega |D_i u_\tau(t)|^4 . \tag{26}$$

Hence, we have

$$\frac{1}{2} \int_0^t \int_\Omega \chi_\tau \frac{d}{dt} |\nabla \chi_\tau|^2 = \frac{1}{2} \int_\Omega \chi_\tau(t) |\nabla u_\tau(t)|^2 - \frac{1}{2} \int_0^t \int_\Omega \chi_{\tau_t} |\nabla u_\tau|^2$$
$$= \frac{1}{2} \int_\Omega \chi_\tau(t) |\nabla u_\tau(t)|^2 + \int_0^t \int_\Omega |\chi_{\tau_t}|^2 + \frac{1}{2} \int_\Omega |\nabla \chi_\tau(t)|^2 - \int_0^t \int_\Omega w\chi_{\tau_t} \tag{27}$$

In particular, in (27) we have used the fact that

$$\int_0^t \int_\Omega \xi_\tau \chi_{\tau t} = \int_\Omega I_{[0,1]}(\chi_\tau(t)) - \int_\Omega I_{[0,1]}(\chi_\tau(0)) = 0, \tag{28}$$

as $\chi_\tau \in [0,1]$ a.e. in Ω and $\chi_\tau(0) = 1$. Hence, we set

$$B_\tau(t) = \int_0^t \int_\Omega |\chi_{\tau t}|^2 + \frac{1}{2} \int_\Omega |\nabla \chi_\tau(t)|^2 - \int_0^t \int_\Omega w \chi_{\tau t}. \tag{29}$$

Thus, we can rewrite T_τ as follows

$$T_\tau(t) = A_\tau(t) + B_\tau(t) + \frac{1}{2} \int_\Omega \chi_\tau(t) |\nabla u_\tau(t)|^2. \tag{30}$$

Then, we investigate the limit as $\tau \searrow 0$ of T_τ. By Theorem 41 we can infer that

$$\liminf_{\tau \searrow 0} B_\tau(t) = Q_B(t) - \frac{1}{2} \int_0^t \int_\Omega d\chi_t = Q_B(t) + \mathcal{D}(t), \tag{31}$$

where $Q_B \geq 0$. Analogously, we have that

$$\liminf_{\tau \searrow 0} A_\tau(t) = Q_A(t) + \frac{1}{4} \|d(t)\|_H^2, \tag{32}$$

and $Q_A \geq 0$. Finally, there holds

$$\lim_{\tau \searrow 0} \frac{1}{2} \int_\Omega \chi_\tau(t) |\nabla u_\tau(t)|^2 = \frac{1}{2} \int_\Omega \chi(t) d(t). \tag{33}$$

Hence, we can take a suitable subsequence of τ_j such that $\lim_{\tau_j \searrow 0} T_{\tau_j}(t) = \liminf_{\tau \searrow 0} T_\tau(t)$. Thus, denoting by $T(t) = \lim_{\tau_j \searrow 0} T_{\tau_j}(t)$ we can eventually infer that

$$\lim_{\tau_j \searrow 0} T_{\tau_j}(t) = T(t) = Q(t) + \mathcal{D}(t) + \mathcal{S}(t), \tag{34}$$

where $Q(t) = Q_A(t) + Q_B(t)$, $\mathcal{D}$ is introduced in (31), and

$$\mathcal{S}(t) = \frac{1}{4} \|d(t)\|_H^2 + \frac{1}{2} \int_\Omega \chi(t) d(t).$$

By the above argument, we conclude that the work which is provided to the structure $T(t)$ is divideed between damaging external work $\mathcal{D}(t)$, external source of heat $Q(t)$, and stored energy $\mathcal{S}(t)$. In particular, when no instantaneous damage work is applied at the final time t ($d(t) = 0$), it results $\mathcal{S}(t) = 0$ and, consequently, the work which has been provided is exactly the sum of the damaging work $\mathcal{D}$ and of the heat sources Q resulting from the dissipative phenomena.

References

1. E. Bonetti and M. Frémond, 2002, Some notes on the damage theory: microscopic effects of macroscopic movements, submitted.
2. E. Bonetti and G. Schimperna, 2003, Local existence for Frémond's model for elastic materials, Contin. Mech. Thermodyn., to appear
3. M. Frémond, 2001, Non-smooth Thermomechanics, Springer-Verlag, Heidelberg.
4. M. Frémond, 2002, Damage theory. A macroscopic motion vanishes but its effects remain, *Comp. Appl. Math.*, 21: 1-14.
5. M. Frémond, K.L. Kuttler, B. Nedjar, and M. Shillor, 1998, One dimensional models of damage, *Adv. Math. Sci. Appl.*, 8:541–570.
6. M. Frémond, K.L. Kuttler, and M. Shillor, 1999, Existence and uniqueness of solutions for a dynamic one-dimensional damage model, *J. Math. Anal. Appl.*, 229:271–294.
7. M. Frémond and B. Nedjar, 1996, Damage, gradient of damage, and principle of virtual power, *Internat. J. Solids Structures*, 33:1083-1103.
8. P. Germain, 1973, Mécanique des milieux continus, Masson, Paris.
9. J.L. Lions, 1969, Quelques Méthodes de Résolution des Problèmes aux Limites non Linéaires, Dunod, Gauthier-Villars Paris.
10. J. L. Lions and E. Magenes, 1972, *Non-homogeneous Boundary Value Problems and Applications*, Vol. I, Springer-Verlag, Berlin.
11. J. J. Moreau, 1966, Fonctionnelles convexes, Séminaire sur les équations aux dérivées partielles, Collége de France, Paris.

A Numerical Method for Fracture of Rods

Maurizio Angelillo, Enrico Babilio, Antonio Fortunato

Dipartimento di Ingegneria Civile
Università degli Studi di Salerno
I-84084 Fisciano (SA), Italy

Abstract. We consider fracture problems for one-dimensional bodies, such as rods, through energy minimization. For an elastic-brittle body $\mathcal{B}$ occupying in the reference configuration the domain Ω, the energy depends on both some unknown closed crack set K and a displacement field u, smooth on $\Omega \setminus K$. On $\Omega \setminus K$ a bulk energy density and on K an interface energy density are defined. As a result of this assumption the total potential energy is a nonconvex functional. We propose a new numerical method for the search of minimal states of rods by minimizing the energy both on K and u. To seek a minimizer $(\tilde{K}; \tilde{u})$ a nonlinear gradient iterative procedure is employed. During the process of minimization both the displacement field u and the set K evolve toward local minima. To illustrate our approach, some examples are reported.

1 Introduction

A new numerical method for simulating brittle fracture of one-dimensional bodies through energy minimization is introduced. The stimulus for the present work came from some recent unpublished notes of Del Piero [13] in which he considers a material characterized by an interface energy defined on some *a priori* unknown set K plus a bulk elastic energy defined outside K, and seeks the equilibrium states of the solid by minimizing the potential energy in convenient function sets.

The study of variational problems where both volume and surface energy densities are considered, initiated in [12] by De Giorgi, who conied the name of "free discontinuity problems", is rather challenging both from the analytical and the numerical point of view.

Indeed, for $n-$dimensional problems, the energy depends on both a closed $(n-1)-$dimensional set K and a displacement field $\mathbf{u}$, smooth over the complement of K relative to some $n-$dimensional domain Ω. Notice that K is not known *a priori* and its regularity properties are difficult to establish. A lucid mathematical exposition of this kind of problems can be found in the recent book "Functions of bounded variation and free discontinuity problems" by Ambrosio, Fusco and Pallara [6]. The beauty of De Giorgi's theory is that it gives (based essentially on the results of De Giorgi and Ambrosio [12] and Ambrosio [1]) precise information on the qualitative shape that bulk and surface energies must have in order to ensure K is sufficiently regular to be approximated using sets composed by regular arcs, such as edge elements.

From a computational point of view it is essential to know that the K which minimizes the energy, if indeed it exists, is closed, one dimensional and sufficiently regular to be "segmented".

The answer to this question is essentially linked to the proved existence of a solution to the minimum problem within proper function sets. A rough account of this mathematical question is given in Section 2.

From a theoretical point of view, the problem has been solved under very general assumptions, but in any case, explicit solutions can only be computed in very few cases, even for primitive forms of energies. Furthermore, in the case of fracture, no account has been given, up to now, for unilateral constraints on crack lips. Therefore numerical solutions are not only interesting from a practical viewpoint, but also provide useful insight into the possibility of finding a solution.

It should also be pointed out that if one chooses to approximate the problem through standard finite element codes with gap elements, even if the interface energy on gap elements has the required properties, fixed discretizations will not work. Indeed the skeleton of a fixed discretization cannot Γ–converge to arbitrary fracture patterns K. The use of adaptive or "variable" discretizations is then compulsory.

Here a finite element variable discretization is employed. To perform the minimization the nonlinear Polak–Ribière gradient method is applied.

The potential energy is the sum of a quadratic term (elastic energy) and an interface Griffith-like energy. Guided by the regularity and density result, the energy densities we adopt are endowed with all the ingredients that seem to be necessary to get convergence in SBV with the known theorems [6]. In particular the model interface energy we consider:

is subadditive with respect to the jump [u];

its slope at [u] = 0 is infinite.

Section 2 is devoted to a brief survey about known theorems in SBV. Then, the numerical method for one-dimensional problems is presented in Sect. 3.

2   Preliminaries on known mathematical results on free discontinuity problems

Many mathematical models arising from Liquid Crystals Theory, Image Segmentation, Computer Vision, Folding of Thin Shells (see the book [6] and references therein), are based on the minimization of two competing energy forms, these energies being of volume and surface type. In two dimensional fracture mechanics (see for example the models proposed by Francfort and Marigo in [16] and Dal Maso and Toader in [11]) this kind of variational formulation leads to the minimization (under displacement boundary conditions) of a functional of the form

$$F(K;\mathbf{u}) = \int_{\Omega\setminus K} \varphi(\nabla\mathbf{u})\mathrm{d}\mathbf{x} + \int_{K} \vartheta([\mathbf{u}])\mathrm{d}\mathcal{H}^{n-1}, \tag{1}$$

$\mathcal{H}^{n-1}$ being the $(n-1)$-dimensional Hausdorff measure, Ω the n-dimensional region occupied by the body, K a closed $(n-1)$-dimensional set contained in the closure of Ω, φ and ϑ the elastic and the interface energies, $\mathbf{u}$ the displacement vector field, $[\mathbf{u}] = \mathbf{u}^+ - \mathbf{u}^-$ the jump of $\mathbf{u}$ across K. Notice that K is not known *a priori* and no known topology of the closed subsets K of Ω ensures at the same time compactness for minimizing sequences and lower semicontinuity of the energy, even in the simple case $\vartheta = 1$.

Compactness and lower semicontinuity are indeed crucial to show existence via the direct method of the calculus of variations, and to get them it is necessary to give a weak formulation to the problem.

The idea of De Giorgi is to interpret K as the set of discontinuous points of $\mathbf{u}$, this justifies the name "free discontinuity problems" for the minimization of functionals of the form (1), since one looks for minimizing functions whose discontinuities are not known in advance.

The most natural choice for the work space is $BV(\Omega)$, the set of functions of bounded variation, i.e. the set of functions $\mathbf{u}$ whose gradient is a finite measure.

In doing this two difficulties arise in fracture problems. The first one is that the bulk energy depends on the symmetric part of $\nabla\mathbf{u}$ only. The second, more profound, is that the distributional derivative $\nabla\mathbf{u}$ of $\mathbf{u} \in BV$ consists of three parts: the absolutely continuous part $\nabla^a\mathbf{u}$, the jump part $\nabla^j\mathbf{u}$, and the Cantor part $\nabla^c\mathbf{u}$. $\nabla^a\mathbf{u}$ is a "volume" term but only $\nabla^j\mathbf{u}$ has the structure of a surface term.

In other words the support of the Cantor part can be geometrically nasty and has not the desired properties to represent an interface energy of fracture type. Putting aside the first difficulty, caused by the dependence of φ on deformation rather than on $\nabla\mathbf{u}$ (of minor concern from the numerical point of view), the second difficulty is overcome by De Giorgi and Ambrosio (see[12]) by setting the problem in $SBV(\Omega)$, that is the subset of the functions $\mathbf{u} \in BV(\Omega)$ such that $\nabla^c\mathbf{u} = 0$.

Functions $\mathbf{u} \in SBV(\Omega)$ seem to have the desired properties to model fracture, through minimization of the functional

$$\mathcal{E}(\mathbf{u}) = F(\mathbf{u}, S_{\mathbf{u}}), \tag{2}$$

$S_{\mathbf{u}}$ being the set of discontinuous points of $\mathbf{u}$. The variational problem is then formulated as follows:

"find $\mathbf{u}^\circ \in SBV(\Omega)$ such that

$$\mathcal{E}(\mathbf{u}^\circ) = \min_{\substack{\mathbf{u}\in SBV(\Omega)\\ \mathbf{u}=\bar{\mathbf{u}} \text{ on } \partial\Omega_{\mathrm{D}}}} \mathcal{E}(\mathbf{u}) \qquad (3)$$

".

To obtain existence in $SBV(\Omega)$ for this variational problem it is essential to have lower semicontinuity of $\mathcal{E}(\mathbf{u})$ and closure and compactness properties for the minimizing sequences.

Compactness is obtained by Ambrosio in [1] by giving sufficient conditions on weakly convergent sequences in $SBV(\Omega)$ from which suitable uniform bounds on the approximate gradient and on the jump are derived, that prevent the appearance of a Cantor part in the limit.

In particular in order to use the Ambrosio's theorems on closure and compactness, the bulk energy φ must have suitable growth properties and the interface energy ϑ must satisfy the condition

$$\lim_{\|[\mathbf{u}]\|\to 0} \frac{\vartheta\left([\mathbf{u}]\right)}{\|[\mathbf{u}]\|} = +\infty \ , \qquad (4)$$

that is infinite stresses are needed to open up a crack. Examples of 1D interface energies that have such properties are depicted in Fig. 1. In order for

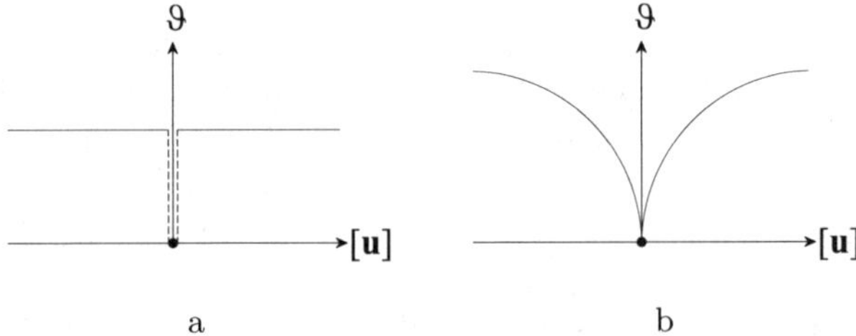

Fig. 1. Typical interface energies with the property (4): a) Griffith , b) Barenblatt.

$\mathcal{E}(\mathbf{u})$ to be lower semicontinuous it is sufficient to assume that φ be convex and ϑ be subadditive functions of their arguments. Subadditivity of ϑ means that:

$$\vartheta(\alpha + \beta) \leq \vartheta(\alpha) + \vartheta(\beta) \ , \qquad (5)$$

therefore in dimension one, if ϑ is subadditive, we can say that it is more convenient for the material, on energy grounds, to get a fracture of a certain size with just a single crack rather than with a large number of cracks. Notice that if ϑ is concave then is subadditive and that both Griffith and Barenblatt energies depicted in Fig. 1 are endowed with all the properties sufficient to ensure compactness and semicontinuity of $\mathcal{E}(\mathbf{u})$.

3   The one-dimensional problem

Consider a one-dimensional body $\mathcal{B}$ occupying in the original reference configuration the interval $\Omega = \{x \in \mathbb{R} \mid 0 \le x \le L\}$, let

$$y = x + \mathrm{u}(x) \, , \tag{6}$$

be its deformation and

$$\varepsilon(x) = \mathrm{u}'(x) \, , \tag{7}$$

the local strain.

We admit that the displacement u can be discontinuous and denote $[\mathrm{u}] = \mathrm{u}^+ - \mathrm{u}^-$ the jump of u. The jump set of u, K, is assumed to belong to an admissible set of cracks $\mathbb{K}$

$$\mathbb{K} = \left\{ K \subset \bar{\Omega}, \ \#(K) < \infty \right\} , \tag{8}$$

that is K is the collection of a finite number of isolated points. We assume u to be regular outside K, say

$$\mathrm{u} \in H^1(\Omega \setminus K) \, . \tag{9}$$

The material is assumed to be homogeneously elastic, that is characterized by a convex elastic energy density $\varphi(\varepsilon(x))$, defined over $\Omega \setminus K$. For $|\varepsilon|$ sufficiently small (small deformations) the energy density φ is well approximated by the quadratic function

$$\varphi = \frac{1}{2}\beta\varepsilon^2 \, , \tag{10}$$

where β is the elastic modulus.

Moreover, the material is assumed to be brittle in the sense of Griffith, that is characterized by an interface energy density $\vartheta([\mathrm{u}](x), x)$, defined over K, of the form

$$\vartheta([\mathrm{u}], x) = \begin{cases} 0, & [\mathrm{u}] = 0 \, , \\ \gamma_x, & [\mathrm{u}] > 0 \, , \\ +\infty, & [\mathrm{u}] < 0 \, . \end{cases} \tag{11}$$

where γ_x represents the fracture thoughness of the material. The dependence of γ_x on x is considered to allow for localized defects.

Notice that with the definition of interface energy density (11), the fracture set K represents only potential points of crack, actual cracks arise in the subset of K where the interface energy is activated ($[\mathrm{u}] > 0$) . We consider that displacements are given at the ends, that is

$$\mathrm{u}(0) = 0 \, , \qquad \mathrm{u}(L) = \bar{\mathrm{u}} \, . \tag{12}$$

A way to impose the boundary conditions (12) for discontinuous u is

$$\int_{\Omega \setminus K} \mathrm{u}'(x)dx + \sum_{x \in K} [\mathrm{u}](x) = \bar{\mathrm{u}} \, . \tag{13}$$

3.1   Stress and potential energy

We restrict to quasi-static processes, that is assume the final state of displacement, stress, strain and fracture is reached through a sequence of equilibrium states continuous in the time interval, except possibly at some points of jump or bifurcation. The equilibrium states are searched through minimization of the potential energy

$$F(K;\mathbf{u}) = \int_{\Omega \setminus K} \varphi(\mathbf{u}'(x))dx + \sum_{x \in K} \vartheta([\mathbf{u}](x), x) \; , \tag{14}$$

over the set

$$\mathbb{K} \times H^1(\Omega \setminus K) \; , \tag{15}$$

under constraint (13). On introducing the notation $a = [u]$, the energy of a homogeneous bar Ω for a single "potential" crack at $x^\circ \in (0, L)$ is represented in Fig. 2, where the notation $a = [u]$ is introduced. Notice that the system,

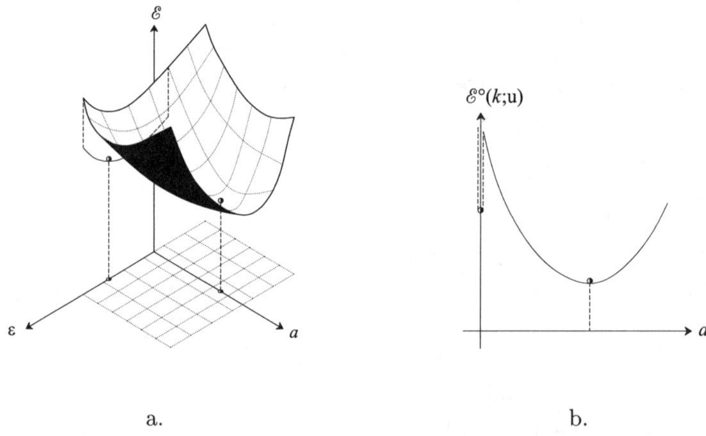

Fig. 2. Energy pocket. Three-dimensional (a) and lateral (b) views of the energy.

in order to get out from the elastic pocket and reach the absolute minimum, must sormount an energy barrier. This steep barrier arises due to the infinite cohesion assumption (4).

Indeed the stress σ inside Ω can be obtained either in the bulk part as

$$\sigma = \frac{\partial \varphi}{\partial \varepsilon} \; , \tag{16}$$

or at the interfaces

$$\sigma = \frac{\partial \vartheta}{\partial a} \; . \tag{17}$$

On adopting the elastic energy (10) in $\Omega \setminus K$ and the Griffith energy (11) in K we have

$$\sigma = \beta\varepsilon \text{ in } \Omega \setminus K \, , \tag{18}$$

$$\sigma = \begin{cases} \sigma^* \in \mathbb{R}, & \text{if } a = 0 \, , \\ 0 \, , & \text{if } a > 0 \, , \end{cases} \tag{19}$$

Action–reaction at interfaces K gives

$$\left. \frac{\partial\varphi}{\partial\varepsilon} \right|_{K+} = \left. \frac{\partial\varphi}{\partial a} \right|_{K} = \left. \frac{\partial\varphi}{\partial\varepsilon} \right|_{K-} \tag{20}$$

where $\frac{\partial\varphi}{\partial\varepsilon}\big|_{K+}$ and $\frac{\partial\varphi}{\partial\varepsilon}\big|_{K-}$ represent the extensions for continuity of $\frac{\partial\varphi}{\partial\varepsilon}$ on the positive and negative sides of the interface.

The material is characterized by an infinite cohesion and infinite stress is needed to create a crack.

We see from (19) that where the crack forms inside Ω the stress must be zero and, therefore, in equilibrium, between two successive active cracks $x^1, x^2 \in K$ the strain must be zero. In other words, for equilibrium, in absence of body forces, if cracks form, the bar breaks into undeformed pieces. This is a peculiarity of one-dimensional problems.

3.2   Discretization

To discretize the problem we split the interval Ω into M equally spaced subintervals $\widetilde{\Omega}^m$ (see Fig. 3). This is the initial mesh: $\widetilde{\Omega}^m$ are elastic elements and cracks can form at interfaces among elements

$$K = \left\{ x(n) = n\frac{L}{M}, \quad n \in \{1, 2, \ldots, M - 1\} \right\} . \tag{21}$$

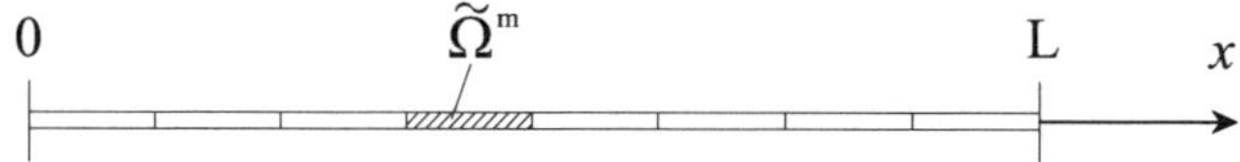

Fig. 3. Discretization of the domain Ω.

To duplicate the interface nodes $x(n)$ and better visualize fracture elements, we generate a new mesh by introducing a fictitious dimension t and assuming

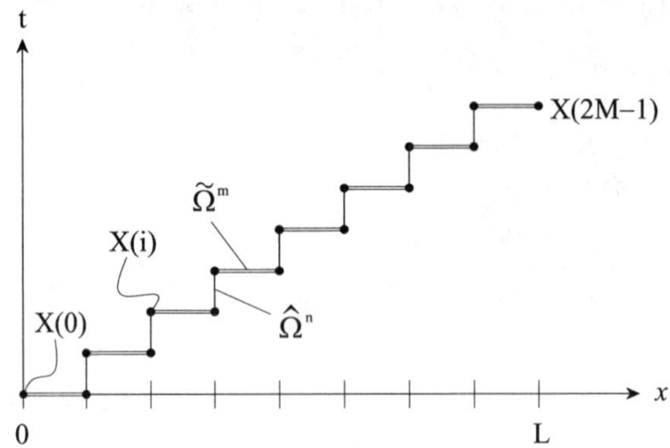

Fig. 4. New discretization of the domain with fictitious interface elements.

that each element $\widetilde{\Omega}^{m}$ occupies a different "t" level (see Fig. 4). In this way we create new fictitious elements $\widehat{\Omega}^{n}$, $n \in \{1, 2, \ldots, M - 1\}$. Each node $X(j) = (x(j), t(j))$, $j \in \{0, 1, \ldots, 2M - 1\}$, of the new mesh has a displacement degree of freedom u(j). Consistent linear shape functions inside each element Ω^{i}, $i \in \{1, 2, \ldots, 2M - 1\}$, are considered to represent the displacement field.

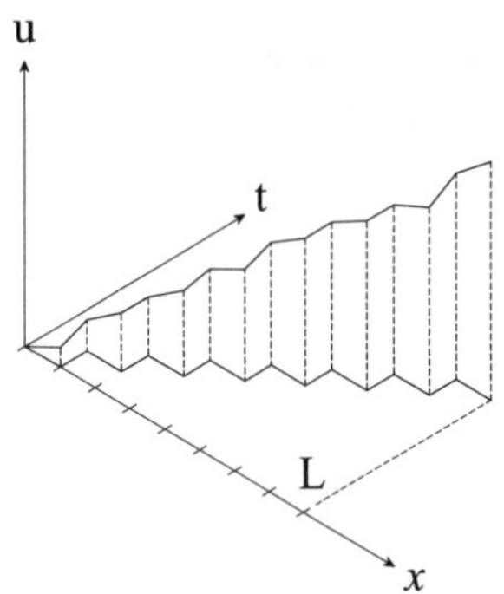

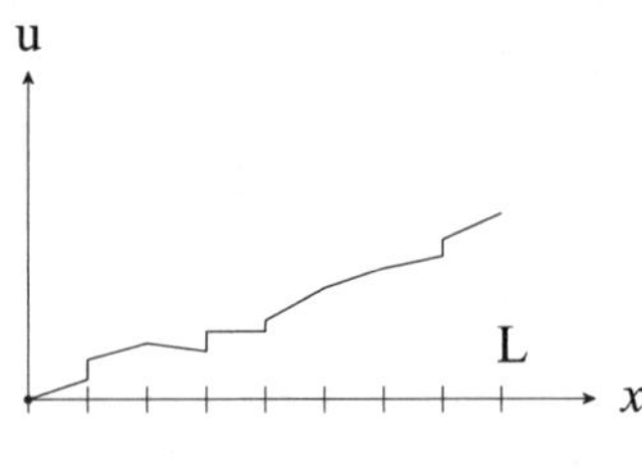

a.   Displacement in 3D space.

b.   Projection of S in the plane t=0: displacement plot.

Fig. 5. Displacement of B.

An admissible displacement of the bar is then described by a continuous one-dimensional piecewise linear point set S embedded in a 3D space (see Fig. 5.a).

The $\widetilde{\Omega}^{m}$ elements carry the elastic energy $\varphi(\mathrm{u}')$, the $\widehat{\Omega}^{n}$ elements carry the fracture energy $\vartheta(a)$.

In particular, calling $x(i,0), x(i,1), \mathrm{u}(i,0), \mathrm{u}(i,1)$ the end positions and the displacements of the generic oriented interval element Ω^i, we take for the energy densities the expressions

$$\varphi(i) = \frac{1}{2}\,\beta\,\left(\frac{\mathrm{u}(i,1) - \mathrm{u}(i,0)}{x(i,1) - x(i,0)}\right)^2\,,\tag{22}$$

$$\vartheta(i) = \,\vartheta\,(\mathrm{u}(i,1) - \mathrm{u}(i,0))\,,\tag{23}$$

and apply the elastic energy (22) on the $\widetilde{\Omega}^m$ elements and the interface energy (23) on the $\widehat{\Omega}^n$ elements.

3.3   Descent minimization

To perform the minimization we employ the program of descent minimization *Surface Evolver* [9]. We quote from the introduction of the Surface Evolver manual:

> The Surface Evolver is a interactive program for the study of surfaces shaped by surface tension and other energies. A surface is implemented as a simplicial complex, that is a union of triangles. The user defines an initial surface in a datafile. The Evolver evolves the surfaces toward minimum energy by a gradient descent method. The energy in the Evolver can be a combination of several energies.

In our one-dimensional problem, the simplicial S is represented by a continuous piecewise linear curve (polyline) embedded in a 3D space. The shape of the polyline S is determined by $2M$ nodal positions $Y(j)$ in the 3D space, defined by the three coordinates $\{x, \mathrm{u}, t\}$ (see Fig. 5.a)

$$Y(j) = \{x(j), \mathrm{u}(j), t(j)\}\,,\quad j \in \{0, 1, \ldots, 2M - 1\}\,.\tag{24}$$

The potential energy is assumed to be the sum of the bulk elastic energy (22) defined on the intervals $\widetilde{\Omega}^m$ plus the interface energy (23) defined on the intervals $\widehat{\Omega}^n$ and can be explicitly expressed in terms of the nodes $Y(j)$.

The search for the minimum proceeds from an initial polyline S° and evolves it through the descent gradient algorithm. The minimization is constrained by the conditions

$$x(j) = x(j+1),\quad j \in \{1, 3, \ldots, 2M - 3\}\,,\tag{25}$$

$$t(j) = t^\circ(j),\qquad j \in \{0, 1, \ldots, 2M - 1\}\,.\tag{26}$$

Actually the energy is independent of t and the constraints (26) play no role.

3.4   Approximate interface energy

The numerical handling of the function ϑ defined in (11) and plotted in Fig. 6.a is rather awkward, therefore we approximate the exact Griffith energy ϑ with the smooth function

$$\bar{\vartheta}(a) = \gamma(c_d + \mathrm{f}_{d,l}(a) + \mathrm{g}_{d,l}(a)) \tag{27}$$

where c_d is a constant, $\mathrm{f}_{d,l}(a)$ and $\mathrm{g}_{d,l}(a)$ are shape functions that approximate the barrier at $a = 0^-$ and the step function at $a = 0$. The form used for $\mathrm{f}_{d,l}(a)$ and $\mathrm{g}_{d,l}(a)$ and the effect of the parameters d and l are described in the Appendix A.

A plot of $\bar{\vartheta}$ is reported in Fig. 6.b. As depicted in Fig. 6.b the unilateral

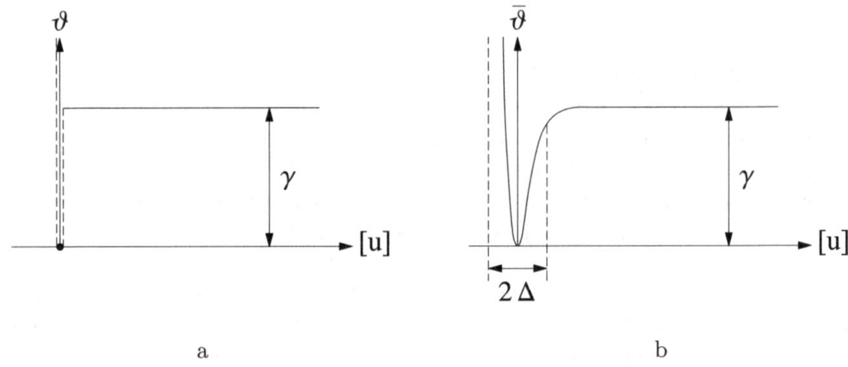

Fig. 6. Exact (a) and approximate (b) interface energy.

constraint on fracture is essentially imposed through a penalty term. The scaling of the parameters d, l is obviously rather delicate and *ad hoc* tuning is required.

We already observed that with our definition of interface energy density ϑ, the fracture pattern K represents only potential lines of fracture, actual cracks arise in the subset of K where the interface energy is activated. In this approximate context active cracks are those for which $a \in [\Delta, +\infty)$, 2Δ being the band width, shown in Fig. 6.b. The band width depends on the parameters d and l controlling the sharpness of the well and the steepness of the barrier. For convergence the band width must be mesh–dependent, in particular $\Delta \to 0$ for $h \to 0$, h being the mesh parameter. Therefore the parameters d and l are mesh–dependent too (see Appendix A).

Notice that, since the original mesh is variable, adjusting d and l during the minimization process could be necessary.

3.5   Numerical example

Three benchmark examples are presented here. We consider a bar of unit length L with a neck. The bar is discretized into $2M-1$ Ω^i elements, $M = 10$. The boundary conditions

$$u(1,0) = 0 \;, \quad u(2M-1,1) = \bar{u} \;, \tag{28}$$

are also considered. We attack the case in which the neck is located on a skeleton point, first (examples (a) and (b)). Then, in the example (c), the neck position is no more on a skeleton point. Notice that in such a case a standard fixed discretization does not work. Indeed the neck location can be found through an adaptation of discretization.

Notice that, in our discretization, the element Ω^i $i = \{1, 2, \ldots, 2M-1\}$ is labeled in a such a way that $\Omega^i \equiv \widetilde{\Omega}^{(2i-1)}$ if i is odd, $\Omega^i \equiv \widehat{\Omega}^{(i/2)}$ otherwise.

In the actual elements $\widetilde{\Omega}^m$ we put for the Young modulus $k = 1$. For the fictitious elements $\widetilde{\Omega}^n$ we put $d = 1.5$ and $l = 10^5$.

In examples (a) and (b) the thoughness parameter of $\widehat{\Omega}^5$ located at $0.5\,L$ is set to a value of $\gamma = 0.01$ to simulate the presence of the neck (actually, in such a way, one can consider the neck just as a defect).

In the example (a) the boundary conditions 28 are fixed and additional initial displacement conditions

$$u(i+1,0) = u(i,1) = 0 \;, \quad i = \{1, 3, 5, \ldots, 2M-3\} \;. \tag{29}$$

are set on the internal nodes.

In Fig. 7 the initial conditions and the results of the calculation are shown.

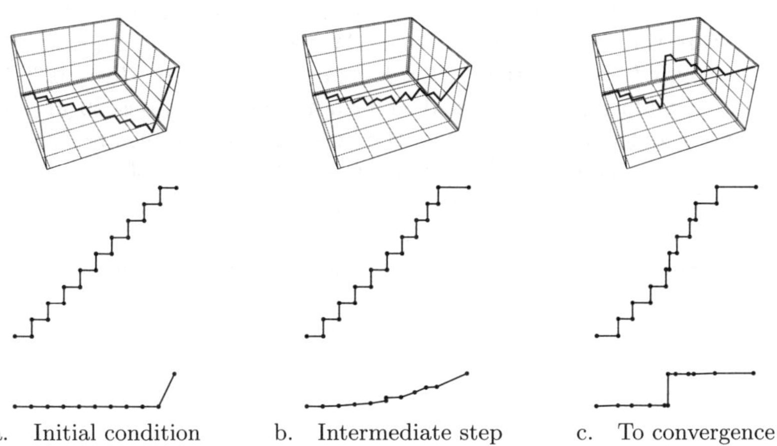

a. Initial condition	b. Intermediate step	c. To convergence

Fig. 7. Plot of polyline in the 3D space $\{x, u, t\}$, mesh variation in (x, t)-plane and diplacement plot, for a displacement applied on the right end of the rod.

In the example (b) the boundary conditions 28 and initial nodal displacements

$$u(i,0) = \frac{\bar{u}}{L} X(i-1)\,, \quad u(i,1) = \frac{\bar{u}}{L} X(i)\,, \qquad i = \{1,3,5,\dots,2M-3\}\,,(30)$$

are also considered.

In Fig.8 the initial conditions and the results of the calculation are shown.

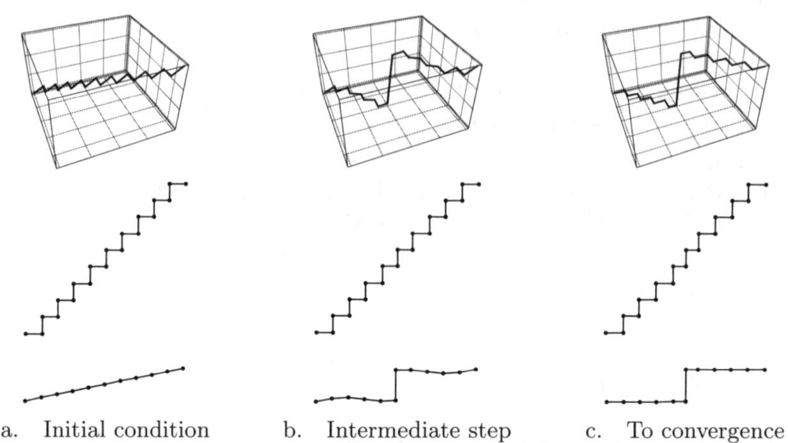

a.   Initial condition   b.   Intermediate step   c.   To convergence

Fig. 8. Plot of polyline in the 3D space $\{x, u, t\}$, mesh variation in (x,t)-plane and diplacement plot, for a linear initial displacement.

In the example (c), the neck is located at 0.36 L from the left end in between the third and the fourth interface(see Fig. 9). Because of the neck, the fracture threshold γ depends on x. We used for γ_x the expression $\gamma_x = \gamma^\circ(1 - k_1 \exp^{-k_2(k_3+x)})$. Here $\gamma^\circ = 1$, $k_1 = 0.5$, $k_2 = 10^4$, $k_3 = 0.36$. The boundary conditions 28 are fixed and additional conditions 29 are set. In Fig. 9 the geometry of the rod and the boundary conditions are shown.

In Figs. 10, 11 the initial condition and the results of the calculation are shown. Optimization of energy through descent minimization locate the interface $i = 4$ at the neck position, the bar breaks at $x = 0.36\,L$ and split into two undeformed pieces. Finally, in Fig. 12 the evolution of the minimization is reported.

A   Appendix

On introducing the constant

$$r_d = \sqrt{2d - 1} + 2d \arccot\sqrt{2d - 1}\,, \tag{31}$$

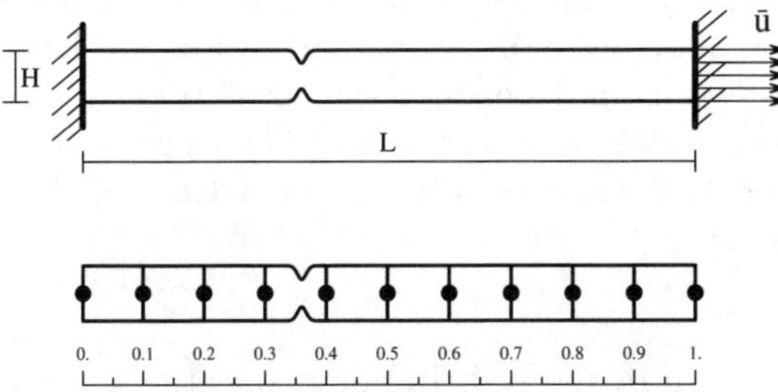

Fig. 9. Geometry of the bar with location of the neck, boundary coditions. Initial discretization.

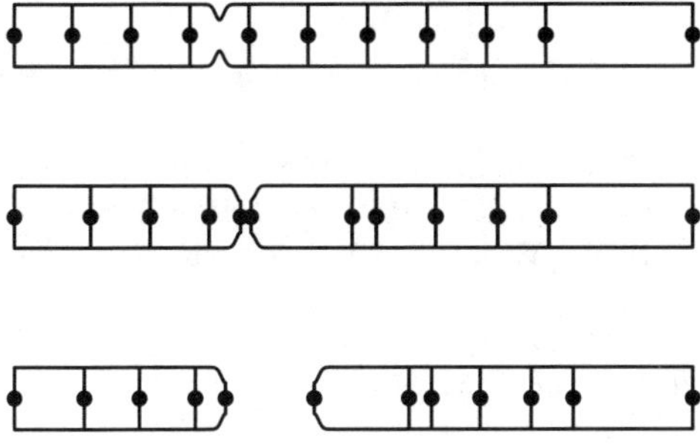

Fig. 10. Evolution of mesh and deformation through descent minimization from initial state to convergence.

the constant c_d and the shape functions in (27) are

$$c_d = \frac{r_d}{r_d - d\pi} \, , \tag{32}$$

$$f_{d,l}(a) = -e^{-la}\sqrt{2d-1}\, r_d\, c_d^{-1} \, , \tag{33}$$

$$g_{d,l}(a) = -2d \arccot(-e^{-la}\sqrt{2d-1})\, r_d\, c_d^{-1} \, , \tag{34}$$

The parameters d and l control the steepness of the barrier at $a < 0$ and the sharpness of the well at $a = 0^+$. Increasing l and decreasing d the steepness of barrier and the sharpness of the well increase. Notice that we assume $d > 1/2$ and $l > 0$.

By choosing to define active cracks those for which the fracture energy reaches 99% of γ, the half band width Δ is determined by the value of the abscissa for which

$$\frac{\bar{\vartheta}(a)}{\gamma} = 0.99 \, . \tag{35}$$

The band width 2Δ, shown in Fig. 6, depends on the parameters l and d. The plot l–Δ, for several values of d, is reported in Fig. 13. For convergence the band width must be mesh-dependent and, in particular, $\Delta \to 0$ for $h \to 0$, h being the length of the smallest element. Therefore the parameters l and d are mesh-dependent. We assume a linear dependence of Δ on h

$$\Delta \leq \frac{h}{L} \tag{36}$$

where L is the length of the rod. In our applications, we set $d = 1.5.$ and $l = 1.0 \ 10^5$.

Plots of $f_{d,l}(a)$ and $g_{d,l}(a)$, for particular values of the parameters, are depicted in Fig. 14.

References

1. Ambrosio, L. (1989) Compactness theorem for a special class of functions of bounded variation. *Boll. Un. Mat. Ital.***3**–B, 857–881.
2. Ambrosio, L. (1989) Variational problems in *SBV* and image segmentation. *Acta Appl. Math.* **17**, 1–40.
3. Ambrosio, L. (1990) Existence theory for a new class of variational problems. *Arch. Rat. Mech. Anal.* **111**, 291–322.
4. Ambrosio, L. (1995) A new proof of the *SBV* compactness theorem. *Calc. Var.* **3**, 127–137.
5. Ambrosio, L., Braides, A. (1997) Energies in *SBV* and variational models in fracture mechanics. *Homogenization and Applications to Material Sciences (D.*

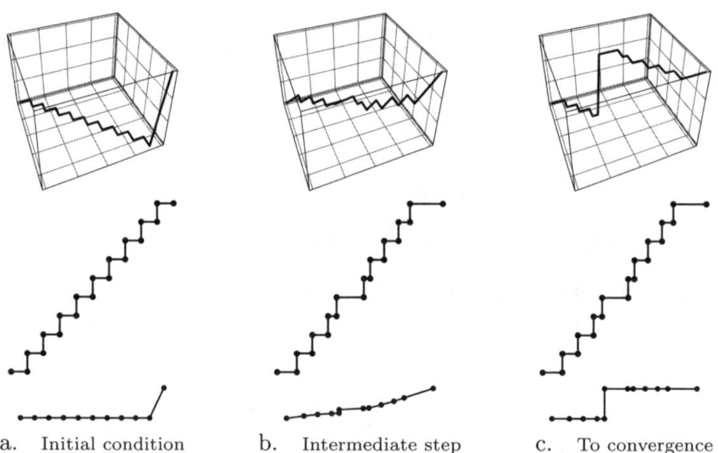

Fig. 11. Plot of the polyline in the 3D space $\{x, u, t\}$, mesh variation in (x, t)-plane and diplacement plot, for the given relative displacement considered.

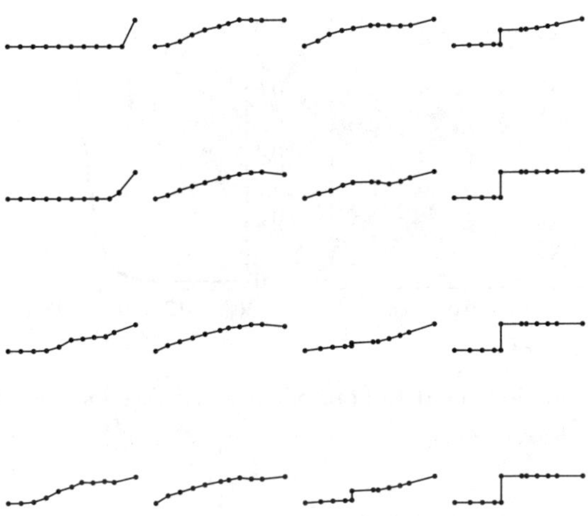

Fig. 12. Evolution of deformation through descent minimization from initial state to convergence.

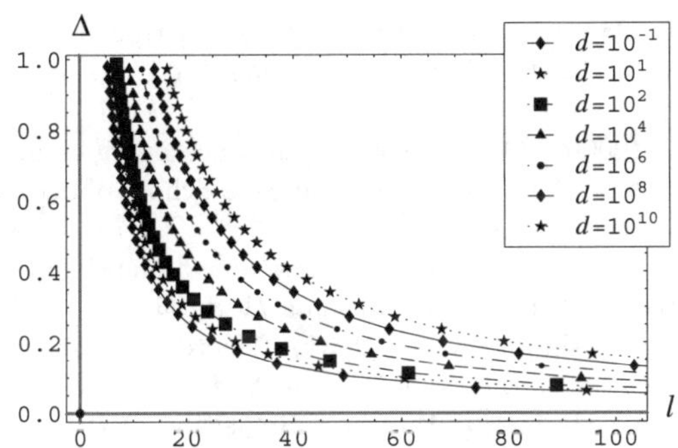

Fig. 13. Plots l–Δ for several values of d.

Cioranescu, A. Damlamian, P. Donato eds.), *GAKUTO, Gakkōtosho, Tokio, Japan*, 1–22.

6. Ambrosio, L., Fusco, N., Pallara, D. (2000) Functions of bounded variation and free discontinuity problems. *Clarendon Press*, Oxford.

7. Bourdin, B. (2000) Francfort, G. A., Marigo, J. J., Numerical experiments in revisited brittle fracture. *J. Mech. Phys. Solids*, **48** (4), 797–826.

8. Braides, A. (1998) *Approximation of Free–Discontinuity Problems.* Lecture Notes in Mathematics, Vol. **1694**, Springer, Berlin.

9. Brakke, K. A. (1999) *Surface Evolver Manual.* Mathematics Department, Susquehanna University, Selinsgrove.
Available at `http://www.susqu.edu/facstaff/b/brakke/evolver`.

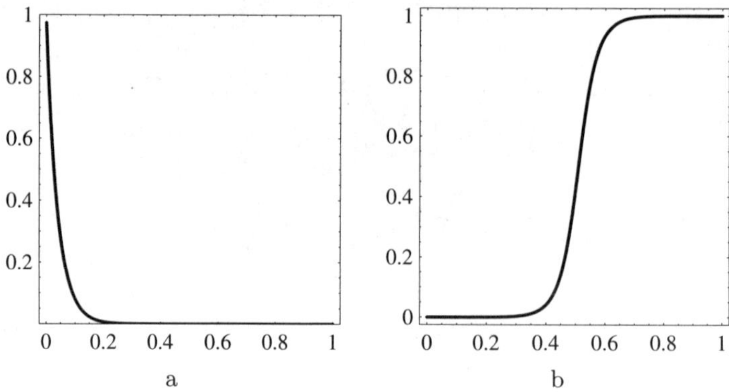

Fig. 14. Shape functions used to approximate interface energies $\bar{\vartheta}(a)$: a) plot of $f_{d,l}(a)$, b) plot of $g_{d,l}(a)$.

10. Dal Maso, G., Toader, R. (2002) A model for the quasi-static growth of brittle fractures: existence and approximation results. *Arch. Rational Mech. Anal.* **162**, 101-135.

11. Dal Maso, G., Toader, R. (2002) A model for the quasi–static growth of brittle fractures based on local minimization. *Math. Models Methods Appl. Sci.*, 12 (**12**), 1773–1799.

12. De Giorgi, E., Ambrosio, L. (1988) Un nuovo tipo di funzionale del calcolo delle variazioni. *Atti Accad. Naz. Lincei Rend. Cl. Sci. Fis. Mat. Nat. s. 8* **82**, 199–210.

13. Del Piero, G. (2000) On the role of interface energies in the description of material behaviour. Lecture notes, Graduate School of Structural Engineering, *Dip. Ing. Civ.*, Università di Salerno.

14. Del Piero, G., Truskinovsky, L. (1998) A one–dimensional model for localized and distributed fracture. *J. de Physique IV* **8**, 95–102.

15. Del Piero, G., Truskinovsky, L. (2001) Macro– and micro–cracking in one–dimensional elasticity. *Int. J. Solids and Structures*, **38**, 1135–1148.

16. Francfort, G. A., Marigo, J. J. (1998) Revisiting brittle fracture as an energy minimization problem. *J. Mech. Phys. Solids*, **46**(8), 1319–1342.

17. Negri, M. (2003) A finite element approximation of the Griffith's model in fracture mechanics. *Numerische Mathematik*, **95**, 653–687.

18. Negri, M. (2001) Numerical methods for free–discontinuity problems based on approximations by Γ–convergence. PhD Theses.

Softening Behavior of Reinforced Cementitious Beams

Sonia Marfia, Elio Sacco

Dipartimento di Meccanica Strutture A. & T.
Università di Cassino
03043 - Cassino, Italy

Abstract. In the present paper, a one-dimensional elastoplastic-damage model for the analysis of the mechanical response of beams constituted by cementitious materials, eventually reinforced by classical steel bars and/or advanced composite laminates, is developed. The proposed model is thermodynamically consistent and it takes into account the different behavior in tension and in compression of the cohesive materials. The governing equations are derived and a numerical procedure is proposed. Numerical applications are developed in order to analyze the axial and the bending response of reinforced concrete and masonry beams and to assess the efficiency of the proposed procedure. Comparisons with analytical solutions are reported.

1   Introduction

Cementitious materials are characterized by softening response with different strength in compression and in tension. In particular, experimental results show that they present brittle behavior in traction and inelastic deformations accompanied by damage effects in compression.

The mechanical modeling of cementitious materials can be based on the use of the concepts of the fracture mechanics, or of the continuum damage mechanics neglecting or considering the plasticity effects. In the framework of continuum damage mechanics, different models, which account not only for the damage effects but also for the inelastic, i.e. plastic, deformations have been proposed in the literature (Dragon and Mróz 1979, Lotfi and Shing 1991, Carol and Bažant 1997, de Borst et al., 1999, Addessi et al., 2001).

Aim of the paper is the definition of a simple and effective structural model in order to investigate the behavior of cementitious (concrete or masonry) beams, eventually reinforced by classical steel bars, or/and by advanced composite laminates glued on the top or on the bottom of the beams.

A one-dimensional thermodynamically consistent model for cementitious materials, which takes into account the damage and the plasticity effects is proposed. It is assumed that the plasticity can be activated only when the material is subjected to a compressive stress; on the contrary, the material behavior in tension is governed only by damage. The constitutive equations as well as the damage and plasticity evolutive equations are written in explicit form. The damage and plastic evolutions result coupled since the damage is governed by the elastic strain.

Because of the damage in tension and of the damage and plasticity in compression, highly nonlinear equilibrium equations are deduced. In order to solve the nonlinear problem, a numerical procedure is developed. It is based on the arc-length method, with a proper choice of the control parameters. Then, the numerical procedure is implemented in a computer code.

Some applications are presented. The axial and the bending response of the reinforced concrete and masonry beams is investigated. Moreover, several loading histories are considered. Some comparisons between analytical and numerical solutions are developed in order to assess the efficiency of the procedure.

2   Constitutive model

A thermodynamically consistent one-dimensional constitutive relation is addressed. The free energy is assumed to be:

$$\psi = \frac{1}{2}\left(1 - D\right) E \left(\varepsilon - \varepsilon_p\right)^2 + \eta\, g^+(\xi) + (1 - \eta)\, g^-(\xi) + k(\beta) \tag{1}$$

where: E is the Young modulus of the material; D is the damage parameter satisfying the classical inequalities $0 \leq D \leq 1$, with $D = 0$ for the virgin material and $D = 1$ for the completely damaged material; ε_p is the plastic strain, so that $\varepsilon - \varepsilon_p = \varepsilon_e$ represents the elastic strain; ξ is the internal parameter governing the damage softening; β is the internal parameter governing the plasticity hardening; η is the stepwise function of the elastic strain ε_e, such that $\eta = 1$ if $\varepsilon_e \geq 0$ and $\eta = 0$ if $\varepsilon_e < 0$ and the superscript $^+$ corresponds to the case $\eta = 1$, i.e. $\varepsilon_e \geq 0$, and the superscript $^-$ corresponds to the case $\eta = 0$, i.e. $\varepsilon_e < 0$.

The functions $g^\pm(\xi)$ and $k(\beta)$ are defined as:

$$g^\pm(\xi) = \frac{1}{2}E\frac{\left(\varepsilon_c^\pm\right)^2}{\left(1 + \alpha^\pm\xi - \xi\right)\left(\alpha^\pm - 1\right)} \qquad\qquad k(\beta) = \frac{1}{2}K\beta^2 \tag{2}$$

where $\varepsilon_c^\pm$ represents the starting damage threshold strain and $\alpha^\pm = \varepsilon_c^\pm/\varepsilon_u^\pm$ is the threshold ratio, with $\varepsilon_u^\pm$ the final damage threshold strain. The quantity K is the plastic hardening parameter. The threshold strains $\varepsilon_c^\pm$ and $\varepsilon_u^\pm$ and the plastic hardening quantity K are material parameters.

The state laws result:

$$\sigma = \frac{\partial\psi}{\partial\varepsilon} = (1 - D)\,E\,(\varepsilon - \varepsilon_p) \tag{3}$$

$$Y = -\frac{\partial\psi}{\partial D} = \frac{1}{2}E\,(\varepsilon - \varepsilon_p)^2 = \frac{1}{2}E\varepsilon_e^2$$

$$\zeta = -\frac{\partial\psi}{\partial\xi} = \frac{1}{2}E\frac{\left(\varepsilon_c^\pm\right)^2}{\left(1 + \alpha^\pm\xi - \xi\right)^2}$$

$$\tau = -\frac{\partial \psi}{\partial \varepsilon_p} = (1 - D)\, E\, (\varepsilon - \varepsilon_p)$$

$$\vartheta = -\frac{\partial \psi}{\partial \beta} = -K\beta$$

where σ is the stress, Y is the damage energy release rate, ζ is the thermodynamical force associated to ξ, τ is the thermodynamical force associated to the plastic strain and, indeed, it results $\tau = \sigma$; finally ϑ is the hardening plastic force.

Considering isothermal processes, the Clausius-Duhem inequality can be written in terms of the free energy as:

$$\mathcal{D} = -\dot{\psi} + \sigma\dot{\varepsilon} = Y\dot{D} + \zeta\dot{\xi} + \sigma\dot{\varepsilon}_p + \vartheta\dot{\beta} \geq 0 \tag{4}$$

where $\mathcal{D}$ represents the mechanical dissipation.

The evolutionary equations of the internal state variables $(D, \xi, \varepsilon_p, \beta)$ are deduced introducing a damage yield function f^D and a plastic yield function f^p. In particular, regarding the damage evolutions, it results:

$$f^D(Y, \zeta) = Y - \zeta \leq 0 \qquad \dot{\gamma} \geq 0 \qquad f^D\dot{\gamma} = 0 \tag{5}$$

$$\dot{D} = \frac{\partial f^D}{\partial Y}\dot{\gamma} = \dot{\gamma} \qquad\qquad \dot{\xi} = -\frac{\partial f^D}{\partial \zeta}\dot{\gamma} = \dot{\gamma} \tag{6}$$

so that $\dot{D} = \dot{\xi} = \dot{\gamma}$, i.e. the parameter ξ coincides with the damage internal state variable D.

The evolution of the damage parameter $\dot{D} = \dot{\xi} = \dot{\gamma}$ is deduced by the consistency equations:

$$\dot{D} = \frac{\varepsilon_c^{\pm}}{(1 - \alpha^{\pm})\,\varepsilon_e^2}\dot{\varepsilon}_e \qquad \text{with} \quad f^D = 0 \quad \text{and} \quad \dot{D} > 0 \tag{7}$$

which gives the damage rate as function of the elastic strain rate.

By integrating equation (7), considering a monotonic loading history, the damage parameter is computed as:

$$D = \frac{\varepsilon_c^{\pm} - \varepsilon_e}{(\alpha^{\pm} - 1)\,\varepsilon_e} = \varepsilon_u \frac{\varepsilon_c^{\pm} - \varepsilon_e}{\left(\varepsilon_c^{\pm} - \varepsilon_u^{\pm}\right)\varepsilon_e} \tag{8}$$

Thus, it results $D = 0$ for $\varepsilon_e = \varepsilon_c^{\pm}$ and $D = 1$ for $\varepsilon_e = \varepsilon_u^{\pm}$; moreover, substituting the deduced relation (8) into the expression of the stress, given by the first equation of the state laws (3), a stress-strain linear softening is obtained when no plastic evolution is considered.

For what concerns the plasticity evolution, it is assumed:

$$f^p(\sigma, \vartheta) = -\frac{\sigma}{1 - D} + \vartheta - \sigma_y \leq 0 \qquad \dot{\mu} \geq 0 \qquad f^p\dot{\mu} = 0 \tag{9}$$

$$\dot{\varepsilon}_p = \frac{\partial f^p}{\partial \sigma}\dot{\mu} = -\frac{\dot{\mu}}{1 - D} \qquad\qquad \dot{\beta} = \frac{\partial f^p}{\partial \vartheta}\dot{\mu} = \dot{\mu} \tag{10}$$

such that, it results:

$$\dot{\beta} = -(1 - D)\dot{\varepsilon}_p \tag{11}$$

The quantity $\sigma/(1 - D) = \tilde{\sigma}$ in equation (9) is the effective stress.

The consistency condition for the plastic process leads to:

$$0 = \dot{f}^p(\sigma, \vartheta) = \frac{\partial f^p}{\partial \sigma}\dot{\sigma} + \frac{\partial f^p}{\partial \vartheta}\dot{\vartheta} = E\left(\dot{\varepsilon} - \dot{\varepsilon}_p\right) + K(1 - D)\dot{\varepsilon}_p \tag{12}$$

Solving equation (12) with respect to $\dot{\varepsilon}_p$, the evolutionary equations of the plastic process result:

$$\dot{\varepsilon}_p = H\dot{\varepsilon} \qquad\qquad \dot{\beta} = -(1 - D)H\dot{\varepsilon} \tag{13}$$

where

$$H = \frac{E}{E + K(1 - D)} \tag{14}$$

Finally, the damage rate can be computed in terms of the total strain rate as:

$$\dot{D} = \frac{\varepsilon_c^{\pm}}{(1 - \alpha^{\pm})\,\varepsilon_e^2}\left(\dot{\varepsilon} - \dot{\varepsilon}_p\right) = \frac{\varepsilon_c^{\pm}}{(1 - \alpha^{\pm})\,\varepsilon_e^2}\left(1 - H\right)\dot{\varepsilon} \tag{15}$$

The tangent constitutive modulus E_T is obtained by differentiating the stress-strain relationship:

$$\dot{\sigma} = (1 - D)\,E\,(\dot{\varepsilon} - \dot{\varepsilon}_p) - \dot{D}\,E\,(\varepsilon - \varepsilon_p) \tag{16}$$

$$= \left\{ E\left[(1 - D) - (\varepsilon - \varepsilon_p)\frac{\varepsilon_c^{\pm}}{(1 - \alpha^{\pm})\,\varepsilon_e^2}\right](1 - H)\right\}\dot{\varepsilon} = E_T\dot{\varepsilon}$$

3   Cross-section beam equations

The one-dimensional elastoplastic-damage constitutive law, developed in the previous section, is adopted to study the behavior of softening beams made of concrete or masonry. In particular, beam cross-sections, presenting a symmetry axis y, are considered. Note that the study can be extended to other geometries of the cross-section.

The possibility to account for the presence of elastic reinforcements is considered. Reinforcements could be internal, as steel bars in concrete beams, and/or external, as the ones adopted to rehabilitate damaged concrete or masonry beams by using fiber-reinforced plastic materials.

The response of the damage-plastic cross-section beam is derived considering the elongation e and the bending curvature χ deformations, such that the strain at a typical point of the beam is:

$$\varepsilon = e + y\,\chi \tag{17}$$

The resultants in the softening beam are:

$$N^B(e, \chi) = \int_A \sigma(e, \chi) \, dA \ , \ M^B(e, \chi) = \int_A y\sigma(e, \chi) \, dA \tag{18}$$

where A is the cross-section area of the beam. Moreover, the resultants in the elastic reinforcements are:

$$N^R = A^R \, e + B^R \, \chi \quad , \quad M^R = B^R \, e + D^R \, \chi \tag{19}$$

where:

$$A^R = \sum_{i=1}^{n_R} E_i^R S_i \quad , \quad B^R = \sum_{i=1}^{n_R} E_i^R S_i h_i \quad , \quad D^R = \sum_{i=1}^{n_R} E_i^R S_i h_i^2 \tag{20}$$

with n_R the number of reinforcements, E_i^R, S_i and h_i the Young modulus, the area and the position of the $i-$th reinforcement, respectively.

The total resultant axial forces and the bending moments in the reinforced beam are:

$$N = N^B + N^R \quad , \quad M = M^B + M^R \tag{21}$$

Finally, the behavior of the cross-section softening beam is governed by the equation:

$$\begin{aligned}
R_N(e, \chi, \lambda) &= N(e, \chi) - \lambda N_{ext} = N^B(e, \chi) + N^R(e, \chi) - \lambda N_{ext} = 0 \\
R_M(e, \chi, \lambda) &= M(e, \chi) - \lambda M_{ext} = M^B(e, \chi) + M^R(e, \chi) - \lambda M_{ext} = 0
\end{aligned} \tag{22}$$

where λ is the loading multiplier, introduced with the aim of developing an arc-length procedure. Equations (22) can be rewritten in vectorial form as:

$$\mathbf{R}(\mathbf{u}, \lambda) = \mathbf{S}(\mathbf{u}) - \lambda\omega = \mathbf{0} \tag{23}$$

where $\mathbf{u}$ is the beam elongation and curvature vector, $\mathbf{S}(\mathbf{u})$ is the internal stress resultant vector and ω is the external load vector; in particular, it is set:

$$\mathbf{u} = \begin{bmatrix} e \\ \chi \end{bmatrix} \qquad \mathbf{S}(\mathbf{u}) = \begin{bmatrix} N(e, \chi) \\ M(e, \chi) \end{bmatrix} \qquad \omega = \begin{bmatrix} N_{ext} \\ M_{ext} \end{bmatrix} \tag{24}$$

4 Numerical applications

The nonlinear equilibrium equation (23) is solved developing a numerical procedure, based on the arc-length method, with a proper choice of the control parameters. The evolutionary equations of the damage and plasticity are integrated with respect to the time, developing an implicit Euler algorithm. The finite step nonlinear problem is solved adopting a predictor-corrector procedure. In particular, the coupled damage and plasticity evolutive equations are solved adopting a Newton-Raphson algorithm. The two strains, evaluated in the cross-section where the elastic strains are equal to the tensile and compressive final damage thresholds, are assumed as control parameters. Then, the numerical procedure is implemented in a computer code.

Applications are developed to investigate the behavior of reinforced and concrete and masonry rectangular cross-sections.

Initially, a masonry element reinforced by FRP sheets is analyzed (Triantafillou 1996, Luciano and Sacco 1998, Marfia and Sacco 2001). The axial and bending behavior is studied. Moreover, comparisons between the numerical and the analytical solutions are made in order to assess the efficiency of the procedure.

The analyses are developed to study the mechanical response of a reinforced masonry element, characterized by the following material and geometrical data:

$$
\begin{aligned}
E &= 5000\text{MPa} & \sigma_y &= 3\text{MPa} & K &= 500\text{MPa}\\
\varepsilon_c^+ &= 0.0001 & \varepsilon_c^- &= -0.00067\\
\varepsilon_u^+ &= 0.0004 & \varepsilon_u^- &= -0.00085\\
E^R &= 200000\text{MPa}\\
b &= 130\text{mm} & h &= 250\text{mm}\\
n^R &= 2 & S_1 &= 10\text{mm}^2 & S_2 &= 10\text{mm}^2\\
& & h_1 &= 125\text{mm} & h_2 &= -125\text{mm}
\end{aligned}
$$

where b is the width of the cross-section.

Note that the material and geometrical data correspond to a possible masonry obtained using blocks with a rectangular cross-section of 130x250 mm^2, reinforced by carbon-epoxy composite material at the top and at the bottom of the cross-section.

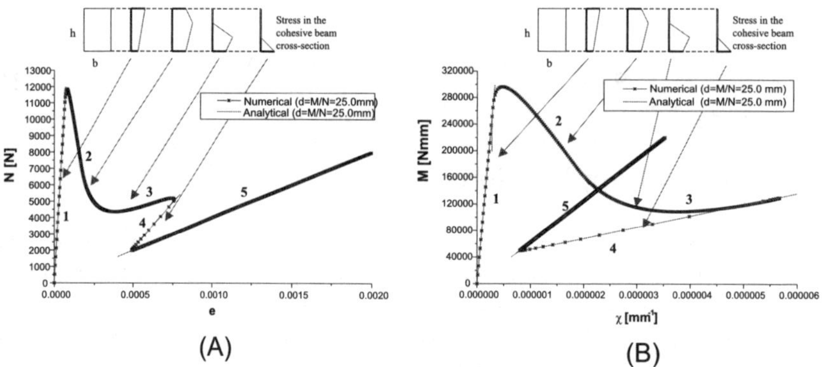

Fig. 1. Comparisons between analitycal and numerical results: (A) Axial force versus elongation; (B) bending moment versus curvature.

Initially, the problem concerning the beam subjected to a tensile axial force N_{ext} and to a bending moment M_{ext}, characterized by a prescribed eccentricity $d = M_{ext}/N_{ext} = 25$mm, is investigated. For the considered problem, the whole cross-section is in tension, thus only damage and no plasticity can occur; the analytical solution is also determined.

In figures 1(A) and 1(B) the comparison between the analytical and the numerical solutions is shown. In particular, in figure 1(A) the axial force versus the elongation is reported; while in figure 1(B) the bending moment is plotted versus the curvature. It can be emphasized that the numerical procedure is able to compute a very satisfactory solution in terms of axial deformation as well as of curvature. In fact, the numerical solution is able to predict also sharp snap-back branches. These results point out that the choice of the control parameters in the arc-length method is appropriate. In figures 1(A) and 1(B), the distributions of tensile stress along the masonry cross-section during the whole loading history are also schematically reported. It can be noted that the mechanical response of the reinforced section both in terms of axial force-elongation, reported in figure 1(A) , and bending moment-curvature, reported in figure 1(B), is characterized by five different branches: a linear elastic behavior when the whole section still behaves elastically; a softening branch after the peak that corresponds to the damage evolution in the upper part of the masonry section; a hardening branch when the upper part of the section is completely damaged, the middle part is partially damaged and the lower one still behaves elastically; a severe snap-back when the greater part of the masonry section is completely damaged and only a little part, is partially damaged and a linear elastic branch that occurs when the whole masonry section is completely damaged and only the reinforcements are active.

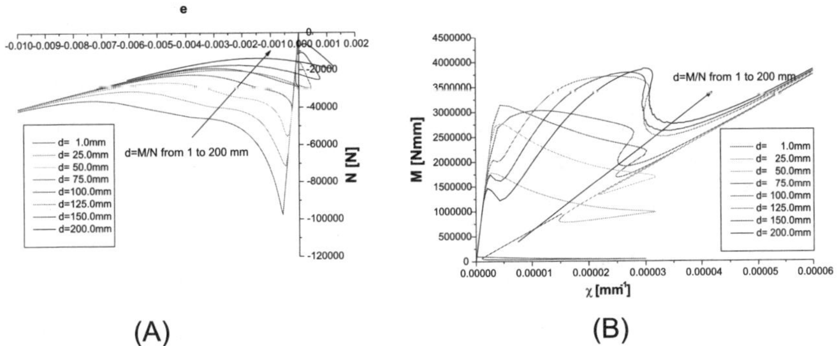

Fig. 2. Mechanical response for different values of the eccentricity: (A) Compressive axial force versus elongation; (B) bending moment versus curvature.

In figures 2(A) and 2(B), the compressive axial force versus the axial strain and the bending moment versus the curvature are plotted, respectively. In these cases, the response of the reinforced beam appears complex as it is significantly influenced not only by the damage evolution but also by the plastic flow even for reduced values of the eccentricity d. The response of the beam for any loading history is characterized by steep softening branches and by severe snap-backs. As it can be seen in figure 2(A), for high values of d in the first part of the loading history a positive elongation occurs although the section is subjected to compressive axial force. Finally, it can be pointed out that, also in these cases, the mechanical response tends to the linear elastic behavior of the reinforcements for any value of d, when the masonry is completely damaged.

A concrete element reinforced by classical steel bars is analyzed. In particular, it is studied a rectangular cross-section characterized by the following geometrical data:

$$b = 300\text{mm} \quad h = 600\text{mm}$$
$$n^R = 2 \qquad S_1 = 462\text{mm}^2 \quad S_2 = 226\text{mm}^2$$
$$h_1 = 270\text{mm} \quad h_2 = -270\text{mm}$$

Note that the data of the steel reinforcements corresponds to 4 bars $\phi 16$ at the bottom and 2 bars $\phi 12$ at the top of the cross-section. The reinforcement is modelled as a linear elastic material with the Young modulus $E^R = 210000\text{MPa}$.

Three different concrete behaviors are considered.

Case 1 considers a concrete response strictly following the Eurocode 2 (1991) prescriptions. Hence, an elastoplastic behavior, with no hardening, is adopted in compression until the uniaxial deformation ε reaches the limit value $\varepsilon_f = -0.0035$; moreover no strength in tension is assumed. The Case 1 concrete response is recovered within the proposed elastoplastic-damage model setting the parameters as:

- compression

$$E = 19050\text{MPa} \quad \sigma_y = 13.226\text{MPa} \quad K = 0\text{MPa}$$
$$\varepsilon_c^- \to \infty \qquad \varepsilon_u^- \to \infty$$

- tension

$$E = 19050\text{MPa}$$
$$\varepsilon_c^+ \to 0 \qquad \varepsilon_u^+ \to 0$$

where the symbols $\to 0$ and $\to \infty$ indicate the assumption of very small and very high numerical values, respectively. Note that the Young modulus E and the yield stress σ_y are determined following the prescriptions of the Eurocode 2.

Case 2 considers the same concrete response of Case 1 in compression; a finite strength, evaluated according to the Eurocode 2 (1991), and a linear

softening response are introduced in tension. Case 2 concrete tensile response
is obtained within the proposed model setting the parameters as:

$$E = 19050\mathrm{MPa}$$
$$\varepsilon_c^+ = 0.000136 \quad \varepsilon_u^+ = 0.00136$$

Case 3 considers the damage response of Case 2 in tension; an elastoplastic-
damage response is assumed in compression in order to better approximate
the experimental behavior of concrete. Case 3 compressive response is recovered within the proposed model, setting the parameters as:

$$E = 19050\mathrm{MPa} \quad \sigma_y = 13.226\mathrm{MPa} \quad K = 870\mathrm{MPa}$$
$$\varepsilon_c^- = -0.00072 \quad \varepsilon_u^- = -0.00082$$

The bending behavior of the reinforced concrete section is analyzed.

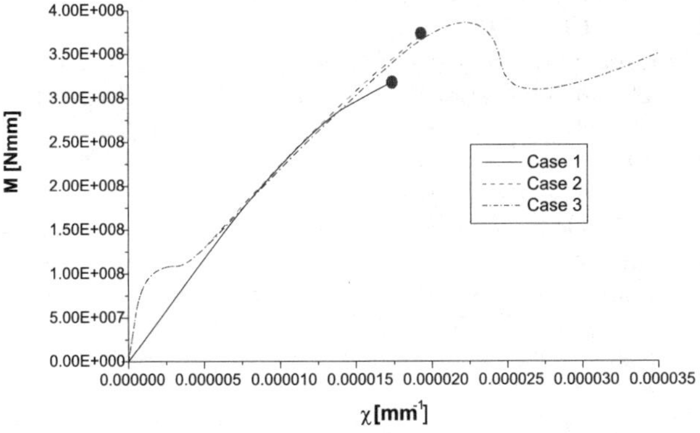

Fig. 3. Mechanical response of a reinforced concrete element in terms of bending
moment versus curvature.

In figure 3, the bending moment versus the curvature is plotted for the
three different proposed concrete responses. It can be noted that there are
significative differences, in the first part of the analyses, between the response
obtained in Case 1 and in Case 2 and 3. In fact, the latter two stress-strain
relationships are characterized by finite tensile strength, while a no tension
material is assumed in Case 1. In the second part of the analyses, when the
material considered in Case 2 and 3 is completely damaged in tension, the
behavior computed with the three stress-strain laws present only light differences. In Case 1 and 2 the analyses are stopped when the axial deformation
reaches the limit value ε_f in a point of the section, since the compressive
concrete behavior is not defined after that point for Case 1 and 2. In Case
3 plastic and damage evolutions in compression are defined also after the
limit deformation ε_f. In fact, the softening branch of the response is due to
the damage and plastic evolutions. When the concrete material is completely

damaged, only the steel reinforcements are able to bear bending loading increments; thus, the response becomes linear with a slope corresponding to the bending stiffness of the reinforceme

References

1. Dragon A., Mróz Z., 1979. A continuum model for plastic-brittle behavior of rock and concrete. Int. J. Engng. Sci. **17**, 121–137.
2. Lotfi H.R., Shing P.B., 1991. An appraisal of smeared crack models for masonry shear wall analysis. Computers & Structures, **41**(3), 413–425.
3. Carol I., Bažant Z., 1997. Damage and plasticity in microplane theory. Int. J. Solids and Structures 34 **29**, 3807–3835.
4. de Borst R., Pamin J., Geers M.G.D., 1999. On coupled gradient dependent plasticity and damage theories with a view to localization analysis. Eur. J. Mech. A/Solids **18**, 939-962.
5. Addessi D., Marfia S., Sacco E., 2002. A plastic nonlocal damage model. Comput. Methods Appl. Mech. Engrg. **191**, 1291–1310.
6. Triantafillou, T.C. (1996) Innovative strengthening of masonry monuments with composites. Proceedings of 2nd International Conference Advanced Composite Materials in Bridges and Structures. Montréal (Canada), August 11-14.
7. Luciano, R., Sacco, E. (1998) Damage of masonry panels reinforced by FRP sheets. International Journal of Solids and Structures **35**, 1723–1741.
8. Marfia S., Sacco E., 2001. Modeling of reinforced masonry elements. International Journal of Solids and Structures **38**, 4177–4198.
9. Eurocode 2, 1991. Design of concrete structures - UNI ENV 1992 -1-1.

An Experimental and Numerical Investigation on the Plating of Reinforced Concrete Beams with FRP Laminates

L. Ascione, V. P. Berardi, E. Di Nardo, L. Feo, G. Mancusi

Dipartimento di Ingegneria Civile
Università degli Studi di Salerno
I-84084 Fisciano (SA), Italy

Abstract. This work deals with some numerical and experimental results on the plating of Reinforced Concrete (RC) beams externally strengthened with FRP laminates. In particular, they concern the tangential interaction distributions along the reinforced boundaries of the strengthened beams. The comparisons between numerical and experimental results show the accuracy of the proposed mechanical model in evaluating the stresses distributions at the interface between concrete core and composite laminates.

1   Introduction

The use of Fiber Reinforced Polymer (FRP) laminates in Civil Engineering applications represents an innovative and versatile technique for the strengthening of reinforced concrete (RC) structures.

In recent years this system is supplanting the classical Hermitian technique concerning the application of steel plates, due to several advantages: high strength-to-weight ratio, low maintenance cost, corrosion resistance and lightness.

An important topic arising in the study of strengthened beams is the evaluation of the interactions at the concrete-FRP interface. These interactions, in fact, make possible the correct transmission of stresses from core to reinforcement and they may be the cause of premature failure.

Several researchers [1-9] have studied this problem and have proposed different analytical models for analysing FRP systems and based on local equilibrium close to the cut-off section. Such models have been validated by comparisons with both FEM solutions in 2D and experimental data.

Recently, the authors have developed numerical and experimental studies on the mechanical behaviour of RC beams strengthened with composite laminates [10-26]. In particular, they have formulated two kinematical models able to analyse the strengthening of RC elements both in bending and in shear [13,18,19].

The present work summarizes the results of an experimental investigation on RC beams strengthened in flexure by the external bonding of carbon fiber reinforced polymer composite (CFRPC) laminates. In particular, the influence of the plate length on the shear stresses transmission near the cut-off cross-sections of the plated beams has been investigated.

The experimental investigation has been carried out at the Testing Laboratory of Material and Structures of the Department of Civil Engineering of the University of Salerno. Comparison of experimental and theoretical results has shown a good agreement.

2   Mechanical Model

Let us consider a straight RC beam strengthened with an external composite plate (Fig.1).

Following the mechanical model presented by the authors in [21], the natural configuration (B) of the strengthened beam is modelled as the assembly of two one-dimensional components corresponding to the concrete core $(B^{(1)})$ and to the FRP overlay $(B^{(2)})$ (Fig.1).

2.1   Kinematical model of the concrete core $(B^{(1)})$

The kinematical model of the concrete core $(B^{(1)})$ is driven from a suitable power expansion of the displacement field components with respect to the coordinates x and y starting from a given point $P_0^{(1)}(x_0^{(1)}, y_0^{(1)})$ of its cross-section $\Sigma^{(1)}$ [13].

This model can be expressed in the following form:

$$u^{(1)}(x, y, z) = u_0^{(1)}(z) - \theta^{(1)}(z)(y - y_0^{(1)}) +$$
$$+ \sum_{i,j} u_{ij}^{(1)}(z)(x - x_0^{(1)})^i(y - y_0^{(1)})^j , \tag{1a}$$

$$v^{(1)}(x, y, z) = v_0^{(1)}(z) + \theta^{(1)}(z)(x - x_0^{(1)}) +$$
$$+ \sum_{m,n} v_{mn}^{(1)}(z)(x - x_0^{(1)})^m(y - y_0^{(1)})^n , \tag{1b}$$

$$w^{(1)}(x, y, z) = w_0^{(1)}(z) + \phi^{(1)}(z)(y - y_0^{(1)}) - \psi^{(1)}(z)(x - x_0^{(1)}) +$$
$$+ \sum_{h,k} w_{ij}^{(1)}(z)(x - x_0^{(1)})^h(y - y_0^{(1)})^k , \tag{1c}$$

where: $u_0^{(1)}$, $v_0^{(1)}$ and $w_0^{(1)}$ are the displacement components of $P_0^{(1)}$; $\phi^{(1)}$, $\psi^{(1)}$, $\theta^{(1)}$ are the rotation vector components of $\Sigma^{(1)}$ and the quantities $u_{ij}^{(1)}$, $v_{mn}^{(1)}$ and $w_{hk}^{(1)}$ are terms of higher order in the power expansion. The terms in $u^{(1)}$ and $v^{(1)}$, related to $u_{ij}^{(1)}$ and $v_{mn}^{(1)}$, model an in-plane deformation of the cross-section $\Sigma^{(1)}$; the terms in $w^{(1)}$, related to $w_{hk}^{(1)}$, model an out-of-plane warping of $\Sigma^{(1)}$.

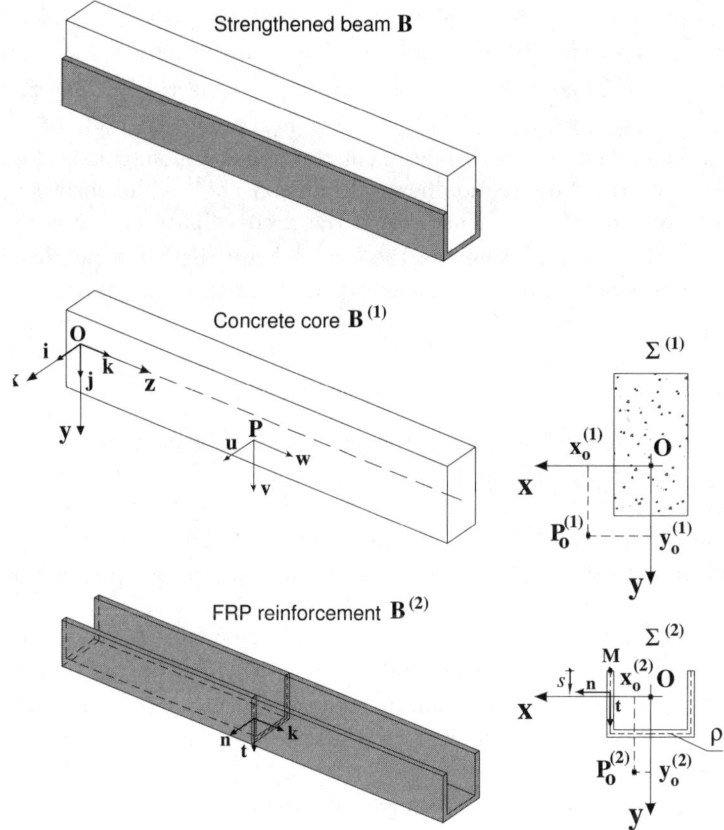

Fig. 1. RC beam strengthened with a FRP overlay: cross section and local reference frame.

2.2 Kinematical model of the FRP laminate ($\mathrm{B}^{(2)}$)

A generalization of Vlasov's theory has been adopted for the external reinforcement ($\mathrm{B}^{(2)}$).

The kinematics of the FRP overlay is characterized by a rigid transformation of the cross-section $\Sigma^{(2)}$ in its own plane and by a warping out of the same plane.

Consequently, the following displacement field is assumed:

$$u^{(2)}(s,z) = u_0^{(2)}(z) - \theta^{(2)}(z)(y(s) - y_0^{(2)}) \,, \tag{2a}$$

$$v^{(2)}(s,z) = v_0^{(2)}(z) - \theta^{(2)}(z)(x(s) - x_0^{(2)}) \,, \tag{2b}$$

$$w^{(2)}(s,z) = w_0^{(2)}(z) - \psi^{(2)}(z)x(s) + \phi^{(2)}(z)y(s) +$$

$$-\dot{\theta}^{(2)}(z)\omega(s) + \sum_{i=1}^{N_s} \gamma_i^{(2)}(z)\omega_i(s) \,, \tag{2c}$$

where: $x_0^{(2)}$, $y_0^{(2)}$ are the coordinates of an arbitrary point $P_0^{(2)}$ on the x-y plane assumed as the pole of the above mentioned rigid transformation; $u_0^{(2)}$ and $v_0^{(2)}$ are the displacement components of the point $P_0^{(2)}$ respectively along x, y and z axis: $w_0^{(2)} = w_M - \phi^{(2)} y_M + \psi^{(2)} x_M$, M (x_M, y_M) being the origin of the curvilinear abscissa s and w_M its displacement components along z axis (Fig. 1); $\phi^{(2)}$ and $\psi^{(2)}$ are the cross-section flexural rotations; $\dot{\theta}^{(2)}$ is the derivative of the twisting rotation $\theta^{(2)}$ with respect to the z coordinate and ω is the sectorial area of the classical Vlasov's theory; $\gamma_i^{(2)}$ are further generalized displacement components and ω_i are geometrical quantities defined as :

$$\omega_i(s) = \int_M^Q f_i^*(s)ds \; , \tag{3}$$

$f_i^*(s)$ being suitable shape functions of the curvilinear coordinate s.

In Eq. (3) Q is a generic point of ρ (Fig. 1). The term $\dot{\theta}^{(2)}(z)\omega(s)$ corresponds to a warping of the cross-section $\Sigma^{(2)}$ without shear deformations of the middle surface; the terms $\gamma_i^{(2)}(z)\omega_i(s)$ corresponds to a warping of the cross-section in presence of shear deformations of the middle surface, depending on s and z.

The kinematical model of the FRP reinforcement coincides with Vlasov's one if functions $\gamma_{xz}^{(2)} = \dot{u}_0^{(2)} - \psi^{(2)}$, $\gamma_{yz}^{(2)} = \dot{v}_0^{(2)} + \phi^{(2)}$ and $\gamma_i^{(2)}$ are equal to zero. The further terms $\gamma_i^{(2)}$ can be used to obtain a more accurate simulation of the behaviour of the FRP overlay with regard to the effects of shear deformability.

2.3 Modelling of Composite/Concrete Interface

The proposed model assumes that the external composite overlay $(B^{(2)})$ is bonded to the concrete core $(B^{(1)})$ by continuous distributions of bilateral elastic springs, arranged along the directions n, t, k of the cartesian local reference system (Fig. 1). The spring reactions give an approximation of the interfacial stresses. In particular, as the stiffness of the springs tends toward infinite, the model approximates the perfect adhesion condition between the two components.

A relevant topic in studying composite plating is the concentration of shear stresses and pealing forces near the ends of the FRP laminate (cut-off cross-sections). Previous theoretical and experimental studies, developed by the authors [18], have suggested to simulate the diffusion of the inter-face bond-stresses within the end region of composite overlay by assuming a fictitious variation of the thickness $t_p(z)$ from zero to actual value t_p, corresponding to a cubic law over an appropriate length l_d (Fig. 2), as presented in [18]. This quantity represents a known term of the problem and is assumed to have the following expression:

$$l_d = \chi \sqrt{\frac{E_p}{G_g} t_p t_g} \; , \tag{4}$$

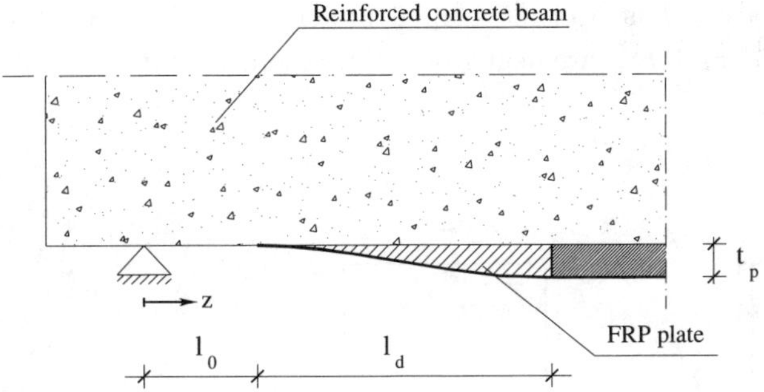

Fig. 2. Lenght l_d.

where E_p is the Young longitudinal modulus of the composite plate, t_p the nominal thickness of the laminate, t_g the thickness of the adhesive layer at the interface, G_g the shear modulus of the adhesive and χ an appropriate factor. This kind of approach has been already adopted by the authors in [18] and provides good agreement with the experimental results presented by Swamy and Roberts in [2].

2.4 Constitutive Laws

The behaviour of the concrete core has been assumed elasto-plastic in compression and not resistant in tension. The internal steel rebars and the FRP reinforcements have been assumed to be isotropic elastic-perfectly plastic and orthotropic linearly elastic, respectively.

2.5 The Numerical Model

The above presented mechanical model has been discretizated by using a finite element approach.

The finite element model uses Hermitian cubic polynomials to approximate all the kinematical unknowns. The corresponding discrete problem, which is not-linear because of the constitutive assumptions, has been solved by using the Newton-Raphson method. The integrations concerning the generic cross-section Σ of the strengthened beam (B), required for assembling the stiffness matrix, have been performed by a numerical technique (Gauss numerical integration).

In order to improve numerical stability and convergence, the cross-section $\Sigma^{(1)}$ of the concrete core (B$^{(1)}$) has been divided into several layers. By this way it is possible to test the strain field in the concrete core in many Gauss points at any iteration. This technique also permits to take into account the contribution of steel rebars to the flexural capacity of the strengthened beam. In fact, the steel rebars are modelled by using an appropriate layer (Fig. 3). More details are available in [13-18].

The numerical analysis has been developed under the assumption that the quantities $u_{ij}^{(1)}$, $v_{mn}^{(1)}$ and $w_{hk}^{(1)}$ are negligible (Timoshenko's model for $B^{(1)}$ component).

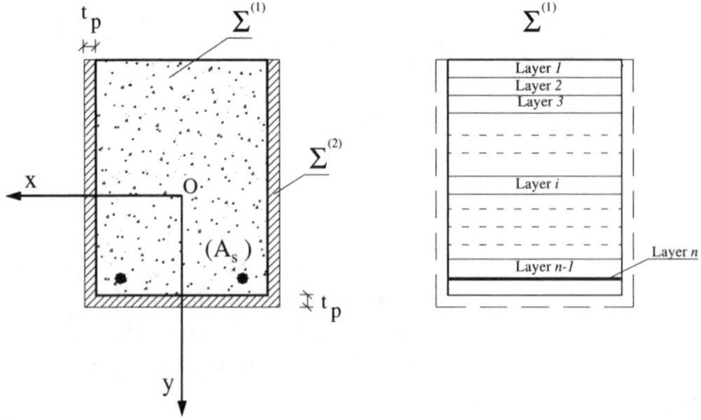

Fig. 3. Partition of cross-section by layers.

3   Experimental tests

Twenty-four rectangular RC beams were cast in four sets. All beams were of size 150 (B) mm x 250 (H) mm x 2300 (L) mm. Each beam of set 1 and set 2 was reinforced with two 12 mm diameter longitudinal steel bars, while each beam of set 3 and set 4 with three 20 mm diameter steel bars. Details of steel reinforcements are given in Fig. 4. Four RC beams strengthened in flexure and in shear by using four or six layers of carbon fabrics (Table 1) (SB1/01/6s, SB2/02/4s, SB3/01/4s and SB3/01/6s) were tested to failure. Beams were tested in four point bending test over a span of 2000 mm (Fig. 5). Details of all the strengthened beams and their experimental ultimate loads (F_u) are summarized in Table 1.

Table 1. Details of the strengthened beams.

Beam	Set	Number of layers	t_p (mm)	F_u (kN)
SB1/01/6s	1	6	2.5	150
SB2/01/4s	2	4	2.0	118
SB3/01/4s	3	4	2.0	205
SB3/01/6s	3	6	2.3	214

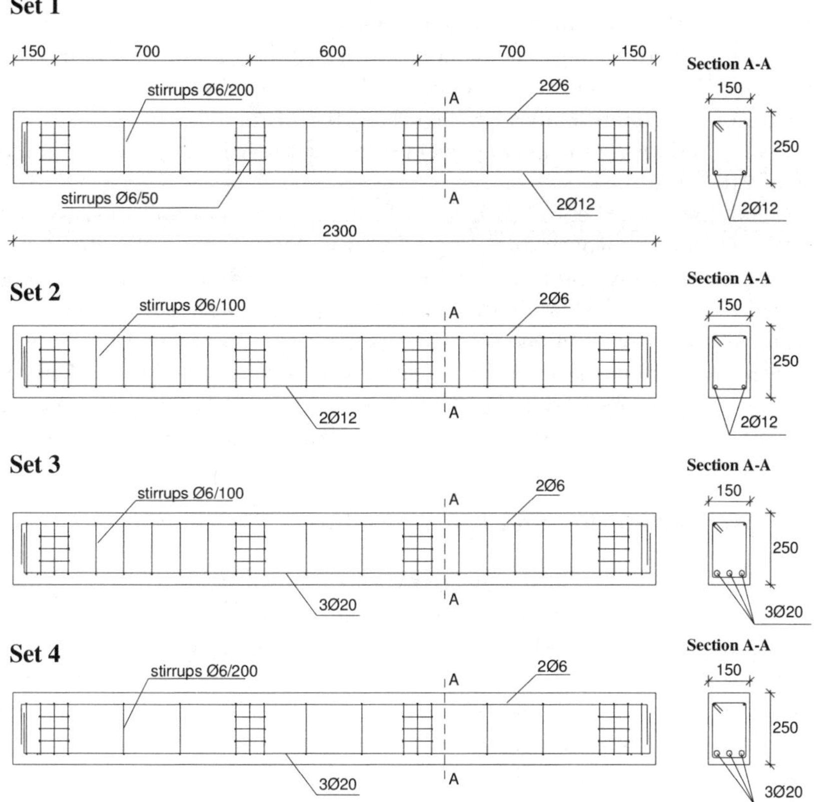

Fig. 4. Details of steel rebars (dimensions in mm).

3.1   Materials

<u>Concrete</u>: Ordinary Portland cement, locally available sand and crushed basalt rock were used for making concrete. The maximum size of coarse aggregate

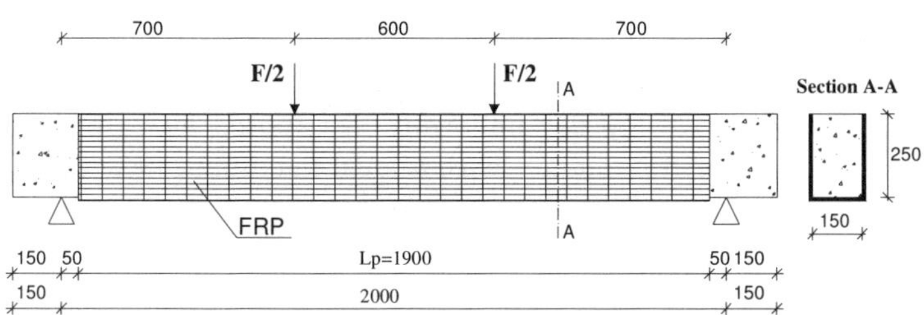

Fig. 5. Statical scheme and loading configuration of the strengthened RC beam (dimensions in mm).

used was 25.4 mm. The cement-sand-gravel ratios in the concrete mix were 1:2:3 by weight. The water-cement ratio was 0.55. Thirty concrete cubes of size 150 mm were cast at the same time in order to evaluate the mechanical properties of the hardened concrete. The average measured compressive strength of the concrete after 28 days was 19.4 MPa.

<u>Steel</u>: steel had an elastic average yield stress of 435 MPa and an elastic modulus of 210 GPa.

<u>CFRP laminates and epoxy adhesive</u>: the composite material consisted of bi-directional carbon woven textiles (SikaWrap-160C 0/90(VP)) bonded together with an epoxy adhesive (Sikadur 330). The values of the mechanical properties of the CFRP laminates and of the epoxy adhesive were supplied by the manufacturer or obtained by experimental tests. In particular, the tensile strength and the tensile modulus of SikaWrap-160C textiles were 3500 MPa and 235 GPa, respectively; the flexural strength and the flexural modulus (E_g) of the Sikadur 330 adhesive were 30 MPa and 3.8 GPa, respectively.

3.2   Instrumentation and test procedure

Beams were extensively instrumented with electrical strain gages in order to evaluate material and structural response during tests.

Strains, deflections and applied load were recorded by a data acquisition system, which was customized through software developed by one of the authors under Lab-Windows CVI environment [23]. The experimental values of the tangential interactions along the beam axis were calculated starting from the measured strains [15].

4   Results and conclusions

Figure 6 shows typical numerical distributions of tangential interactions along the axis of the beam SB1/01/6s and experimental results.

In particular, the numerical predictions have been obtained by assuming: t_g=1 mm, l_d=32 mm ($\chi = 4$), and two values of the parameter N_s=0, 7. Figure 7 shows theoretical and experimental shear interaction along the curvilinear abscissa s of the beam cross-section at z=150 mm. As it can be seen the shear stresses (τ_{nk}) are not uniformly distributed along both the beam axis and the reinforced boundary with a local maximum near the cut-off cross-sections. Futhermore, the numerical results agree closely with the experimental ones.

Table 2 summarizes the theoretical and experimental maximum values of shear interactions in SB1/01/6s and SB3/01/6s beams at z=150 mm for different values of the applied load. In particular, $(\tau_{max}^{(0)})$ and $(\tau_{max}^{(7)})$ denote the maximum values of the interactions corresponding to N_s=0 and N_s=7,

Table 2. Comparison between the theoretical and experimental maximum values of the longitudinal shear interactions at z=150 mm.

Beam	F (kN)	$\tau_{max,exp}$ (N/mm^2)	$\tau_{max}^{(0)}$ (N/mm^2)	$\tau_{max}^{(7)}$ (N/mm^2)	τ_{avg} (N/mm^2)
SB1/01/6s	50	0.216	0.187	0.194	0.080
SB1/01/6s	70	0.253	0.211	0.274	0.120
SB3/01/6s	50	0.066	0.053	0.055	0.025
SB3/01/6s	70	0.087	0.073	0.095	0.035
SB3/01/6s	180	0.312	0.295	0.308	0.090

$(\tau_{max,exp})$ is the maximum experimental value and (τ_{avg}) is the average value, evaluated through Jourawski's formula.

The peak values of the longitudinal shear interactions are localized at the tensile side of the strengthened beam near the cut-off cross-sections (at z=66 mm) and they are greater than the average ones (τ_{avg}), computed in the same cross-section (Table 3).

Similar remarks have been obtained from the experimental tests performed on the other beams (SB2/01/4s, SB3/01/4s,SB3/01/6s). The numerical and experimental results are in good agreement (Table 2).

Based on the findings of this study, the following conclusions can be made:
- high stress concentration near the cut-off cross-sections exist and the abserved experimental values are considerably higher than those given by simple elastic theory;
- the distributions of tangential interactions and their peak values are affected by the plating ratio;

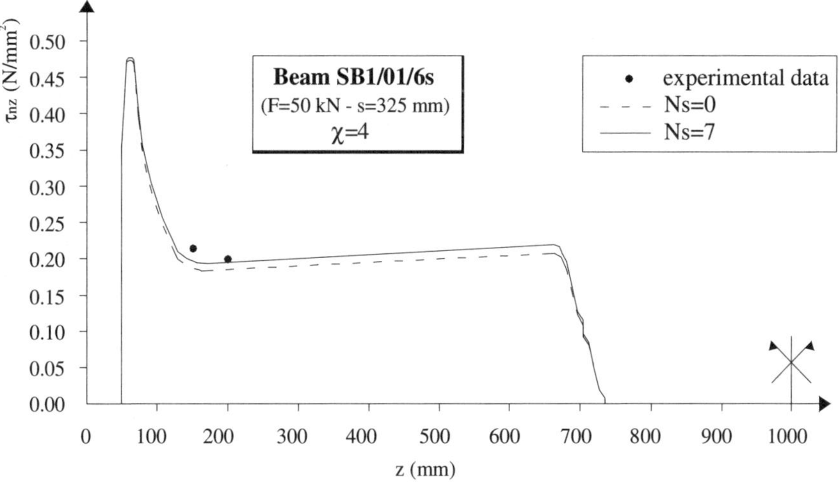

Fig. 6. Shear interactions along the beam axis.

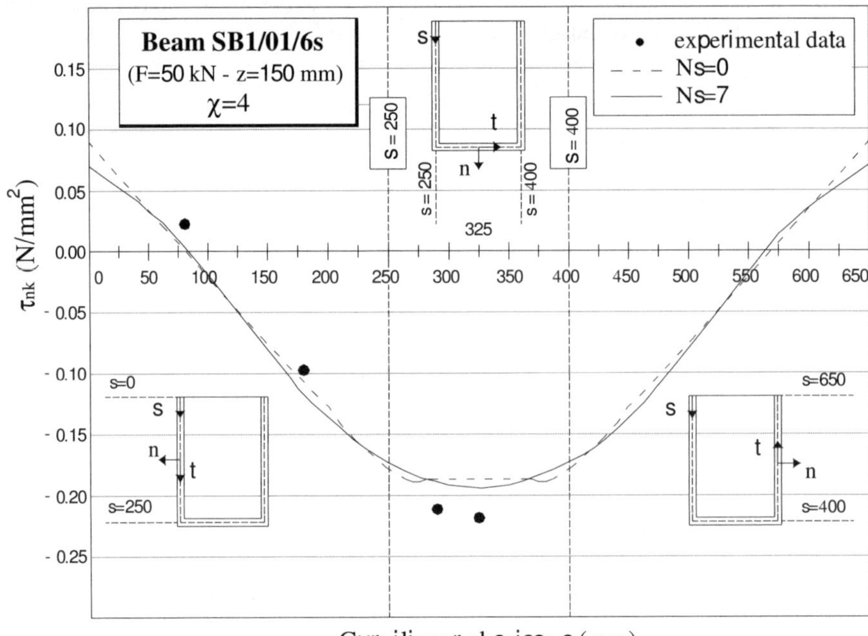

Fig. 7. Shear interactions along the curvilinear abscissa of cross section at $z=150$ mm.

Table 3. Comparison between the theoretical and experimental peak values of the longitudinal shear interactions.

Beam	F (kN)	$\tau_{peak}^{(0)}$ (N/mm²)	$\tau_{peak}^{(7)}$ (N/mm²)	τ_{avg} (N/mm²)	$\tau_{peak}^{(0)}/\tau_{avg}$	$\tau_{peak}^{(7)}/\tau_{avg}$
SB1/01/6s	50	0.469	0.473	0.080	5.863	5.913
SB1/01/6s	70	0.705	0.707	0.120	5.875	5.892
SB3/01/6s	50	0.125	0.127	0.025	5.000	5.080
SB3/01/6s	70	0.181	0.183	0.035	5.171	5.229
SB3/01/6s	180	0.511	0.516	0.090	5.678	5.734

- the tangential interactions play a fundamental role in the static behaviour of the plated beam, because they can produce a sudden and premature failure.

References

1. Swamy R. N., Jones R., Bloxham J. W. (1987) Structural Behaviour of Reinforced Concrete Beams Strengthened by Epoxy Bonded Steel Plates. The Structural Engineer, **65(2)**, 59-68.

2. Jones R., Swamy R. N., Charif A. (1988) Plate Separation and Anchorage of Reinforced Concrete Beams Strengthened by Epoxy-Bonded Steel Plates. The Structural Engineer, **66(5/1)**, 85-94.

3. Swamy R. N., et al. (1989) The Effect of External Plate Reinforcement on the Strengthening of Structurally Damaged RC Beams. The Structural Engineer, **67(3/7)**, 45-54.

4. Roberts T. M. (1990) Approximate Analysis of Shear and Normal Stress Concentrations in the Adhesive Layer of Plated RC Beams. The Structural Engineer, **67(12/20)**, 37-46.

5. Quantrill R. J., et al. (1996) Experimental and Analytical Investigation of FRP Strengthened Beam Response: Part I. Magazine of Concrete Research, **48(177)**, 331-341.

6. Quantrill R. J., et al. (1996) Predictions of Maximum Plate End Stresses of FRP Strengthened Beams: Part II. Magazine of Concrete Research, **48(177)**, 343-351.

7. Malek A.A., Saadatmanesh H., Ehsani M.R. (1996) Shear and Normal Stress Concentrations in RC Beams Strengthened with FRP Plates. Second International Conference on Advanced Composite Materials in Bridges and Structures, Montral, Qubec, Canada, 629-638.

8. Arduini M., Di Tommaso A., Manfroni O. (1995) Fracture Mechanisms of Concrete Beams Bonded with Composite Plates. Second International Symposium on: Non-Metallic Reinforcement for Concrete Structures FRPRCS-2, Ghent, 483-531.

9. Arduini M., Di Tommaso A., Manfroni O., Nanni A. (1996) Failure Mechanisms of Concrete Beams Reinforced with FRP Flexible Sheets. Second International Conference on Advanced Composite Materials in Bridges and Structures, Montral, Qubec, Canada, 253-260.

10. Ascione L., Fraternali F. (1990) On the Mechanical Behaviour of Curved Composite Beams: A Simple Model which Takes into Account the Warping Effects. Rendiconti dell'Accademia Nazionale dei Lincei, **IX(I)**, 223-233.

11. Ascione L., Bilotti G., Fraternali F. (1991) Sul Calcolo delle Tensioni Interlaminari in Travi Curve Composite. XX AIAS National Conference, Palermo, 441-451.

12. Ascione L., Fraternali F. (1992) A Penalty Model for the Analysis of Composite Curved Beams. Computers & Structures, **45(5/6)**, 985-999.

13. Ascione L., Feo L., Fraternali F. (1997) The Wrapping of Reinforced Concrete Beams with FRP Plates: A Mechanical Model. Advancing with Composites '97, Milano, 155-170.

14. Feo L., D'Agostino G., Tartaglione D. (1997) Sul Comportamento Meccanico di Travi Rinforzate con Placche in FRP: uno Studio Numerico. XXVI AIAS National Conference, Catania, 511-518.

15. Feo L., D'Agostino G., Tartaglione D. (1998) On the Statical Behaviour of Reinforced Concrete Beams Wrapped with FRP Plates: an Experimental Investigation. ECCM-8, European Conference On Composite Materials, Napoli, **2**, 189-196.

16. Ascione L., Feo L., Fraternali F. (1998) Stress Analysis of Reinforced Concrete Bemas Wrapped with FRP Plates. ECCM-8, European Conference On Composite Materials, Napoli, **2**, 197-204.

17. Ascione L., Feo L., D'Agostino G., Tartaglione D. (1998) Sul Rinforzo di Travi in Conglomerato Cementizio Armato Mediante Placche in FRP: Alcuni Confronti Teorico-Sperimentali. XXVII AIAS National Conference, Perugia, 101-113.

18. Ascione L., Feo L. (2000) Modeling of Composite/Concrete Interface of R/C Beams Strengthened with Composite Laminates. Composites Part B, **31(6/7)**, 535-540.

19. Ascione L., Feo L., Mancusi G. (2000) On the Statical Behaviour of FRP Thin-Walled Beams. Composites Part B, **31(8)**, 643-654.

20. Ascione L., Feo L. (2000) On the Plating of Reinforced Concrete Beams with FRP: a Numerical Simulation. ECCOMAS 2000, Barcellona.

21. Ascione L., Feo L. (2000) Interaction Stresses in Reinforced Concrete Beams Strengthened with FRP plates. Advancing with Composites 2000, Milano, 259-268.

22. Ascione L., Feo L. (2000) Disuniformit e concentrazioni tensionali all'interfaccia nelle strutture placcate: modellazione strutturale e confronti sperimentali. XXIX AIAS National Conference, Lucca, 709-720.

23. Feo L., Rapuano S. (2000) Un Sistema di Misura per il Rilievo delle Tensioni Interlaminari nelle Strutture Placcate con FRP. XXIX AIAS National Conference, Lucca, 759-768.

24. Ascione L., Berardi V.P., Feo L., Mancusi G. (2001) Il calcolo delle interazioni nel placcaggio di strutture in c.a. mediante lamine in FRP. XV AIMETA National Congress, Taormina.

25. Ascione L., Feo L., Berardi V.P. (2001) Analisi numerico-sperimentale di strutture in c.a. placcate con lamine di materiale composito. XXX AIAS National Conference, Alghero.

26. Ascione L., Feo L., Mancusi G. (2001) Progetto di rinforzi in FRP per strutture in conglomerato cementizio armato. Convegno Nazionale Crolli e Affidabilit delle Strutture Civili, Venezia, 517-533.

Reliability of CFRP Structural Repair for Reinforced Concrete Elements

Antonio Bonati, Giovanni Cavanna

Istituto per le Tecnologie della Costruzione
Consiglio Nazionale delle Ricerche
via Lombardia, 49
20098 San Giuliano Milanese, Italy

Abstract. This paper presents the outcomes of load tests performed on concrete beams reinforced to withstand bending stress through the application of a carbon fibres-based compound. The innovative nature of the proposed experiment lies in the choice of reinforced elements, that is artificially deteriorated concrete beams.

1 Introduction

The use of a new material, the technology for use, and its combination with other traditional materials, must be considered as a "SYSTEM" and, as such, its performances are to be evaluated avoiding the mistake of considering only the evaluation of its individual constituents which is nevertheless essential. Control systems adopted for both process and materials are to be necessarily stepped up as the used technologies and components are becoming more and more complex and sophisticated, in order to obtain reliable performances. Except for some isolated cases mainly related to restoration or recovery interventions of particular historical-cultural value, building yards can not certainly be compared with aeronautic yards where composite materials have been greatly acknowledged. The highest risk is these new materials being used by untrained operators who are familiar with other kinds of working "tolerances"; what's more, the risk is broadened by the seemingly extraordinary simplicity of application techniques, whose outcomes can discredit an innovative system with big potentials. In confirmation of what stated above, as far as the experiment in question is concerned, the execution of the works for the cortical rehabilitation of the beams to be repaired was assigned to and performed by operators highly skilled in the field of volumetric and structural rehabilitation of concrete structures who had not yet reached the necessary experience in the field of composite materials. Fibres were instead applied by workers with unquestionable skills in the sector of composite materials which have been recently introduced in the applications of the building field. Another factor having influenced the development of the described work is related to the main goal of these new reinforcing systems. They are used on structures to improve their strength capacity against new or previously underestimated loads, or to restore their original load-bearing capacities, lost due to exceptional events (earthquakes, fire, etc.) or to a natural decay caused by chemical-physical phenomena. It is evident that in nearly all cases it is necessary to act on existing products having undergone a natural ageing process. Depending on their original intrinsic characteristics and the type of agents (both exceptional and ordinary) they have to endure, each of the structures to be reinforced will be in a more or less acceptable preservation state. This is the reason why, when concrete is in good conditions, all currently employed reinforcing systems envisage at least an accurate phase of cleaning of the surfaces where the reinforcing element will be applied, that has to be performed by means of metal brushes or sandblasting. National and international research have proved the effectiveness of reinforcing interventions on structural elements using composite materials, when applied on sound products, not considering that, actually, these systems nearly always envisage a phase for the geometrical and structural recovery of the support. These remarks induced us to assess the system applied on a decayed support, that is to say on damaged concrete structures where large parts of the concrete, more or less superficial, are missing and where the reinforcements are exposed and extremely oxidized.

2   Experimental survey

The experiment was carried out by submitting restored and bending-reinforced concrete beams to loading tests (bending on four points, load being applied at 1/3 and 2/3 of the span).

The covering concrete on the intrados surface of the beam was almost completely missing and the exposed reinforcement irons were clearly oxidized. The restoration took place after removing the loose concrete, cleaning and treating the exposed reinforcements with anticorrosive mortar and final application of an amount of fibre-reinforced mortar allowing to recreate the original volume of the structure, see figure 1. Usually, the volume reconstruction of beams serves the purpose of recreating the cover layer having all the specific functional characteristics as the original one. Whenever a further phase of application of composite materials is foreseen, the cover layer must ensure a proper flatness of the surface on which the fabric is to be applied and merely perform a structural function, since, through it, stresses will be transferred from the beam to the composite.

After being restored, the beams were reinforced by applying to their intrados surface a composite consisting of three layers of 400 g/m^2 one-way carbon dry fabric and epoxy resin matrix, applied with the manual impregnation technique (wet lay up).

Two beams were taken as reference in order to compare the results obtained on the restored-reinforced beams. The two beams had the same dimensions (15x15x160 cm, symmetric longitudinal reinforcement 2+2 ϕ 12

Fig. 1. Restoration phase, anti-corrosion treatment of the irons (a) and volume reconstruction (b)

and brackets ϕ 8 with 15 cm pitch), one of them was not reinforced and the other was reinforced but not restored.

During the loading tests, the deformations of the strained steel, the compressed concrete and CFRP of all the tested beams were recorded at mid-section by means of strain gauges, while the deflection was measured by means of a potentiometric transducer. The load was created by a hydraulic jack and measured by means of a load cell. All obtained data were then recorded with a computer-aided system.

3   Analysis of results

The results of loading tests are hereafter schematised and analysed.

As-delivered beam: after the initiation of the first cracking (corresponding to the tensile strength of concrete in the lower area) the curve's slope varies and keeps decreasing until complete choking of the section. The diagram shows the ductility of the beam starting from the steel's yield. Failure occurred by concrete's compression.

CFRP-repaired beam: following a first section corresponding to the section of the as-delivered beam and the first cracking having been reached, the curve's slope is considerably steeper, which means a higher stiffness. Failure occurs without any warning with a 22% increase of the ultimate load with respect to the as-delivered beam.

Restored and repaired beam: this type of beam undoubtedly shows the most critical behaviour. Thanks to the increase of bending stiffness, which is similar to that obtained for the other repaired beam, the reinforcing action may be deemed successful, but only by reaching failure it was possible to verify that there was no meaningful increase of the ultimate load with respect to the reference beam. The introduction of a further element, that is the restoration mortar, rendered the system more complex and proved to be the weakest link in the sequence.

	As-delivered beam	Reinforced beam	Restored-reinforced beam
First cracking load kN	800	1200	1000
Steel yield load kN	4100	Not reached	Not reached
Ultimate load kN	4600	5600	4800
Concrete max def. mm/mm	$-3,641*10^{-3}$ (at 4600 daN)	$-1,034*10^{-3}$ (at 5600 daN)	$-1,016*10^{-3}$ (at 4800 daN)
Steel max def. mm/mm	$15,899*10^{-3}$ (at 4600 daN)	$1,676*10^{-3}$ (at 5600 daN)	$1,412*10^{-3}$ (at 4800 daN)
CFRP max. def. mm/mm	-	$2,425*10^{-3}$ (at 5600 daN)	$1.717*10^{-3}$ (at 4800 daN)
Deflection at 4000 daN mm	7,11	3,08	3.91
Failure mode	Crushing (o buckling) of concrete	Sudden detachment of CFRP layer	Sudden detachment of the restoration mortar with fibre

Table 1. Results of loading tests

The failure of reinforced beams must be considered of frail type, having occurred due to a sudden detachment of the reinforcement at one of its ends (peeling) for the reinforced beam and due to a detachment of the composite and restoration mortar (concrete ripping) for the restored and reinforced beam, see figure 2.

Fig. 2. Breaking mode of reinforced (a) and restored + reinforced beams (b)

The "concrete ripping" failure, already outlined in other researches, is caused by a first cracking of concrete at 45 (due to shear stresses) at the ends of application ends of the reinforcement (cut off). As the applied load increases, cracks propagate progressively until they reach the lower longitudinal reinforcement to proceed along a horizontal line, causing this way the detachment of the concrete where the composite is applied. In case of a restored beam, horizontal propagation resulted to coincide with the mortar-concrete interface surface, outlining that in this beam the adhesion value between the two materials was smaller than the shear value of the same materials, which justifies the low ultimate load value reached.

It is remarked that the breaking mode of the restored beam is particularly dangerous; in fact, any loading test, to be executed as a control, might have outlined an increase in the stiffness similar to that of the just reinforced beam that could lead to assume the success of the action without making predict the actual behaviour under ultimate load.

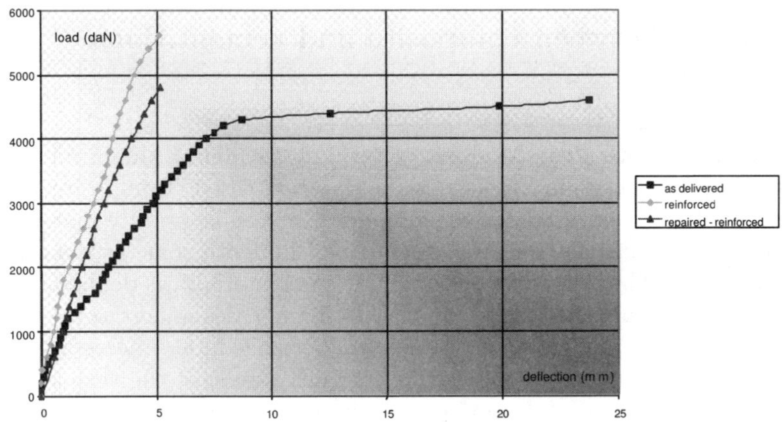

Fig. 3. Load-displacement diagrammes of tested beams

4   Some remarks

The experimental results of this first phase show some essential aspects. " The test performed on the restored beam showed that during the design phase (modelling and subsequent prediction of the ultimate load) it is not possible to set aside the state of the support under test that in many real cases does not only need to be simply "cleaned" but requires heavier actions, introducing new variables in the reinforcing system. Moreover, if the mechanical characteristics of the support are bad, either the laminated layers or the cohesive mortar layer might happen to detach (concrete ripping), always too early as to the value of ultimate moment referred to a bending failure mechanism. " The failure modes observed in all reinforced beams were started by a peeling phenomenon and, as a consequence, the attainable ultimate load depends on the composite/cementitious material adhesion. Other international experiences confirm what is stated above. We were urged analyse thoroughly the aspects connected with the adhesion between composite and adhering elements in the light of the latter remark and following the outcome of different researches showing how the normal stress values and shear values are concentrated in the composite-support interface at the final point of the reinforcing lamination (see the work by Malek, Saadatmanesh and Ehsani in "prediction of failure load of R/C beams strengthened with FRP plate due to stress concentration at the plate end" ACI STRUCTURAL JOURNAL 1998 - Bjorn Taljsten "shear and peeling stresses at the end of concrete beams strengthened with CFRP laminates", proceedings of the congress held in Bologna in 2000 - Advanced FRP Material for civil structures).

5   Adhesion between composite and cementitious materials

The collaboration with Prof. M. Rink of the Polytechnic of Milan, on the occasion of a degree thesis in Engineering of Materials titled "Composites for structural reinforcement: study of the adhesion between composite material and cementitious materials", allowed us to study in depth some aspects related to adhesion. The definition of "cementitious materials" is deliberately generic since the composite materials must in many cases adhere to a support other than concrete. Most of the times it becomes in fact necessary to apply a more or less important mortar layer which serves, at the very least, to recreate the flattest possible surface and, in other cases, to geometrically reconstruct the original section of the products on which the work is done, which may have changed over the time (mainly on account of accidental impacts or more frequently owing to the expulsion of the cover layer due to the corrosion of the metal reinforcement). In particular, in the course of this work, it was possible to outline some factors that remarkably affected the quality of gluing and must therefore be taken into account during the designing phase:

" State of the cementitious surface. The surface must be suitably prepared, possibly by roughing the surface in order to see the aggregates and remove the weak layers (adhesions are better on concrete than on mortar). The surface must also be cleaned from any possible contaminating material (grease, oil, powder).

" Surface porosity. With particular regard to the wet lay up technique, where the resin used for impregnation also is also used as gluing agent, it is important to take into account the right amount of resin that can and must seep into the uneven and porous areas of the substrate to avoid an incomplete impregnation of the fabric (which would imply a bad distribution of stresses and exposes the fibres to the environmental aggression).

" Adhesive thickness. To obtain an even distribution of stresses within the reinforcement, the adhesive thickness must be kept as constant as possible.

" Utilization of primers. Primers are often used to prepare the surfaces before the application of the composite. Primers have the same composition of gluing resins but with a lower viscosity. If, on the one hand, the utilization of primers is advantageous since they better cover the surface, seeping into the uneven areas, thanks to their lower viscosity, on the other hand it introduces a new variable. Primer can become a critical part of the joint since it is weak from a mechanical point of view when applied in too many layers and if it is fully reticulated before the application of the resin. It becomes therefore necessary to define and control application times and temperatures.

" Application temperature: Tresin ¿ Tsurface to avoid the formation of Skin due to thermal shock. During the curing Tresin also affects the Tg of the obtained composite. Ambient RH and adhering elements, during reticulation heat develops and this may take the humidity absorbed by adhering elements out of the adhering parts, close to the gluing surface, thus creating some gaps. Surfaces must be all the more dry the higher is the reaction peak temperature.

In particular, by comparing two calorimetric tests, it is possible to notice that during the reticulation phase, temperature may affect the glass transition temperature Tg. On top of that, it must be considered that other factors, such as for instance water absorption, can further bring down the value of Tg and that the mechanical characteristics of the resin (in particular the elastic modulus) quickly decay when this temperature value is exceeded. The above remarks may lead to question not only the fire resistance of reinforced structures, which is one of the essential requirements for construction materials but also implies a particular context such as a fire, but also not extremely severe environmental conditions (high temperature and humidity values).

However, it must be stated that the variation of mechanical characteristics of the composite related to the temperature value are reversible. That's why it was decided to plan a new experimental campaign in order to assess the performances of FRP reinforcements not only under more or less severe conditions but also making tests under particular environmental conditions (loading tests under thermal effect, hight temperature).

References

1. A. Nanni *Fibre-reinforced-plastic (FRP) reinforcement for concrete structures: properties and applications.* Elsevier, Amsterdam, 1993.
2. G. C. Mays and A. R. Hutchinson *Adhesives in Civil Engineering.* Cambridge University Press, Cambridge, 1997.
3. H. Toutanji and W. Gomez Durability of Concrete Beams Externally Bonded with FRP Sheets in Aggressive Environments. *Cement and Concrete Composites Journal,* 19(4), 1997.

4. J. M. Berthelot *Composite Materials. Mechanical Behavior and Structural Analysis.* Springer Verlag, Berlin, 1999.
5. V. S. GangaRao and P. V. Vijay Draft design guidelines for concrete beams externally strengthened with FRP. In D. Duthinh, editor, *NIST Workshop on Standards Development for the Use of Fiber Reinforced Polymers for the Rehabilitation of Concrete and Masonry Structures*, Tucson, Arizona, 3:113–120, 1999.
6. P.C. Varelidis, C.D. Papaspyrides, and R.L. McCullough The effect of temperature on the single-fiber fragmentation test with coated carbon fibers. *Composites Science and Technology*, 58(9):1487–1496, 1998.

Elastic Plates with Weakly Incoherent Response

G. Lancioni

Dipartimento di Ingegneria Civile
Università di Roma "Tor Vergata",
via del Politecnico, 1
00133 Roma, Italia

Abstract. A two-dimensional dynamic theory is formulated for transversely isotropic plates whose material-symmetry axis slightly deviates from the normal to the plate cross section (this type of transverse isotropy is said to be *weakly incoherent*). The model is based on a representation of the admissible motions capable of capturing obliqueness in response. Some solutions in the form of progressive waves are studied and some incoherency tests via wave propagation are proposed.

1   Introduction

Standard linear elastic plate theories deals with materials which are either isotropic, or characterized by a preferred response direction $\mathbf{c}$ parallel to the normal $\mathbf{z}$ to the plate's cross section. For this kind of plates, which we call *coherent*, the admissible motions are customarily chosen of a form consistent with both the plate shape and the constitutive geometry, being constrained along the direction $\mathbf{z} \equiv \mathbf{c}$: for example, the first-order polynomial in the thickness coordinate ζ

$$\mathbf{u}(x, \zeta, t) = \mathbf{u}_0(x, t) + \zeta \mathbf{u}_1(x, t), \tag{1}$$

with x the place in the middle plane and t the time, allows for a generalization of the classical Reissner-Mindlin theory [1]–[3], in which the additional condition $\mathbf{u}_1 \cdot \mathbf{z} = 0$ forbids changes in the plate thickness.

When $|\mathbf{c} \times \mathbf{z}| = \sin\theta \in\ ]0, 1[$, plates are *incoherent*. For these plates, which we here study, the representation (1) is unable to reproduce the obliqueness of the response. To compensate for this, a standard remedy is to use higher-order polynomial in the thickness (up to order three, and more; see R.C. Batra and S. Vidoli [4]). At variance with this, we propose and then study the following kinematical representation:

$$\tilde{\mathbf{u}}(y, \xi, t) = \tilde{\mathbf{u}}_0(y, t) + \xi \tilde{\mathbf{u}}_1(y, t), \tag{2}$$

where ξ is a coordinate taken along $\mathbf{c}$ starting from the place y in the middle plane. If compared with the representation (1), (2) has the advantage of reproducing the constitutive obliqueness by constraining the motion just in the direction $\mathbf{c}$, and the disadvantage of not allowing for an equally simple description of the displacement field of points belonging to the lateral mantle of the right cylinder occupied by the plate.

The motions (1) and (2) are prescribed in [8]–[10] to deduce two simple two-dimensional theories for the free vibrations of incoherent, transversely isotropic slabs. The two models are tested by comparing their predictions as to wave propagation with those of the parent three-dimensional theory. It turns out that the model based on kinematics (2) gives more accurate results when incoherence is weak ($\sin\theta$ close to 0) and the material's stiffness in the transverse direction is greater than in the isotropy plane.

Bearing these results in mind, in the present work we formulate a plate theory for *weakly incoherent, transversely isotropic* plates by prescribing admissible motions of the form (2). Incoherency weakness allows us to adopt the approximation $\sin\theta \sim \theta$, which is used to obtain an explicit formula for the elasticity tensor as a function of the incoherency angle θ (Section 2) and to perform some manipulations on the kinematics (2) (Section 3). In Section 4, a set of evolution equations and boundary conditions defined on the plate middle plane is deduced. The dimensional reduction is performed by thickness integration of the three-dimensional virtual work equation.

Within the dynamic plate theory we arrive at, we study the free propagation of harmonic waves (Section 5). We find that incoherence couples the vibrational modes in all possible ways: separation between symmetric and antisymmetric waves is lost; moreover, dilatational and isochoric oscillations at cut-off are coupled.

Since incoherence, however weak, destroys the orthogonality in energy of the membrane and flexure regimes, it can be seen as a defect. In Section 6 we propose an easy coherence test capable to detect such undesirable inchherent zones in a plate by propagating harmonic waves with certain assigned frequencies.

2   Weakly Incoherent Transversally Isotropic Plates

In the three-dimensional Euclidean space let us consider a plate-like body which we identify with its reference configuration $\Omega = \{p = x + \zeta\mathbf{z} \mid x \in \mathcal{P},\ \zeta \in \mathcal{I}\}$, where $\mathcal{P}$ is a two-dimensional flat region orthogonal to the unit vector $\mathbf{z}$ with boundary $\partial\mathcal{P}$, and $\mathcal{I} = (-\varepsilon,\ \varepsilon)$ an interval. We itroduce the orthonormal Cartesian frame $(o;\ \mathbf{c}_1,\ \mathbf{c}_2,\ \mathbf{z})$, with $o \in \mathcal{P}$, so that the position vector of the typical point $x \in \mathcal{P}$ with respect to the origin is $x - o = x_\alpha \mathbf{c}_\alpha,\ \alpha = 1, 2$.

The body Ω is supposed to be made of a homogeneous linearly elastic material being transversely isotropic with respect to an axis $\mathbf{c}$ whose orientation (the same considered in [8]–[10]) is obtained by rotating $\mathbf{z}$ around $\mathbf{c}_2$ counterclockwise by the angle θ: $\mathbf{c} = \mathbf{Q}(\theta)\mathbf{z}$, with $\mathbf{Q}(\theta) = \mathbf{I} + \sin\theta\mathbf{W} + (1 - \cos\theta)\mathbf{W}^2$, and $\mathbf{W} = 2\,\mathrm{skw}(\mathbf{c}_1 \otimes \mathbf{z})$ (in fig. 1 a geometrical draft is shown).

For this plate the *elasticity tensor* is

$$\mathbb{C}(\theta) = \mathbb{Q}(\theta)\mathbb{C}_0\mathbb{Q}^T(\theta), \tag{3}$$

obtained by rotating the elasticity tensor $\mathbb{C}_0 = \mathbb{C}(0)$ for a transversely isotropic material coherently oriented[1] by means of the fourth order tensor $\mathbb{Q}(\theta)$ which operates on any second-order tensor $\mathbf{A}$ as follows: $\mathbb{Q}[\mathbf{A}] = \mathbf{Q}\mathbf{A}\mathbf{Q}^T$. The procedure to arrive at (3) is described in details in [8] or [10] and an extended version of the elasticity tensor is there given.

We suppose that incoherence is weak, so that, the function $\mathbb{C}(\theta)$ can be developed in Taylor series around the configuration $\theta = 0$ up to the first order and $o(\theta)$ terms neglected:

$$\mathbb{C}_{lin}(\theta) = \mathbb{C}(0) + \theta\, \partial_\theta\mathbb{C}(\theta)|_{\theta=0}. \tag{4}$$

By developing the derivative in the second term we obtain

$$\partial_\theta\mathbb{C}(\theta)|_{\theta=0} = \mathbb{W}\mathbb{C}_0 - \mathbb{C}_0\mathbb{W},$$

where $\mathbb{W} = \partial_\theta\mathbb{Q}(\theta)|_{\theta=0}$ is a fourth-order tensor which operates on any symmetric tensor $\mathbf{B}$ as follows $\mathbb{W}[\mathbf{B}] = 2\,\mathrm{sym}(\mathbf{W}\mathbf{B})$, and the linearized version of the elasticity tensor is

$$\mathbb{C}_{lin}(\theta) = \mathbb{C}_0 + \theta(\mathbb{W}\mathbb{C}_0 - \mathbb{C}_0\mathbb{W}). \tag{5}$$

We again refer the reader to [8] for an extend representation of $\mathbb{C}_{lin}$. Here we just point out that both $\mathbb{C}_{lin}$ and $\mathbb{C}$ have the same structure and they reflect the response of a special monoclinic material characterized by the symmetry transformation $\mathbf{R}_{\mathbf{c}_2}^\pi$ (the rotation of π around $\mathbf{c}_2$) with each of its 13 material moduli a function of the 5 moduli of $\mathbb{C}_0$ and of the incoherence angle θ.

3   Motion

Kinematical assumptions are crucial to deduce plates theories. To successfully get two-dimensional balance equations from the three-dimensional virtual work equation by thickness integration, admissible motions must have the general form

$$\mathbf{u}(x, \zeta, t; \theta) = \sum_{i=0}^{n} \varphi_i(\zeta; \theta)\, \mathbf{u}_i(x, t), \tag{6}$$

where $\varphi_i(\zeta; \theta)$ are $(n+1)$ given functions which depend on the coordinate ζ and on the incoherence angle θ, and $\mathbf{u}_i(x, t)$ are $(n+1)$ unknown functions of the position $x \in \mathcal{P}$ and of time t.

By setting $\varphi_i(\zeta; \theta) = \zeta^i$ and $n = 1$ in (6), we get the representation (1). The kinematics (1), typical of coherent plates, can also be prescribed in the

---

[1] The tensor $\mathbb{C}_0$ is represented in the form given in §16 of [7]

326   G. Lancioni

incoherent case (see [8,9] and [4] for the elastic and electroelastic cases respectively) although it does not take into account the constitutive obliqueness.

In [8,9] free-waves propagation tests are performed on the conjectures (1) and (2), and since they reveal that the representation (2) is to prefer when incoherence is weak and the transversal isotropy strong (that is the material's stiffness is greater in the transversal direction $\mathbf{c}$ than in the isotropy plane), in the theory we are going to deduce (for weakly incoherent plates) we assume the kinematics (2) which we rewrite here:

$$\tilde{\mathbf{u}}(y, \xi, t) = \tilde{\mathbf{u}}_0(y, t) + \xi\tilde{\mathbf{u}}_1(y, t). \tag{7}$$

The pair (y, ξ) defines univocally the position of a typical point $p \in \Omega$ which can be represented in the following alternative way $p = y + \xi\mathbf{c}$ (see fig. 1), and the one-to-one map $(y, \xi) \leftrightarrow (x, \zeta)$ is

$$y = x - \zeta \tan\theta\, \mathbf{c}_1, \quad \xi = \zeta/\cos\theta. \tag{8}$$

Keeping in mind that our goal is to build a suitable kinematics of the form (6), we perform two operations on the kinematics (7). First we express it in terms of the cartesian coordinates (x, ζ) by means of the map (8):

$$\hat{\mathbf{u}}(x, \zeta;\, \theta) = \tilde{\mathbf{u}}_0(y(x, \zeta;\, \theta)) + \xi(\zeta;\, \theta)\tilde{\mathbf{u}}_1(y(x, \zeta;\, \theta)). \tag{9}$$

Then we make explicit the dependence of the descriptors $\tilde{\mathbf{u}}_0$ end $\tilde{\mathbf{u}}_1$ on ζ and θ: fixed the position (x, ζ), we develop (9) in Taylor series of θ up to the first order and we neglect the $o(\theta)$ terms by virtue of the incoherence weakness:

$$\hat{\mathbf{u}}(\theta) = \hat{\mathbf{u}}(\theta)\,|_{\theta=0} + \theta\partial_\theta\hat{\mathbf{u}}(\theta)\,|_{\theta=0}, \tag{10}$$

which, being

$$\partial_\theta\hat{\mathbf{u}}(\theta)\,|_{\theta=0} = -\zeta[\nabla\tilde{\mathbf{u}}_0(x)\mathbf{c}_1 + \zeta\nabla\tilde{\mathbf{u}}_1(x)\mathbf{c}_1], \tag{11}$$

turns into the following expression

$$\hat{\mathbf{u}}(x, \zeta, \theta) = \tilde{\mathbf{u}}_0(x) + \zeta[\tilde{\mathbf{u}}_1(x) - \theta\nabla\tilde{\mathbf{u}}_0(x)\mathbf{c}_1] - \zeta^2\theta\nabla\tilde{\mathbf{u}}_1(x)\mathbf{c}_1, \tag{12}$$

whose shape is of type (6). We point out that a first-order kinematics in the ξ coordinate is a second-order kinematics in the ζ coordinate. In the following, the two-dimensional theory based on the *Ansatz* (7) for a weakly incoherent plate is deduced and some free-waves solutions are evaluated.

4   Plate Equations

The balance principle we use is the *virtual work principle* which states that

$$\int_\Omega \mathbf{S} \cdot \nabla\mathbf{v} + \int_\Omega \varrho\ddot{\mathbf{u}} \cdot \mathbf{v} - \int_\Omega \mathbf{b}^{ni} \cdot \mathbf{v} - \int_{\partial\Omega} \mathbf{s} \cdot \mathbf{v} = 0, \tag{13}$$

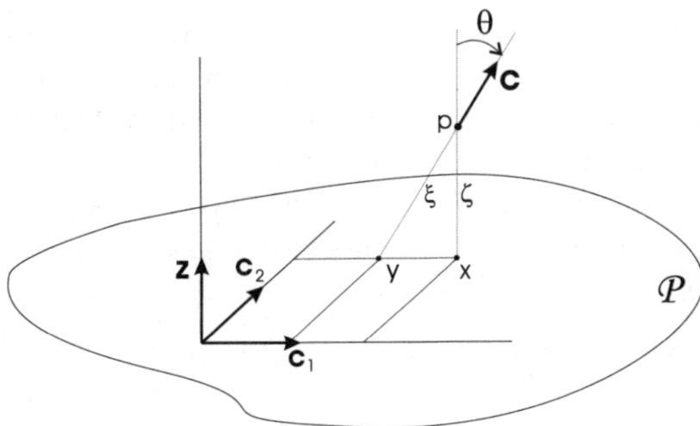

Fig. 1. Oblique geometry.

for all admissible test field $\mathbf{v}$, where $\mathbf{S}$ is the stress, ϱ the mass density, $-\varrho\ddot{\mathbf{u}}$ and $\mathbf{b}^{ni}$ are the inertial and non-inertial body loads and $\mathbf{s}$ is the surfaces external load.

Since we are going to investigate free vibration, we assume $\mathbf{b}^{ni} = \mathbf{s} = \mathbf{0}$. The application of the virtual work principle with test fields of the form (12) gives the following balance equations

$$\operatorname{div}\mathbf{M}_0 + \theta\nabla(\operatorname{div}\mathbf{M}_1 - \mathbf{M}_0\mathbf{z})\mathbf{c}_1 = 2\varepsilon\varrho\ddot{\tilde{\mathbf{u}}}_0 - \theta^2\frac{2}{3}\varepsilon^3\varrho\nabla((\nabla\ddot{\tilde{\mathbf{u}}}_0)\mathbf{c}_1)\mathbf{c}_1,$$

$$\operatorname{div}\mathbf{M}_1 - \mathbf{M}_0\mathbf{z} + \theta\nabla(\operatorname{div}\mathbf{M}_2 - 2\mathbf{M}_1\mathbf{z})\mathbf{c}_1] =$$
$$= \frac{2}{3}\varepsilon^3\varrho\ddot{\tilde{\mathbf{u}}}_1 - \theta^2\frac{2}{5}\varepsilon^5\varrho\nabla((\nabla\ddot{\tilde{\mathbf{u}}}_1)\mathbf{c}_1)\mathbf{c}_1,$$

$$(14)$$

where

$$\mathbf{M}_i(x,t;\theta) = \int_{-\varepsilon}^{\varepsilon}\zeta^i\mathbf{S}(x,\zeta,t;\theta), \qquad i = 0,...,2 \tag{15}$$

are dynamical descriptors defined in $\mathcal{P}$, and the boundary conditions either of Dirichlet type

$$\tilde{\mathbf{u}}_i = \tilde{\mathbf{w}}_i, \quad \partial_\nu\tilde{\mathbf{u}}_i = \partial_\nu\tilde{\mathbf{w}}_i, \quad i = 0, 1, \tag{16}$$

or of Neumann type

$$\mathbf{M}_0\nu + \theta(\mathbf{c}_1\cdot\nu)(\operatorname{div}\mathbf{M}_1 - \mathbf{M}_0\mathbf{z} - \frac{2}{3}\varepsilon^3\varrho\ddot{\tilde{\mathbf{u}}}_1)+$$
$$+\theta\partial_\tau[(\tau\cdot\mathbf{c}_1)\mathbf{M}_1\nu] + \theta^2\frac{2}{3}\varepsilon^3\varrho(\nu\cdot\mathbf{c}_1)(\nabla\ddot{\tilde{\mathbf{u}}}_0)\mathbf{c}_1 = 0,$$

$$\mathbf{M}_1\nu + \theta(\mathbf{c}_1\cdot\nu)(\operatorname{div}\mathbf{M}_2 - 2\mathbf{M}_1\mathbf{z} - \frac{2}{3}\varepsilon^3\varrho\ddot{\tilde{\mathbf{u}}}_0)+$$
$$+\theta\partial_\tau[(\tau\cdot\mathbf{c}_1)\mathbf{M}_2\nu] + \theta^2\frac{2}{5}\varepsilon^5\varrho(\nu\cdot\mathbf{c}_1)(\nabla\ddot{\tilde{\mathbf{u}}}_1)\mathbf{c}_1 = 0,$$

$$(\nu\cdot\mathbf{c}_1)\mathbf{M}_1\nu = 0, \quad (\nu\cdot\mathbf{c}_1)\mathbf{M}_2\nu = 0,$$

$$(17)$$

defined on $\partial\mathcal{P}$, where ν and τ are the unit outer normal and the unit tangent to $\partial\mathcal{P}$ respectively and $\partial_\nu(\cdot)$ and $\partial_\tau(\cdot)$ the derivatives in the ν and τ directions. We point out that the Dirichlet conditions (16) are equivalent to assign a displacement $\hat{\mathbf{w}}$ with the shape (9) on the mantle $\partial\mathcal{P}\times\mathcal{I}$.

If we introduce the constitutive relation $\mathbf{S}=\mathbb{C}_{lin}[\mathrm{sym}(\nabla\hat{\mathbf{u}})]$, with $\mathbb{C}_{lin}$ defined in (5), into the dynamical descriptors (15), the system (14) turn into the following evolution equations

$$2\varepsilon\{\mu\Delta\bar{\mathbf{u}}_0+(\lambda+\mu)\nabla\operatorname{div}\bar{\mathbf{u}}_0+\tau_2\nabla w_1+\theta[-(\lambda+\mu)\nabla(\nabla w_0\cdot\mathbf{c}_1)+$$
$$+(\tau_2-\lambda)\nabla(\bar{\mathbf{u}}_1\cdot\mathbf{c}_1)+(\eta-\mu)\Delta w_0\mathbf{c}_1+(\eta-\mu)(\operatorname{div}\bar{\mathbf{u}}_1)\mathbf{c}_1-\mu(\nabla\bar{\mathbf{u}}_1)\mathbf{c}_1]\}=$$
$$=2\varepsilon\varrho\ddot{\bar{\mathbf{u}}}_0,$$

$$2\varepsilon\{\eta(\Delta w_0+\operatorname{div}\bar{\mathbf{u}}_1)+\theta[-(\tau_2+2\eta)(\nabla w_1\cdot\mathbf{c}_1)-(\lambda+\mu)\operatorname{div}((\nabla\bar{\mathbf{u}}_0)\mathbf{c}_1)+$$
$$+(\eta-\mu)\Delta(\bar{\mathbf{u}}_0\cdot\mathbf{c}_1)]\}=2\varepsilon\varrho\ddot{w}_0,$$

$$\tfrac{2}{3}\varepsilon^3[\mu\Delta\bar{\mathbf{u}}_1+(\lambda+\mu)\nabla\operatorname{div}\bar{\mathbf{u}}_1]-2\varepsilon\eta(\nabla w_0+\bar{\mathbf{u}}_1)+$$
$$+\theta\{\tfrac{2}{3}\varepsilon^3[-(\lambda+\mu)\nabla(\nabla w_1\cdot\mathbf{c}_1)+(\eta-\mu)\Delta w_1\mathbf{c}_1]+$$
$$+2\varepsilon[(\tau_2+2\eta-\tau_1)w_1\mathbf{c}_1+(\lambda-\tau_2)(\operatorname{div}\bar{\mathbf{u}}_0)\mathbf{c}_1+\mu(\nabla\bar{\mathbf{u}}_0)\mathbf{c}_1+\mu(\nabla\bar{\mathbf{u}}_0)\mathbf{c}_1+$$
$$-(\eta-\mu)\nabla(\bar{\mathbf{u}}_0\cdot\mathbf{c}_1)]\}=\tfrac{2}{3}\varepsilon^3\ddot{\bar{\mathbf{u}}}_1,$$

$$\tfrac{2}{3}\varepsilon^3\eta\Delta w_1-2\varepsilon(\tau_1 w_1+\tau_2\operatorname{div}\bar{\mathbf{u}}_0)+\theta\{\tfrac{2}{3}\varepsilon^3[-(\lambda+\mu)\operatorname{div}((\nabla\bar{\mathbf{u}}_1)\mathbf{c}_1)+$$
$$+(\eta-\mu)\Delta(\bar{\mathbf{u}}_1\cdot\mathbf{c}_1)]+2\varepsilon[(\tau_2+2\eta)\nabla w_0\cdot\mathbf{c}_1+(\tau_2-\tau_1+2\eta)\bar{\mathbf{u}}_1\cdot\mathbf{c}_1]\}=$$
$$=\tfrac{2}{3}\varepsilon^3\varrho\ddot{w}_1.$$

$$(18)$$

where $w_i=\tilde{\mathbf{u}}_i\cdot\mathbf{z}$, $\bar{\mathbf{u}}_i=\tilde{\mathbf{u}}_i-w_i\mathbf{z}$ $(i=0,\,1)$ and all the $o(\theta)$ terms are neglected. We notice that, if we set $\theta=0$, the system (18) decouples into two sets of equations in the unknowns $(\bar{\mathbf{u}}_0,w_1)$ and $(w_0,\bar{\mathbf{u}}_1)$ which govern the evolution of *membrane* and *flexure* regimes respectively.

5   Free-wave Propagation

We now study solution of system (18) having the form of *plane progressive waves propagating in the direction* $\mathbf{c}_1$, with wavenumber k and frequency f. Accordingly, we assume for the unknowns of system (18) the following representations:

$$\bar{\mathbf{u}}_0=\mathbf{U}_0 g(x,t),\ \bar{\mathbf{u}}_1=\mathbf{U}_1 g(x,t),\ w_0=W_0 g(x,t),\ w_1=W_1 g(x,t),\quad(19)$$

where $\mathbf{U}_i=(U_i,\,V_i)^T$ and W_i $(i=0,\,1)$ are constant amplitudes and $g(x,t)=\exp(i(k\mathbf{c}_1\cdot(x-o)-ft))$ is the *wave shape*. Substitution of (19) into the evolution equations (18) yields two linear algebraic systems, namely,

$$\mathbf{L}(\tilde{k},\tilde{f})\mathbf{a}=\mathbf{0},\quad\mathbf{I}(\tilde{k},\tilde{f})\mathbf{b}=\mathbf{0},\qquad(20)$$

where the unknowns are the vectors $\mathbf{a}=(W_0,\,U_0,\,\varepsilon W_1,\,\varepsilon U_1)^T$ and $\mathbf{b}=(V_0,\,V_1)^T$, and the matrices $\mathbf{L}(\tilde{k},\tilde{f})$ and $\mathbf{I}(\tilde{k},\tilde{f})$, whose detailed form is to be found in [8], depend on the dimensionless parameters $\tilde{k}=\varepsilon k$ and $\tilde{f}=\varepsilon\sqrt{\varrho/\mu}f$.

As a consequence of our choice of the propagation direction $\mathbf{c}_1$, the wave motion splits into an irrotational oscillation in the coordinate plane $x_2 = 0$ determined by system $(20)_1$ and a solenoidal oscillation directed along the x_2-axis solution of system $(20)_2$. For nontrivial solutions to exist, we require that

$$\det \mathbf{L}(\tilde{k}, \tilde{f}) = 0 \quad \text{and} \quad \det \mathbf{I}(\tilde{k}, \tilde{f}) = 0. \tag{21}$$

Let us consider equation $(21)_1$. Since the left-hand side is a polynomial of fourth order both in $\tilde{k}^2$ and in $\tilde{f}^2$, for each fixed positive $\tilde{f}$, four waves with different wavelength $\tilde{k}$ propagate and, on the contrary, for each fixed $\tilde{k} \in \mathbb{R}^+$, four progressive waves propagate with different frequencies $\tilde{f}$. We notice that if $\tilde{k}$ is real (or immaginary), the wave is *progressive* (or *standing*).

In Fig.2 the dispersion curves for a ceramic plate [2] with an incoherence angle $\theta = 15^o$ are drawn in solid line and compared with those of a coherent plate in dashed line.

It is worth pointing out some changes in the wave propagation which are brought by incoherence. First of all, separation of *flexure* and *membrane waves* is lost. Moreover, at cut-off ($\tilde{k} << 1$), incoherence couples dilatational and isochoric waves. In the coherent case, when $\tilde{f} = \tilde{f}_1^c$, an isochoric in-plane shear oscillation of the "deck-of-cards" type takes place ($\tilde{\mathbf{u}} \cdot \mathbf{z} = 0$), and, when $\tilde{f} = \tilde{f}_2^c$, a symmetric dilatational oscillation occurs ($\tilde{\mathbf{u}} \cdot \mathbf{c}_1 = 0$). As soon as incoherence appears, both at frequency f_1^{inc} end at frequency f_2^{inc}, the progressive waves are the sum of an in-plane shear oscillation and a symmetric dilatational oscillation

$$\tilde{\mathbf{u}} = \exp(-i\, f_{1,2}^{inc}\, t)\, \zeta\, (U_1 \mathbf{c}_1 + W_1 \mathbf{z}). \tag{22}$$

At zero frequency a rigid motion in the $x_2 = 0$ plane is predicted both in the coherent and incoherent cases.

In the incoherent case, the no null cut-off frequencies values are

$$\tilde{f}_{1,2}^{inc} = (\sqrt{6(\eta + \tau_1 \mp \sqrt{4\theta^2(2\eta + \tau_2 - \tau_1)^2 + (\eta - \tau_1)^2}\,)/\mu}\,)/2, \tag{23}$$

and they differ from the corresponding coherent ones for the presence of the term $4\theta^2(2\eta + \tau_2 - \tau_1)^2$ whose second factor is a combination of constitutive moduli which characterize the mechanical behavior in the direction $\mathbf{c}$ of transversal isotropy. It follows that the shifts of the cut-off frequencies from the corresponding coherent ones are as great as θ is great and/or as the material stiffness in the direction $\mathbf{c}$ is greater than in the isotropy plane.

The dispersion equation $(21)_2$ for the *twist* waves is independent of θ, and it is satisfied by two parabolic dispersion curves[3]. In [5] and [6] the study of harmonic waves propagating in plates is extended to the electroelastic case.

---

[2] We use the constitutive moduli of lead titanate-zirconate P1-88 taken from Table 1 in [8].

[3] For their analytical expressions and representations we refer the reader to [8].

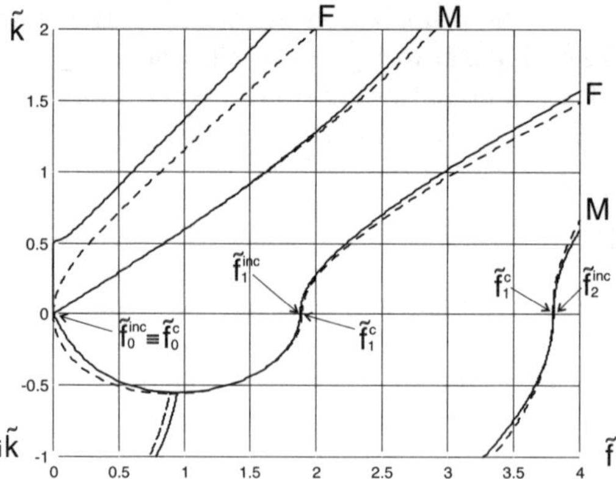

Fig. 2. Dispersion branches for the irrotational wave motions: incoherent slab $\equiv$ solid line, coherent slab $\equiv$ dashed line. The dispersion branches for the coherent case labelled by the letters f and m characterize the flexure and membrane waves respectively.

5.1   Coherence Tests

In some plate problems the coupling of membrane and flexure regimes due to incoherence may be undesired and, as a result, it may be useful to check the presence of incoherent zones in a plate. Here we suggest a test to detect such regions via wave propagation. Looking at fig. 2, we zero in on the two couples of branches with no null cut-off frequencies and small $\tilde{k}$ and the enlargements of these portions are drawn in fig. 3. We point out that the shifts of the branches along the $\tilde{f}$ axis yield two frequencies ranges within which waves with low wave-number are progressive or standing depending on coherence or incoherence and viceversa.

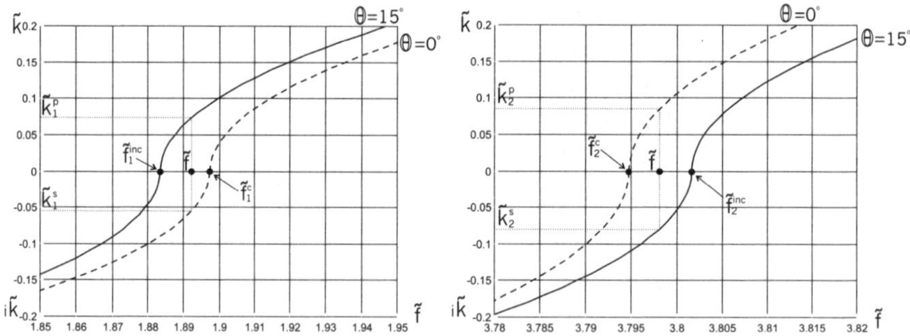

Fig. 3. Dispersion branches near the cut-off frequencies: dashed line $\equiv$ $(\theta = 0)$, solid line $\equiv$ $(\theta = 15^{o})$.

To be more precise, from fig. 3.1 we notice that if we fix the frequency $\tilde{f} \in (\tilde{f}_1^{inc}, \tilde{f}_1^{c})$, an essentially in plane-shear progressive wave propagates in incoherent regions with wavenumber $\tilde{k} = \tilde{k}_1^{p}$ and, as soon as it enters a coherent zone, it becomes standing and it decays in the direction $\mathbf{c}_1$ according to the factor $\exp(-\tilde{k}_1^{s} x_1 / i\varepsilon)$. Looking to fig. 3.2, an essentially symmetric dilatational wave with frequency $\tilde{f} \in (\tilde{f}_2^{c}, \tilde{f}_2^{inc})$ is progressive in coherent areas and standing in the incoherent ones. The wave motions described above can be carried out by easy dynamical experiments with a frequency control and their sudden changes observed to draw a map of the incoherent zones.

References

1. Reissner E. (1946) On the theory of transverse bending of elastic plates Int. J. Solids Structures **12**.
2. Mindlin R.D. (1951) Thickness-Shear and Flexural Vibrations of Cristal Plates. J. of Applied Physics **22** (3).
3. Mindlin R.D. (1951) Influence of rotatory inertia and shear on flexural motions of isotropic, elastic plates. J. Appl. Mech. **31-38**.
4. Batra R.C., Vidoli S. (2000) Derivation of plate and rod equations for a piezo-electric body from a mixed three-dimensional variational principle. J. Elasticity **59** 23-50.
5. Podio-Guidugli P., Tomassetti G. (2001) Thickness waves in electroelastic plates. Wave motion **34** 175-191.
6. Lancioni G., Tomassetti G. (2002) Flexure waves in electroelastic plates. Wave motion **35** 257-269.
7. Podio-Guidugli P. (2000) A Primer in Elasticity. Kluwer.
8. Lancioni G. (2002) Dinamica di piastre incoerenti. Tesi di dottorato.
9. Lancioni G. (2002) Free waves in incoherent slabs. Proceedings Euromech Colloquium 444, Critical Review of the Theories of Plates and Shells and New Applications, Bremen, Germany.
10. Lancioni G., Podio-Guidugli P. (2003) forthcoming.

A Finite Element for the Analysis of Monoclinic Laminated Plates

Ferdinando Auricchio[1], Elio Sacco[2], Giuseppe Vairo[3]

[1] Dipartimento di Meccanica Strutturale
   Università di Pavia, via Ferrata, 1
   27100 Pavia, Italy
[2] Dipartimento di Meccanica, Strutture, A&T
   Università di Cassino, via Di Biasio, 43
   03043 Cassino, Italy
[3] Dipartimento di Ingegneria Civile
   Università di Roma "Tor Vergata",
   via del Politecnico, 1
   00133 Roma, Italy

Abstract. This paper presents a 4-node finite element, based on a First-order Shear-Deformation Theory (FSDT), for the analysis of composite monoclinic laminated plates. A mixed-enhanced variational formulation is adopted. It includes as primary variables the transverse shear as well as enhanced incompatible modes, which are introduced to improve in-plane deformations. Several numerical applications are presented in order to show the effectiveness of the proposed element.

1   Introduction

Composite laminated plates have generally an anisotropic behaviour that involves extension-bending coupling and they usually exhibit non-negligible shear deformation in the thickness. Furthermore, the determination of accurate values for interlaminar normal and shear stresses represents an important task in order to analyse activation and development of delamination mechanisms.

The First-order Shear-Deformation Theory (FSDT) is a 2D laminate theory based on both strain and stress hypotheses and it was originally developed by Yang et al. [16] and by Whitney and Pagano [15] as an extension of the plate theory proposed by Reissner [12] and Mindlin [6]. As specific literature confirms [2–4,15], the FSDT gives the possibility of taking into account shear deformation effects in a simple way and it appears as the best compromise between prediction ability and computational costs for a wide class of laminate problems even for moderately thick laminates. However, it is worth to observe that accurate results are obtained only if proper values for the shear correction factors are adopted.

This paper starts with a review of the FSDT's basic hypotheses and techniques usually adopted for the recovery of transverse shear stresses are also recalled. Hence, a mixed-enhanced variational formulation is used to develop a 4-node finite element for the analysis of laminated plates with monoclinic constitutive behaviour. The element is locking free and it employs enhanced incompatible modes in order to improve the interpolation scheme for in-plane deformations. It is able to provide accurate in-plane/out-of-plane deformations and transverse shear stresses as proved through several numerical applications.

2   First-Order Laminate Theory (FSDT)

A laminated plate Ω refers to a flat body formed by n perfectly bonded layers and it is defined as:

$$\Omega = \left\{ (x_1, x_2, x_3) \in \mathbb{R}^3 : x_3 \in (-h/2, h/2), \mathbf{x} = (x_1, x_2) \in \mathcal{P} \subset \mathbb{R}^2 \right\} \quad (1)$$

where the plane $x_3 = 0$ identifies the middle cross-section $\mathcal{P}$ of the undeformed plate. The kth lamina occupies the region $\mathcal{P} \times \left] x_3^{(k-1)}, x_3^{(k)} \right[$, where the superscript index k discriminates any quantity referred to the kth layer. The thickness h is constant and it is assumed to be small compared to in-plane dimensions. In the following, Greek indices assume values in $\{1, 2\}$ and Einstein summation convention is adopted.

Basic FSDT strain and stress assumptions are: out-of-plane shear stress is null for each layer, i.e. $\sigma_{33}^{(k)} = 0$; straight lines orthogonal to the mid-plane cannot be stretched and they remain straight after the deformation, i.e. $\varepsilon_{33}^{(k)} = 0$, $\varepsilon_{\alpha 3,3}^{(k)} = 0$. It can be observed that, in a general 3D elastic formulation, the simultaneous presence of these statements is formally correct as proved through the concepts of constrained continua [5].

The above-cited kinematical assumptions lead to the following representation form for the displacement field $\mathbf{s}$ as a function of the mid-plane displacements $\mathbf{u} = u_\alpha \mathbf{e}_\alpha$, w and rotations $\boldsymbol{\varphi} = \varphi_\alpha \mathbf{e}_\alpha$:

$$\mathbf{s}(\mathbf{x}, x_3) = \mathbf{u}(\mathbf{x}) + w(\mathbf{x})\mathbf{e}_3 + x_3 \boldsymbol{\varphi}(\mathbf{x}) \quad (2)$$

where $\{\mathbf{e}_\alpha, \mathbf{e}_3\}$ represents an orthonormal basis for a fixed Cartesian coordinate frame.

Hence, the strain tensor $\boldsymbol{\varepsilon}$ can be decomposed as:

$$\boldsymbol{\varepsilon} = \nabla^{(s)}\mathbf{s} = \begin{bmatrix} \bar{\boldsymbol{\varepsilon}} & \frac{1}{2}\boldsymbol{\gamma} \\ \frac{1}{2}\boldsymbol{\gamma}^T & 0 \end{bmatrix} \quad (3)$$

where $\bar{\boldsymbol{\varepsilon}}$ is the in-plane strain tensor and $\boldsymbol{\gamma}$ represents the shear strain vector:

$$\bar{\boldsymbol{\varepsilon}} = \overline{\nabla}^{(s)}\mathbf{u} + x_3 \overline{\nabla}^{(s)}\boldsymbol{\varphi} = \mathbf{E} + x_3 \boldsymbol{\Theta}, \qquad \boldsymbol{\gamma} = \{\gamma_{13}, \gamma_{23}\}^T = \boldsymbol{\varphi} + \overline{\nabla} w. \quad (4)$$

The following notation rules are introduced: symbol ∇ denotes gradient operator, $\overline{\nabla}$ denotes in-plane gradient operator, i.e. with respect to variables x_α, and finally the superscript index (s) denotes the symmetrical part of a tensor.

The laminated plate is assumed to be formed by monoclinic layers with symmetry plane parallel to $\mathcal{P}$ so that it results $C^{(k)}_{\alpha\beta\gamma3} = C^{(k)}_{33\gamma3} = 0$, where $\left(\mathbb{C}^{(k)}\right)_{\alpha\beta\gamma\delta} = C^{(k)}_{\alpha\beta\gamma\delta}$ is the fourth-order elasticity tensor for the kth layer. Taking into account the above-mentioned stress assumptions and introducing the in-plane stress tensor $\overline{\sigma}$ and the transverse shear stress vector $\tau = \{\sigma_{13}, \sigma_{23}\}^T$ the stress-strain relations for the kth layer become:

$$\overline{\sigma}^{(k)} = \tilde{\mathbb{C}}^{(k)}\overline{\varepsilon} = \tilde{\mathbb{C}}^{(k)}\left(\mathbf{E} + x_3\boldsymbol{\Theta}\right), \qquad \tau^{(k)} = \tilde{\mathbf{Q}}^{(k)}\gamma \tag{5}$$

where $\tilde{\mathbb{C}}^{(k)}$ is the reduced in-plane elasticity tensor and $(\tilde{\mathbf{Q}}^{(k)})_{\alpha\beta} = C^{(k)}_{\alpha3\beta3}\chi_{\alpha\beta}$ is the corrected shear elastic one. The quantities $\chi_{\alpha\beta}$ are the so-called shear correction factors and they are assumed to be constant along the laminate thickness direction. It is interesting to recall that the well known values $\chi_{11}(= \chi_{22}) = 5/6$, $\chi_{12}(= \chi_{21}) = 0$ are strictly correct only for homogeneous isotropic plates.

Introducing the resultant stresses as:

$$\mathbf{N} = \int_{-h/2}^{h/2} \overline{\sigma}\,dx_3, \qquad \mathbf{M} = \int_{-h/2}^{h/2} x_3\overline{\sigma}\,dx_3, \qquad \mathbf{S} = \int_{-h/2}^{h/2} \tau\,dx_3, \tag{6}$$

the constitutive equations between resultant stresses and kinematic variables can be put in the form:

$$\begin{bmatrix} \mathbf{N} \\ \mathbf{M} \\ \mathbf{S} \end{bmatrix} = \begin{bmatrix} \mathbf{A} & \mathbf{B} & 0 \\ \mathbf{B} & \mathbf{D} & 0 \\ 0 & 0 & \mathbf{H} \end{bmatrix} \begin{bmatrix} \mathbf{E} \\ \boldsymbol{\Theta} \\ \gamma \end{bmatrix} \tag{7}$$

where the laminate stiffness tensors are obtained by:

$$\mathbf{A} = \sum_{k=1}^{n} \left(x_3^{(k)} - x_3^{(k-1)}\right)\tilde{\mathbb{C}}^{(k)}, \quad \mathbf{B} = \frac{1}{2}\sum_{k=1}^{n} \left(x_3^{2(k)} - x_3^{2(k-1)}\right)\tilde{\mathbb{C}}^{(k)}, \tag{8}$$

$$\mathbf{D} = \frac{1}{3}\sum_{k=1}^{n} \left(x_3^{3(k)} - x_3^{3(k-1)}\right)\tilde{\mathbb{C}}^{(k)}, \quad \mathbf{H} = \sum_{k=1}^{n} \left(x_3^{(k)} - x_3^{(k-1)}\right)\tilde{\mathbf{Q}}^{(k)}. \tag{9}$$

The equations (7) highlight the high extension-bending coupling in the case of monoclinic laminae.

An accurate evaluation for through-the-thickness shear stress can be recovered using the 3D equilibrium equations. If no in-plane loads per unit volume are considered, shear stresses in the kth layer result:

$$\hat{\tau}^{(k)} = \hat{\tau}_o^{(k)} - \int_{x_3^{(k-1)}}^{x_3} \overline{\nabla}\cdot\overline{\sigma}^{(k)}\,d\zeta = \hat{\tau}_o^{(k)} - \int_{x_3^{(k-1)}}^{x_3} \overline{\nabla}\cdot\left[\tilde{\mathbb{C}}^{(k)}\left(\mathbf{E} + \zeta\boldsymbol{\Theta}\right)\right]\,d\zeta \tag{10}$$

where $\hat{\tau}_o^{(k)} = \hat{\tau}^{(k)}\Big|_{x_3=x_3^{(k-1)}}$, with $\hat{\tau}_o^{(1)} = \mathbf{0}$.

3   Mixed-Enhanced Finite-Element Formulation

In order to avoid locking phenomena, the finite element formulation is built up using a Hu-Washizu-like functional, with partial mixed terms for the transverse shear [2]. In the framework of incompatible modes [8,13], in-plane strains are enhanced with a field $\overline{\varepsilon}^{en}$, such that $\mathbf{E} = \overline{\nabla}^{(s)}\mathbf{u} + \overline{\varepsilon}^{en}$, $\boldsymbol{\Theta} = \overline{\nabla}^{(s)}\boldsymbol{\varphi}$. Accordingly, the following mixed functional for monoclinic laminates is obtained [2]:

$$
\Pi(\mathbf{u}, \boldsymbol{\varphi}, w, \mathbf{S}, \overline{\varepsilon}^{en}) = \frac{1}{2} \int_{\mathcal{P}} \left(\overline{\nabla}^{(s)}\mathbf{u} + \overline{\varepsilon}^{en} \right)^T \mathbf{A} \left(\overline{\nabla}^{(s)}\mathbf{u} + \overline{\varepsilon}^{en} \right) dA
$$

$$
+ \int_{\mathcal{P}} \left(\overline{\nabla}^{(s)}\mathbf{u} + \overline{\varepsilon}^{en} \right)^T \mathbf{B}\overline{\nabla}^{(s)}\boldsymbol{\varphi}\, dA + \frac{1}{2} \int_{\mathcal{P}} \left(\overline{\nabla}^{(s)}\boldsymbol{\varphi} \right)^T \mathbf{D}\overline{\nabla}^{(s)}\boldsymbol{\varphi}\, dA
$$

$$
- \frac{1}{2} \int_{\mathcal{P}} \mathbf{S}^T \mathbf{H}^{-1} \mathbf{S}\, dA + \int_{\mathcal{P}} \mathbf{S}^T \left(\boldsymbol{\varphi} + \overline{\nabla} w \right) dA - \Pi_{ext}. \qquad (11)
$$

An iso-parametric 4-node finite-element for laminated plates is built up discretizing the functional (11) and considering the standard iso-parametric map [17]: $\mathbf{x} = \boldsymbol{\Psi}\hat{\mathbf{x}}$, where $\mathbf{x}$ denotes the global coordinate vector in the physical element, $\boldsymbol{\Psi}$ contains the bi-linear shape functions and $\hat{\mathbf{x}}$ is the nodal coordinate vector.

The in-plane displacements are taken bi-linear in the nodal parameters $\hat{\mathbf{u}}$, i.e. $\mathbf{u} = \boldsymbol{\Psi}\hat{\mathbf{u}}$. The interpolation scheme for the rotational field is also bi-linear in the nodal parameters $\hat{\boldsymbol{\varphi}}$, with added internal degrees of freedom $\hat{\boldsymbol{\varphi}}_b$, i.e. $\boldsymbol{\varphi} = \boldsymbol{\Psi}\hat{\boldsymbol{\varphi}} + \boldsymbol{\Psi}_b\hat{\boldsymbol{\varphi}}_b$, where $\boldsymbol{\Psi}_b$ are bubble functions defined as [1]:

$$
\boldsymbol{\Psi}_b = \frac{(1 - \xi_1^2)(1 - \xi_2^2)}{j} \begin{bmatrix} J_{22}^o & -J_{12}^o & J_{22}^o \xi_2 & -J_{12}^o \xi_1 \\ -J_{21}^o & J_{11}^o & -J_{21}^o \xi_2 & J_{11}^o \xi_1 \end{bmatrix} \qquad (12)
$$

being $\boldsymbol{\xi} = \{\xi_1, \xi_2\}^T$ the local coordinate vector in the parent element, $\mathbf{J}_o$ the jacobian of the iso-parametric map evaluated at $\boldsymbol{\xi} = \mathbf{0}$, i.e. $\mathbf{J}_o = \partial\mathbf{x}/\partial\boldsymbol{\xi}|_{\boldsymbol{\xi}=\mathbf{0}}$, and $j = det[\mathbf{J}]$. Furthermore, the transverse displacement interpolation is bi-linear in the nodal parameters $\hat{\mathbf{w}}$, enriched with linked quadratic functions expressed in terms of nodal rotations:

$$
w = \boldsymbol{\Psi}\hat{\mathbf{w}} + \boldsymbol{\Psi}_{w\varphi}\hat{\boldsymbol{\varphi}} = \sum_{i=1}^{4} \boldsymbol{\Psi}^i \hat{w}^i - \sum_{i=1}^{4} \boldsymbol{\Psi}^i_{w\varphi} L^i \left(\hat{\varphi}_n^j - \hat{\varphi}_n^i \right) \qquad (13)
$$

where $L^{(i)}$ is the $i - j$ side length, $\hat{\varphi}_n^i$ and $\hat{\varphi}_n^j$ are the rotational components of the nodes i and j along the direction orthogonal to the $i - j$ side. Moreover, the shape functions $\boldsymbol{\Psi}_{w\varphi}$ are set as:

$$
\boldsymbol{\Psi}_{w\varphi} = \begin{Bmatrix} \Psi_{w\varphi}^1 \\ \Psi_{w\varphi}^2 \\ \Psi_{w\varphi}^3 \\ \Psi_{w\varphi}^4 \end{Bmatrix} = \frac{1}{16} \begin{Bmatrix} (1 - \xi_1^2)(1 - \xi_2) \\ (1 + \xi_1)(1 - \xi_2^2) \\ (1 - \xi_1^2)(1 + \xi_2) \\ (1 - \xi_1)(1 - \xi_2^2) \end{Bmatrix}. \qquad (14)
$$

For that regards the shear interpolation scheme it is assumed also bi-linear and defined locally to each element, such as: $\mathbf{S} = \boldsymbol{\Psi}_S \hat{\mathbf{S}}$, resulting $\hat{\mathbf{S}}$ as local parameters to each element and

$$\boldsymbol{\Psi}_S = \begin{bmatrix} J_{11}^o & J_{21}^o & J_{11}^o \xi_2 & J_{21}^o \xi_1 \\ J_{12}^o & J_{22}^o & J_{12}^o \xi_2 & J_{22}^o \xi_1 \end{bmatrix}. \tag{15}$$

Finally, the vector field of the enhanced strain field is expressed as a function of internal degrees of freedom $\hat{\boldsymbol{\varepsilon}}^{en}$, i.e. $\overline{\boldsymbol{\varepsilon}}^{en} = \boldsymbol{\Gamma} \hat{\boldsymbol{\varepsilon}}^{en}$, where $\boldsymbol{\Gamma}$ is an interpolation matrix which is constructed mapping an interpolation matrix defined on the parent element into the physical one, as reported in [13].

The relevant balance equation is obtained introducing the above-cited interpolation schemes and performing the functional stationary conditions for a single element [2].

Since enhanced strains, bubble rotations and resultant shear stresses are local parameters to each element, they can be eliminated by static condensation. Thus, an element with only 5 global degrees of freedom per node, named MEML4 (Monoclinic Enhanced Mixed Linked 4-node), is built up.

Due to the adopted mixed formulation, an accurate evaluation of the resultant shear stress $\mathbf{S}$ is expected and, as a consequence, this occurrence leads to the possibility of shear stress profiles improving. Hence, by using the equation (10), the shear stresses are numerically computed as [2,3]:

$$\hat{\boldsymbol{\tau}}^{(k)}(x_3) = \begin{bmatrix} b_1 & 0 \\ 0 & b_2 \end{bmatrix} \left\{ \mathbf{a} \left(x_3 + \frac{h}{2} \right) - \int_{x_3^{(k-1)}}^{x_3} \overline{\nabla} \cdot \left[\tilde{\mathbb{C}}^{(k)} \left(\mathbf{E} + \zeta \boldsymbol{\Theta} \right) \right] d\zeta \right\} \tag{16}$$

where the vector $\mathbf{a}$ and the quantities b_1 and b_2 are evaluated enforcing the equilibrium constraints:

$$\int_{-h/2}^{h/2} \hat{\boldsymbol{\tau}} dx_3 - \mathbf{S} = \mathbf{0}, \qquad \hat{\boldsymbol{\tau}}|_{\pm h/2} = \mathbf{0}. \tag{17}$$

It is worth to observe that the ability to obtain accurate stress profiles opens the possibility to use iterative procedures for an accurate evaluation of the shear correction factors, as shown in [2,7].

4   Numerical Applications

Several examples have been investigated in order to asses the performances of the MEML4 laminate element, which has been implemented in FEAP (*Finite Element Analysis Program*) [14].

Square cross-ply laminates with side length L are considered and they are acted upon by a transversal sinusoidal load, i.e. $q = q_o \sin(\phi x_1) \sin(\phi x_2)$, where $\phi = \pi/L$. Layer properties are set as: $E_L/E_T = 25$, $\nu_{TT} = 0.25$, $G_{LT}/E_T = 0.5$, $G_{TT}/E_T = 0.2$, i.e. corresponding to an high modulus orthotropic graphite/epoxy composite material. The subscript index L refers to the along-the-fiber direction, whereas the subscript T refers to any orthogonal-to-fiber direction. The side-to-thickness ratio is assumed to be $\eta = L/h = 10$ and shear factors are set constant and equal to $\chi_{\alpha\alpha} = 5/6$, $\chi_{\alpha\beta} = 0$ $(\alpha \neq \beta)$.

Different cross-ply lamination sequences are analysed: on the first, the symmetrical 0/90/90/0 and the unsymmetrical 0/90/90 laminates are considered, with simply supported boundary conditions

$$s_2 = s_3 = \varphi_2 = 0 \text{ at } x_1 = 0, L \text{ and } s_1 = s_3 = \varphi_1 = 0 \text{ at } x_2 = 0, L. \quad (18)$$

The numerical analyses are performed using regular meshes, that is through square MEML4 elements. Moreover, due to symmetry considerations, only one quarter of laminated plates are discretized.

Table 1 reports the values of dimensionless transversal displacement $w^* = 100 E_T w / q_o h \eta^3$ at the plate center for the 0/90/90/0 laminate. Analogously, Table 2 reports the transversal displacement w^* at the plate center and the dimensionless horizontal displacement $u^* = 100 E_T s_1 / q_o h \eta^3$ at $x_1 = 0$, $x_2 = L/2$, $x_3 = 0$ for the 0/90/90 laminate. The MEML4's numerical results are compared with the FSDT analytical solution, which is obtained as discussed in [9]. The accuracy and the convergence of the MEML4 finite element are proved. Furthermore, it clearly appears that the element is locking free and it does not suffer free energy modes.

Table 1. Dimensionless transversal displacement w^* for the symmetrical laminate 0/90/90/0: comparison between analytical (AS) and finite-element solutions

AS	6.6271			
mesh	3×3	6×6	12×12	24×24
MEML4	6.6381	6.6302	6.6279	6.6273

Table 2. Dimensionless transversal and horizontal displacements w^* and u^* for the unsymmetrical laminate 0/90/90: comparison between analytical (AS) and finite-element solutions

MEML4	w^*	u^*
3×3	10.7109	-0.87993
6×6	10.7067	-0.86545
12×12	10.7055	-0.86189
24×24	10.7053	-0.86101
(AS)	10.7052	-0.86071

The laminated plate is now completely discretized using 11×11 MEML4 square elements in order to investigate about the ability of the element to compute satisfactory through-the-thickness and interlaminar shear stresses. The relevant results, which are obtained with and without enhanced strains, are compared with the analytical solution (AS). Figure 1 shows dimensionless shear stress profiles $t_{13} = 100\hat{\tau}_{13}/q_o\eta$ at $x_1 = L/22$, $x_2 = L/2$ and $t_{23} = 100\hat{\tau}_{23}/q_o\eta$ at $x_1 = L/2$, $x_2 = L/22$, for the 0/90/90/0 laminate.

It appears in evident way that there are not differences between enhanced and non-enhanced solutions. This occurrence is justified observing that transversal loading does not induce horizontal deformations and, as a consequence, the enhanced modes are not activated for this problem.

Figure 2 shows the dimensionless shear stress profiles t_{13} and t_{23} for the 0/90/90 laminate. In this case it can be observed a significant improvement of the enhanced solution with respect to the non-enhanced one. In fact, transversal loading induces horizontal deformations in the unsymmetrical laminate and, as a consequence, the enhanced modes are activated. In this way, a beneficial effect on the numerical solution is clearly produced. The excellent approximation of the shear stresses computed via MEML4 with respect to the FSDT analytical solution is proved.

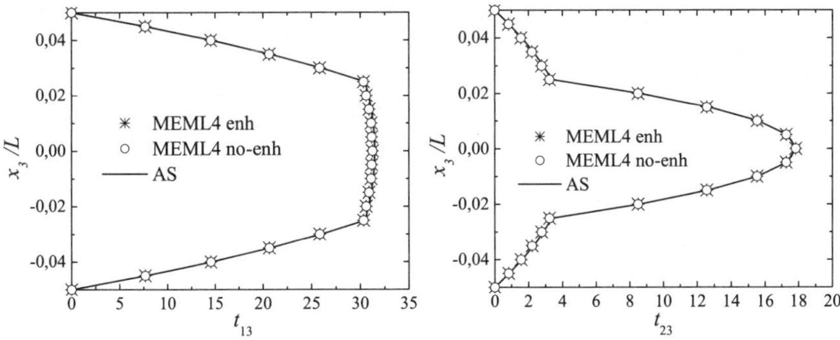

Fig. 1. Dimensionless shear stress profiles t_{13} and t_{23} for the symmetrical 0/90/90/0 laminate: comparison between numerical and analytical (AS) solutions

In order to check the element capability to describe a monoclinic behaviour, the axes x_1 and x_2 of the Cartesian frame, which are assumed to be parallel to the plate edges, are now chosen not to be coincident with the axes of material symmetry into the plane $x_3 = 0$. In this way, each layer exhibits an apparent monoclinic response in the Cartesian frame (O, x_1, x_2, x_3).

Also in this case square laminated plates, subjected to sinusoidal transversal load, are considered. Moreover, they are completely discretized using regular meshes based on MEML4 square elements. Table 3 shows the values of the dimensionless transversal displacement w^* at the plate center in the case of a symmetrical monoclinic-like lamination sequence, whereas Table 4 reports the values of w^* at the plate center and the dimensionless horizontal displacement u^* at $x_1 = L/4$, $x_2 = L/2$, $x_3 = 0$ for the unsymmetrical 30/120

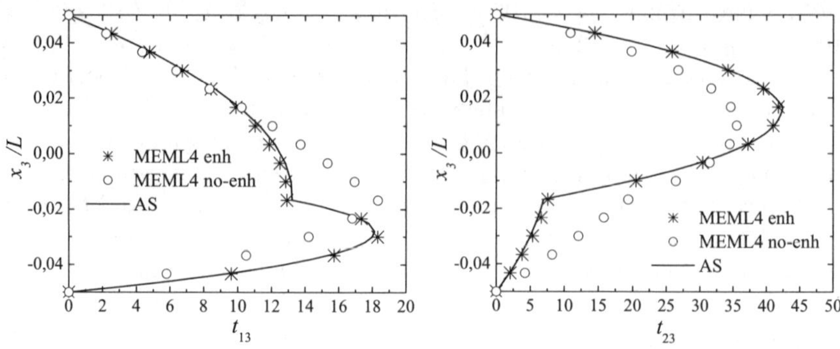

Fig. 2. Dimensionless shear stress profiles t_{13} and t_{23} for the unsymmetrical 0/90/90 laminate: comparison between numerical and analytical (AS) solutions

laminate. The MEML4's numerical results, which are obtained imposing non-homogeneous boundary conditions on the monoclinic-like laminated plates, are compared with the corresponding FSDT analytical solution. In detail, it can be proved that the closed-form FSDT solution for a monoclinic square laminated plate acted upon by a transversal sinusoidal load can be expressed as:

$$
\begin{aligned}
w &= \overline{w}\sin(\phi x_1)\sin(\phi x_2) + \widetilde{w}\cos(\phi x_1)\cos(\phi x_2) \\
\varphi_1 &= \overline{\varphi}_1\cos(\phi x_1)\sin(\phi x_2) + \widetilde{\varphi}_1\sin(\phi x_1)\cos(\phi x_2) \\
\varphi_2 &= \overline{\varphi}_2\sin(\phi x_1)\cos(\phi x_2) + \widetilde{\varphi}_2\cos(\phi x_1)\sin(\phi x_2) \\
u_1 &= \overline{u}_1\cos(\phi x_1)\sin(\phi x_2) + \widetilde{u}_1\sin(\phi x_1)\cos(\phi x_2) \\
u_2 &= \overline{u}_2\sin(\phi x_1)\cos(\phi x_2) + \widetilde{u}_2\cos(\phi x_1)\sin(\phi x_2)
\end{aligned}
\tag{19}
$$

where the amplitudes $(\overline{\cdot})$ and $(\widetilde{\cdot})$ of the unknown fields are computed solving the algebraic system which is obtained through the equilibrium equations. In this way, the external compatibility conditions are satisfied only if the laminated plate is subjected to non-homogeneous boundary conditions. These latter are obtained evaluating the quantities introduced in (19) on the edges $x_1 = 0$, $x_1 = L$ and $x_2 = 0$, $x_2 = L$.

Table 3. Displacement w^* for the monoclinic-like symmetrical 30/120/30 laminate: comparison between analytical (AS) and finite-element solutions

AS	1.3089			
mesh	3×3	6×6	12×12	24×24
MEML4	1.3025	1.3074	1.3086	1.3088

Table 4. Dimensionless displacements w^* and u^* for the monoclinic-like 30/120 laminate: comparison between analytical (AS) and finite-element solutions

MEML4	w^*	u^*
3×3	1.2758	0.02569
6×6	1.2808	0.02675
12×12	1.2818	0.02679
24×24	1.2820	0.02680
(AS)	1.2821	0.02681

Also in this case the convergence is proved and it can be again emphasized that the element does not suffer locking phenomena.

The plate is then completely discretized using 11×11 MEML4 square elements in order to compute the through-the-thickness and interlaminar stresses. The relevant numerical results, which are obtained with and without enhanced strains, are compared with the analytical solutions (AS). Figures 3 and 4 depict the dimensionless shear stress profiles t_{13} at $x_1 = L/22$, $x_2 = L/2$ and t_{23} at $x_1 = L/2$, $x_2 = L/22$ for the symmetrical 30/120/30 and the unsymmetrical 30/120 monoclinic-like laminates, respectively.

The same considerations of the previous cases can be expressed, that is in the case of unsymmetrical configuration the numerical solution is significantly improved by using enhanced modes. Moreover, it can be emphasized that even coarse MEML4-based mesh discretizations produce accurate results not only in terms of displacements but also in terms of shear stress profiles.

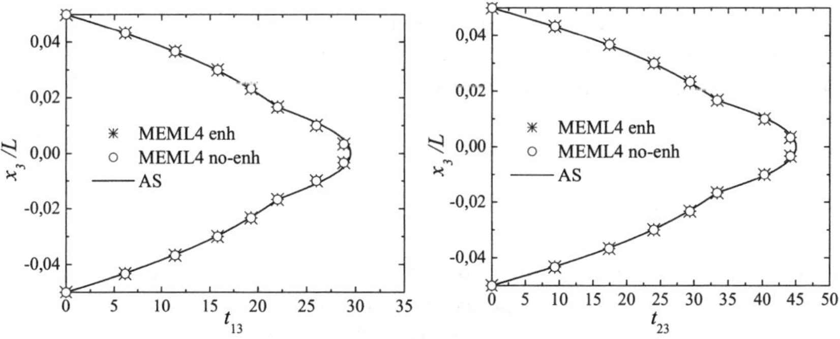

Fig. 3. Dimensionless shear stress profiles t_{13} and t_{23} for the symmetrical monoclinic-like 30/120/30 laminate: comparison between numerical and analytical (AS) solutions

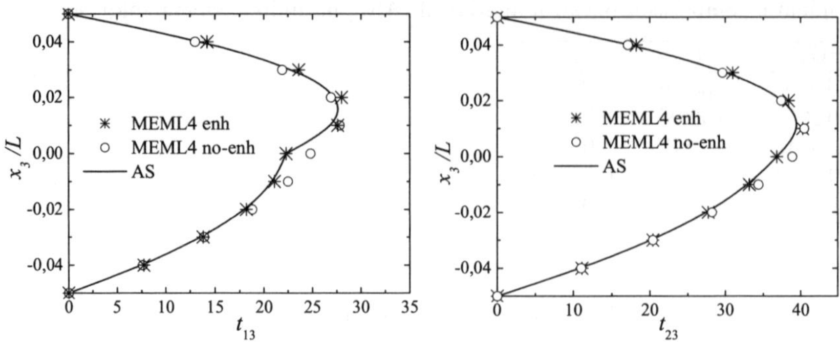

Fig. 4. Dimensionless shear stress profiles t_{13} and t_{23} for the unsymmetrical monoclinic-like 30/120 laminate: comparison between numerical and analytical (AS) solutions

5   Concluding Remarks

An advanced finite element for the analysis of laminated composite plates with monoclinic constitutive behaviour has been developed starting from a First-order Shear Deformation Theory and adopting a mixed-enhanced variational formulation. The element employs bubble functions for rotational degrees of freedom, special functions which link transverse displacements to rotations and enhanced modes in order to enrich in-plane strains.

As proved by several numerical examples, the element is locking free and it provides very accurate in-plane/out-of-plane displacements. Furthermore, the ability to compute very accurate shear stress profiles in the case both of symmetrical and unsymmetrical monoclinic laminated plates is also assessed. Accordingly, the proposed finite element opens the possibility of interesting works concerning the modelling of delamination effects for monoclinic laminates.

Acknowledgements

This research was developed within the framework of Lagrange Laboratory, an European research group between CNRS, CNR, University of Rome "Tor Vergata", University of Montpellier II, ENPC and LCPC.

References

1. Auricchio F., Taylor R. L. (1994) A shear deformable plate element with an exact thin limit, Comp. Meth. Appl. Mech. Engng. **118**, 393-412.

2. Auricchio F., Sacco E. (1999) A mixed-enhanced finite-element for the analysis of laminated composite plates, Int. J. Num. Meth. Engng. **44**, 1481-1504.
3. Auricchio F., Sacco E. (1999) Partial-mixed formulation and refined models for the analysis of composite laminates within an FSDT, Comp. Struct. **46**, 103-113.
4. Auricchio F., Lovadina C., Sacco E. (2001) Analysis of mixed finite elements for laminated composite plates, Comput. Methods Appl. Mech. Engrg. **190**, 4767-4783.
5. Bisegna P., Sacco E. (1997) A rational deduction of plate theories from the three-dimensional linear elasticity, Z. Angew. Math. Mech. **77**, 349-366.
6. Mindlin R. D. (1951) Influence of rotatory inertia and shear on flexural motions of isotropic, elastic plates, J. Appl. Mech. **38**, 31-38.
7. Noor A. K., Burton W. S., Peters J. M. (1990) Predictor-corrector procedures for stress and free vibration analyses of multilayered composite plates and shells, Int. J. Comp. Meth. Appl. Mech. Eng. **82**, 341-363.
8. Pian T. H., Sumihara K. (1984) Rational approach for assumed stress finite elements, Int. J. Numer. Meth. Engng. **20**, 1685-1695.
9. Reddy J. N. (1984) Energy and variational methods in applied mechanics, Wiley, New York.
10. Reddy J. N. (1987) A generalization of two dimensional theories of laminated plates, Commun. Appl. Numer. Meth. **3**, 173-180.
11. Reddy J. N. (1990) On refined theories of composite laminates, Meccanica **25**, 230-238.
12. Reissner E. (1945) The effect of transverse shear deformation on the bending of elastic plates, J. Appl. Mech. **12**, 69-77.
13. Simo J. C., Rifai M. S. (1990) A class of mixed assumed strain methods and the method of incompatible modes, Int. J. Numer. Meth. Engng. **29**, 1595-1638.
14. Taylor R. L. (1997) Manual of user of Finite Element Analysis Program, Univ. of California at Berkeley.
15. Whitney J. M., Pagano N. J. (1970) Shear deformation in heterogeneous anisotropic plates, J. Appl. Mech., Trans. ASME **37**(92/E), 1031-1036.
16. Yang P. C., Norris C. H., Stavsky Y. (1966) Elastic wave propagation in heterogeneous plates, Int. J. Solids Struct. **2**, 665-684.
17. Zienkiewicz O. C., Taylor R. L. (1997) The Finite Element Method, vol. 1, 4th Ed., McGraw-Hill.

A Mixed FSDT Finite-Element Formulation for the Analysis of Composite Laminates Without Shear Correction Factors

Ferdinando Auricchio[1], Elio Sacco[2], Giuseppe Vairo[3]

[1] Dipartimento di Meccanica Strutturale
Università di Pavia, via Ferrata, 1
27100 Pavia, Italy
[2] Dipartimento di Meccanica, Strutture, A&T
Università di Cassino, via Di Biasio, 43
03043 Cassino, Italy
[3] Dipartimento di Ingegneria Civile
Università di Roma "Tor Vergata",
via del Politecnico, 1
00133 Roma, Italy

Abstract. A new 4-node finite element for the analysis of laminated plates is developed starting from a partial-mixed FSDT variational formulation, which considers out-of-plane shear stresses as primary variables. It does not require either the introduction of shear correction factors or post-processing procedures to obtain transverse shear stress profiles. Presented numerical examples show that the element exhibits a quadratic convergence rate and it is locking free.

1 Introduction

Laminated composite plates are nowadays widely used in a variety of complex structures, especially in space, automotive and civil applications. The main reason for this growing use and development is related to the improvement of the ratio between performances and weight for laminated plates with respect to the homogeneous case. However, as a consequence of their anisotropic response, which is characterized by significant shear deformation in the thickness direction and extension-bending coupling, the behaviour of laminated plates is much more complex than homogeneous ones.

Actually, several laminate models are available in literature and some of these are also implemented in commercial finite-element codes. Among the various laminate theories the *First-order Shear Deformation Theory*, denoted as FSDT, appears to be simple and effective in many structural problems. The main features of this approach, which arises from the classical Reissner–Mindlin assumptions [7,11], are related to the possibility of using C^0-conforming methods and the ability to capture shear deformations [4]. In this way, it is possible to consider also the case of moderately thick plates [2].

It is worth to observe that the correct use of the FSDT requires the introduction of the so-called *shear correction factors*, which are defined through the exact characterization of out-of-plane shear stress profiles. Unfortunately, these latter are known a priori only for homogeneous plates or for simple cases, such as cross-ply laminates under cylindrical bending [6,13,14], while no closed-form expression is available for more general cases. On the other hand, shear correction factors have a great influence on the overall structural response. Therefore this aspect represents a clear FSDT's limitation. As a consequence, the most FSDT plate and laminate formulations are able to recover satisfactory values of in-plane stresses, while out-of-plane stresses are usually obtained through post-processing of in-plane solution.

In this work, a new 4-node finite element for laminated plates is developed starting from a partial-mixed FSDT variational formulation, which considers the out-of-plane shear stresses as primary variables. Through-the-thickness shear stress profiles, which are introduced to define a suitable functional, are deduced from the three-dimensional local equilibrium equations, i.e. integrating the first two equilibrium equations with respect to the thickness direction. Accordingly, shear stresses are expressed as functions of in-plane stresses, which can be written as functions either of in-plane strains or of displacement and rotational fields.

The proposed element does not require either the introduction of shear correction factors or post-processing procedures in order to obtain out-of-plane stresses. Moreover, a static condensation procedure leads to a finite element with only five degrees of freedom per node, as it would be the case of a classical displacement-based approach. Several numerical examples are performed for investigating the effectiveness of the proposed formulation and the main features of the new finite element.

2   FSDT Laminate Model

A laminate plate is a three-dimensional flat body Ω, defined as:

$$\Omega = \left\{ (x_1, x_2, z) \in \mathbb{R}^3 : z \in (-h/2, h/2), \ (x_1, x_2) \in \mathcal{P} \subset \mathbb{R}^2 \right\}. \tag{1}$$

The laminate is assumed to be formed by ℓ layers perfectly bonded and whose mechanical properties can be different. The plate thickness h is assumed to be constant and the plane $z = 0$ identifies the mid-plane $\mathcal{P}$ of the undeformed plate. Top and bottom surfaces of Ω are indicated as $\mathcal{P}^+ = \mathcal{P} \times \{h/2\}$ and $\mathcal{P}^- = \mathcal{P} \times \{-h/2\}$, respectively. Moreover, the kth layer (index k assumes values in $\{1, 2, ..., \ell\}$) lies between thickness coordinates z_k and z_{k+1}, such that $z_1 = -h/2$ and $z_{\ell+1} = h/2$.

From here onwards the following notation rules are considered, unless explicitly stated: Greek indices assume values in $\{1, 2\}$, whereas Latin indices assume values in $\{1, 2, 3\}$ with the exception of index n, which assumes values in $\{1, 2, 3, 4\}$. Furthermore, partial derivative of f with respect to the in-plane coordinate x_α is denoted by $f_{,\alpha}$, whereas partial derivative with respect to the thickness coordinate z is indicated with an apex, i.e. f'. Finally, repeated indices are understood to be summed within their ranges, except for the index k, which is used to denote any quantity relative to the kth layer.

The FSDT laminate model is based on the following well-known assumptions on both strain and stress fields [5]: straight lines perpendicular to the mid-plane cannot stretch and they remain straight, i.e. $\varepsilon_{33} = 0$ and $\varepsilon'_{\alpha 3} = 0$; out-of-plane normal stress is zero, i.e. $\sigma_{33} = 0$; out-of-plane shear stresses $\sigma_{\alpha 3}$ are continuous piece-wise quadratic functions of the coordinate z.

The previous strain field assumptions lead to the classical representation form of the displacement field, i.e. as a linear function of the thickness coordinate z:

$$s_\alpha(x_1, x_2, z) = u_\alpha(x_1, x_2) + z\varphi_\alpha(x_1, x_2) \tag{2}$$

$$s_3(x_1, x_2, z) = w(x_1, x_2). \tag{3}$$

where u_α and w represent the mid-plane displacements and φ_α accounts for the rotation in the plane $x_\beta = 0$ (with $\beta \neq \alpha$) of perpendicular to mid-plane lines.

Hence, the strain tensor can be represented as:

$$2\varepsilon_{\alpha\beta} = s_{\alpha,\beta} + s_{\beta,\alpha} = 2(e_{\alpha\beta} + z\kappa_{\alpha\beta}) \tag{4}$$

$$2\varepsilon_{\alpha 3} = s'_\alpha + s_{3,\alpha} = \varphi_\alpha + w_{,\alpha} = \gamma_\alpha \tag{5}$$

$$\varepsilon_{33} = s'_3 = 0 \tag{6}$$

where γ_α is the transverse shear strain vector, $e_{\alpha\beta} = (u_{\alpha,\beta} + u_{\beta,\alpha})/2$ is the in-plane deformation tensor and $\kappa_{\alpha\beta} = (\varphi_{\alpha,\beta} + \varphi_{\beta,\alpha})/2$ accounts for the curvature.

The laminated plate is assumed to be formed by linearly elastic monoclinc layers with symmetry plane parallel to $\mathcal{P}$ so that $C^{(k)}_{\alpha\beta\gamma\delta} = C^{(k)}_{33\alpha 3} = 0$, where $C^{(k)}_{ijlm}$ is the fourth-order elasticity tensor for the kth layer. Taking into account the stress assumptions, the in-plane stress-strain relations for the kth layer can be written as:

$$\sigma^{(k)}_{\alpha\beta} = \overline{C}^{(k)}_{\alpha\beta\gamma\delta}\varepsilon_{\gamma\delta} = \overline{C}^{(k)}_{\alpha\beta\gamma\delta}(e_{\gamma\delta} + z\kappa_{\gamma\delta}) \tag{7}$$

where $\overline{C}^{(k)}$ is the reduced in-plane elasticity tensor. It can be observed that, since each lamina may present different elastic properties, in-plane stresses $\sigma^{(k)}_{\alpha\beta}$ are discontinuous piece-wise linear functions of the coordinate z.

As it is well known [5], evaluation of out-of-plane shear stresses $\tau_\alpha = \sigma_{\alpha 3}$ through local constitutive equation is inappropriate, resulting in a stress field which is not equilibrated at laminae interfaces and which does not generally satisfy boundary conditions on $\mathcal{P}^+$ and $\mathcal{P}^-$. On the other hand, an accurate evaluation for τ_α can be recovered using the three-dimensional equilibrium equations. No in-plane loads per unit volume as well as tangential surface forces on $\mathcal{P}^+$ and $\mathcal{P}^-$ are considered; thus, it results:

$$\tau_\alpha = \sigma_{\alpha 3} = -\int_{-h/2}^{z} \sigma_{\alpha\beta,\beta}\,d\zeta \tag{8}$$

where only the boundary condition $\tau_\alpha(-h/2) = 0$ is automatically enforced, whereas nothing is still said on the condition $\tau_\alpha(h/2) = 0$.

3   Evaluation of the Shear Stress Profile

Taking into account the representation (7) for the in-plane stress $\sigma_{\alpha\beta}^{(k)}$ and the equation (8), the shear stress in the kth lamina can be written as:

$$\tau_\alpha^{(k)} = \tau_\alpha^{(k,0)} - \int_{z_k}^{z} \overline{C}_{\alpha\beta\gamma\delta}^{(k)} \left(e_{\gamma\delta,\beta} + z\kappa_{\gamma\delta,\beta}\right) d\zeta \tag{9}$$

where $\tau_\alpha^{(k,0)}$ is the shear stress evaluated for $z = z_k$:

$$\tau_\alpha^{(k,0)} = -\sum_{i=1}^{k-1} \int \sigma_{\alpha\beta,\beta}^{(i)}\,d\zeta = -\left(\hat{\mathcal{A}}_{\alpha\beta\gamma\delta}^{(k)}e_{\gamma\delta,\beta} + \hat{\mathcal{B}}_{\alpha\beta\gamma\delta}^{(k)}\kappa_{\gamma\delta,\beta}\right) \tag{10}$$

with

$$\hat{\mathcal{A}}_{\alpha\beta\gamma\delta}^{(k)} = \sum_{i=1}^{k-1}(z_{i+1} - z_i)\overline{C}_{\alpha\beta\gamma\delta}^{(i)}; \qquad \hat{\mathcal{B}}_{\alpha\beta\gamma\delta}^{(k)} = \frac{1}{2}\sum_{i=1}^{k-1}(z_{i+1}^2 - z_i^2)\overline{C}_{\alpha\beta\gamma\delta}^{(i)}. \tag{11}$$

Hence, the expression (9) takes the form:

$$\tau_\alpha^{(k)} = -\left(\mathcal{A}_{\alpha\beta\gamma\delta}^{(k)}e_{\gamma\delta,\beta} + \mathcal{B}_{\alpha\beta\gamma\delta}^{(k)}\kappa_{\gamma\delta,\beta}\right) \tag{12}$$

where

$$\mathcal{A}_{\alpha\beta\gamma\delta}^{(k)}(z) = (z - z_k)\overline{C}_{\alpha\beta\gamma\delta}^{(k)} + \hat{\mathcal{A}}_{\alpha\beta\gamma\delta}^{(k)}; \tag{13}$$

$$\mathcal{B}_{\alpha\beta\gamma\delta}^{(k)}(z) = \frac{(z^2 - z_k^2)}{2}\overline{C}_{\alpha\beta\gamma\delta}^{(k)} + \hat{\mathcal{B}}_{\alpha\beta\gamma\delta}^{(k)}. \tag{14}$$

In order to enforce the boundary condition $\tau_\alpha(h/2) = 0$, the expression (12) is enhanced with a linear term in the thickness coordinate z, so that the shear stress function $\tau_\alpha^{(k)}$ can be assumed to have the form:

$$\tau_\alpha^{(k)} = -\left(\mathcal{A}_{\alpha\beta\gamma\delta}^{(k)}e_{\gamma\delta,\beta} + \mathcal{B}_{\alpha\beta\gamma\delta}^{(k)}\kappa_{\gamma\delta,\beta}\right) + a_\alpha(z + h/2) \tag{15}$$

where the vector a_α is solved by the condition $\tau_\alpha(h/2) = 0$, that is:

$$a_\alpha = \frac{1}{h}\left(\mathcal{A}_{\alpha\beta\gamma\delta}e_{\gamma\delta,\beta} + \mathcal{B}_{\alpha\beta\gamma\delta}\kappa_{\gamma\delta,\beta}\right) \tag{16}$$

with

$$\mathcal{A}_{\alpha\beta\gamma\delta} = A^{(\ell)}_{\alpha\beta\gamma\delta}\Big|_{z=\frac{h}{2}} = \sum_{k=1}^{\ell}(z_{k+1} - z_k)\overline{C}^{(k)}_{\alpha\beta\gamma\delta};\tag{17}$$

$$\mathcal{B}_{\alpha\beta\gamma\delta} = B^{(\ell)}_{\alpha\beta\gamma\delta}\Big|_{z=\frac{h}{2}} = \frac{1}{2}\sum_{k=1}^{\ell}(z^2_{k+1} - z^2_k)\overline{C}^{(k)}_{\alpha\beta\gamma\delta}.\tag{18}$$

It is worth to observe that $\mathcal{A}_{\alpha\beta\gamma\delta}$ and $\mathcal{B}_{\alpha\beta\gamma\delta}$ are the classical laminate membrane and membrane-bending coupling fourth-order in-plane elasticity tensors. They do not depend either on the thickness coordinate z or on the specific lamina.

Finally, substituting the expression (16) into the equation (15), shear stresses can be represented putting in evidence the explicit dependency on the thickness coordinate z:

$$\tau^{(k)}_\alpha(z) = -\Bigg[\ \left(\mathcal{A}^{(k,0)}_{\alpha\beta\gamma\delta} + z\mathcal{A}^{(k,1)}_{\alpha\beta\gamma\delta}\right)e_{\gamma\delta,\beta}+$$

$$\tag{19}$$

$$+\left(\mathcal{B}^{(k,0)}_{\alpha\beta\gamma\delta} + z\mathcal{B}^{(k,1)}_{\alpha\beta\gamma\delta} + z^2\mathcal{B}^{(k,2)}_{\alpha\beta\gamma\delta}\right)\kappa_{\gamma\delta,\beta}\Bigg]$$

where the following fourth-order in-plane tensors are introduced:

$$\begin{cases}\mathcal{A}^{(k,0)} = \hat{A}^{(k)} - z_k\overline{C}^{(k)} - \frac{1}{2}\mathcal{A} \\ \mathcal{A}^{(k,1)} = \overline{C}^{(k)} - \frac{1}{h}\mathcal{A}\end{cases} \qquad \begin{cases}\mathcal{B}^{(k,0)} = \hat{B}^{(k)} - \frac{z_k^2}{2}\overline{C}^{(k)} - \frac{1}{2}\mathcal{B} \\ \mathcal{B}^{(k,1)} = -\frac{1}{h}\mathcal{B} \\ \mathcal{B}^{(k,2)} = \frac{1}{2}\overline{C}^{(k)}.\end{cases}\tag{20}$$

4   Variational Formulation

The following mixed functional is considered:

$$\Pi(\varepsilon_{\alpha\beta}, \sigma_{\alpha\beta}, \sigma_{\alpha3}, s_\alpha) = \Pi^{(mb)}(\varepsilon_{\alpha\beta}, \sigma_{\alpha\beta}, s_\alpha) + \Pi^{(s)}(\sigma_{\alpha3}, s_\alpha) - \Pi_{ext}\tag{21}$$

where $\Pi^{(mb)}$ accounts for membrane-bending terms, $\Pi^{(s)}$ accounts for transverse shear terms and Π_{ext} accounts for boundary and loading conditions. In detail, $\Pi^{(mb)}$ has a Hu–Washizu-like form and it is defined as:

$$\Pi^{(mb)} = \frac{1}{2}\int_\Omega \overline{C}_{\alpha\beta\gamma\delta}\varepsilon_{\alpha\beta}\varepsilon_{\gamma\delta}dv + \int_\Omega \sigma_{\alpha\beta}\left[\left(\frac{s_{\alpha,\beta} + s_{\beta,\alpha}}{2}\right) - \varepsilon_{\alpha\beta}\right]dv$$

$$= \frac{1}{2}\int_\Omega \overline{C}_{\alpha\beta\gamma\delta}(e_{\alpha\beta} + z\kappa_{\alpha\beta})(e_{\gamma\delta} + z\kappa_{\gamma\delta})dv\tag{22}$$

$$+ \int_\Omega \frac{\sigma_{\alpha\beta}}{2}\left[(u_\alpha + z\varphi_\alpha)_{,\beta} + (u_\beta + z\varphi_\beta)_{,\alpha} - 2(e_{\alpha\beta} + z\kappa_{\alpha\beta})\right]dv$$

while $\Pi^{(s)}$ has an Hellinger–Reissner-like form and it is defined as:

$$\Pi^{(s)} = \int_\Omega \mathcal{T}_\alpha(w_{,\alpha} + \varphi_\alpha)dv - \frac{1}{2}\int_\Omega \mathcal{T}_{\alpha\beta}\mathcal{T}_\alpha\mathcal{T}_\beta dv \tag{23}$$

where $\mathcal{T}_{\alpha\beta} = \mathcal{C}^{-1}_{\alpha 3\beta 3}$ represents the compliance shear elastic tensor.

Performing the integration along the thickness coordinate z, the membrane-bending part $\Pi^{(mb)}$ of the functional (21) can be written as:

$$\widehat{\Pi}^{(mb)} = \frac{1}{2}\int_\mathcal{P} \left(\mathcal{A}_{\alpha\beta\gamma\delta}e_{\alpha\beta}e_{\gamma\delta} + 2\mathcal{B}_{\alpha\beta\gamma\delta}e_{\alpha\beta}\kappa_{\gamma\delta} + \mathcal{D}_{\alpha\beta\gamma\delta}\kappa_{\alpha\beta}\kappa_{\gamma\delta} \right) dA$$
$$+ \int_\mathcal{P} \mathcal{N}_{\alpha\beta}\left[\left(\frac{u_{\alpha,\beta} + u_{\beta,\alpha}}{2} \right) - e_{\alpha\beta} \right] dA \tag{24}$$
$$+ \int_\mathcal{P} \mathcal{M}_{\alpha\beta}\left[\left(\frac{\varphi_{\alpha,\beta} + \varphi_{\beta,\alpha}}{2} \right) - \kappa_{\alpha\beta} \right] dA$$

where $\mathcal{D}_{\alpha\beta\gamma\delta} = \sum_{i=1}^{\ell}(z_{i+1}^3 - z_i^3)\overline{\mathcal{C}}^{(k)}_{\alpha\beta\gamma\delta}/3$ is the fourth-order in-plane laminate bending elasticity tensor, $\mathcal{N}_{\alpha\beta} = \int_{-h/2}^{h/2}\sigma_{\alpha\beta}dz$ is the in-plane resultant force tensor and $\mathcal{M}_{\alpha\beta} = \int_{-h/2}^{h/2}z\sigma_{\alpha\beta}dz$ is the bending moment one.

Moreover, noting that the quantity $(w_{,\alpha} + \varphi_\alpha)$ is constant with reference to the thickness coordinate z, the first term of $\Pi^{(s)}$ can be written as:

$$\int_\Omega \mathcal{T}_\alpha(w_{,\alpha} + \varphi_\alpha)dv = \int_\mathcal{P} \mathcal{S}_\alpha(w_{,\alpha} + \varphi_\alpha)dA \tag{25}$$

where the resulting shear force $\mathcal{S}_\alpha$ is introduced:

$$\mathcal{S}_\alpha = \int_{-h/2}^{h/2}\mathcal{T}_\alpha dz = \sum_{k=1}^{\ell}\int_{z_k}^{z_{k+1}}\tau_\alpha^{(k)}dz = -\left[\mathcal{Z}^{(e)}_{\alpha\beta\gamma\delta}e_{\gamma\delta,\beta} + \mathcal{Z}^{(\kappa)}_{\alpha\beta\gamma\delta}\kappa_{\gamma\delta,\beta} \right] \tag{26}$$

with

$$\mathcal{Z}^{(e)}_{\alpha\beta\gamma\delta} = \sum_{k=1}^{\ell}\sum_{i=1}^{2}\frac{(z_{k+1}^i - z_k^i)}{i}A^{(k,i-1)}_{\alpha\beta\gamma\delta}; \tag{27}$$

$$\mathcal{Z}^{(\kappa)}_{\alpha\beta\gamma\delta} = \sum_{k=1}^{\ell}\sum_{i=1}^{3}\frac{(z_{k+1}^i - z_k^i)}{i}B^{(k,i-1)}_{\alpha\beta\gamma\delta}. \tag{28}$$

Similarly, recalling the expression (19), formal integration through the thickness for the second term of $\Pi^{(s)}$ leads to:

$$\int_\Omega \mathcal{T}_{\alpha\beta}\mathcal{T}_\alpha\mathcal{T}_\beta dv = \int_\mathcal{P}\sum_{k=1}^{\ell}\int_{z_k}^{z_{k+1}}\mathcal{T}_{\alpha\beta}^{(k)}\tau_\alpha^{(k)}\tau_\beta^{(k)}dz\,dA$$
$$= \int_\mathcal{P}\left(\tilde{\Pi}^{(s,ee)} + 2\tilde{\Pi}^{(s,e\kappa)} + \tilde{\Pi}^{(s,\kappa\kappa)} \right)dA \tag{29}$$

where

$$\tilde{\Pi}^{(s,ee)} = \sum_{k=1}^{\ell}\sum_{i=1}^{3} \frac{\left(z_{k+1}^{i} - z_{k}^{i}\right)}{i} g^{(k,i)};$$

$$\tilde{\Pi}^{(s,e\kappa)} = \sum_{k=1}^{\ell}\sum_{i=1}^{4} \frac{\left(z_{k+1}^{i} - z_{k}^{i}\right)}{i} m^{(k,i)}; \tag{30}$$

$$\tilde{\Pi}^{(s,\kappa\kappa)} = \sum_{k=1}^{\ell}\sum_{i=1}^{5} \frac{\left(z_{k+1}^{i} - z_{k}^{i}\right)}{i} p^{(k,i)}.$$

Introduced quantities in (30) are defined as:

$$g^{(k,i)} = \sum_{q,r} \mathcal{T}_{\alpha\beta}^{(k)} \left(\mathcal{A}_{\alpha\gamma\delta\zeta}^{(k,q)} e_{\delta\zeta,\gamma}\right) \left(\mathcal{A}_{\beta\vartheta\lambda\mu}^{(k,r)} e_{\lambda\mu,\vartheta}\right) \qquad q,\ r \in \mathbb{S}_A$$

$$m^{(k,i)} = \sum_{q,r} \mathcal{T}_{\alpha\beta}^{(k)} \left(\mathcal{A}_{\alpha\gamma\delta\zeta}^{(k,q)} e_{\delta\zeta,\gamma}\right) \left(\mathcal{B}_{\beta\vartheta\lambda\mu}^{(k,r)} \kappa_{\lambda\mu,\vartheta}\right) \quad q \in \mathbb{S}_A,\ r \in \mathbb{S}_B \tag{31}$$

$$p^{(k,i)} = \sum_{q,r} \mathcal{T}_{\alpha\beta}^{(k)} \left(\mathcal{B}_{\alpha\gamma\delta\zeta}^{(k,q)} \kappa_{\delta\zeta,\gamma}\right) \left(\mathcal{B}_{\beta\vartheta\lambda\mu}^{(k,r)} \kappa_{\lambda\mu,\vartheta}\right) \qquad q,\ r \in \mathbb{S}_B$$

where $\mathbb{S}_A = \{0,1\}$; $\mathbb{S}_B = \{0,1,2\}$ and symbol $\sum_{q,r}$ denotes summation on every combination of q and r which satisfies the condition $i = (q+r+1)$, for a fixed value of i.

Therefore, in accordance with the conditions (26) and (29), the functional $\Pi^{(s)}$ can be written as:

$$\widehat{\Pi}^{(s)} = -\int_{\mathcal{P}} \left[\mathcal{Z}_{\alpha\beta\gamma\delta}^{(e)} e_{\gamma\delta,\beta} + \mathcal{Z}_{\alpha\beta\gamma\delta}^{(\kappa)} \kappa_{\gamma\delta,\beta}\right] (w_{,\alpha} + \varphi_\alpha)\, dA$$

$$-\frac{1}{2}\int_{\mathcal{P}} \left(\tilde{\Pi}^{(s,ee)} + 2\tilde{\Pi}^{(s,e\kappa)} + \tilde{\Pi}^{(s,\kappa\kappa)}\right) dA \tag{32}$$

as well as the functional Π introduced in (21) becomes:

$$\Pi = \widehat{\Pi}^{(mb)}(e_{\alpha\beta}, \kappa_{\alpha\beta}, u_\alpha, \varphi_\alpha, \mathcal{N}_{\alpha\beta}, \mathcal{M}_{\alpha\beta}) + \widehat{\Pi}^{(s)}(e_{\alpha\beta}, \kappa_{\alpha\beta}, \varphi_\alpha, w) - \Pi_{ext} \tag{33}$$

where both $\widehat{\Pi}^{(mb)}$ and $\widehat{\Pi}^{(s)}$ are referred to the two-dimensional plate midsurface $\mathcal{P}$. The field and boundary laminate governing equations are deduced performing the stationary conditions of Π with respect to the parameters $e_{\alpha\beta}$, $\kappa_{\alpha\beta}$, u_α, φ_α, w, $\mathcal{N}_{\alpha\beta}$ and $\mathcal{M}_{\alpha\beta}$.

5   The Finite Element

A four-node laminate finite element is obtained performing the discretization of the functional (33) and considering a standard iso-parametric map [15]:

$$x_\alpha = \Psi_n(\widehat{x}_\alpha)_n \tag{34}$$

where x_α is the global in-plane coordinate vector in the physical element, $\Psi_n(\xi_1, \xi_2) = (1 + \widehat{\xi}_{1n}\xi_1)(1 + \widehat{\xi}_{2n}\xi_2)$ are the standard bi-linear shape functions, ξ_α is the parent element coordinate vector and $(\widehat{x}_\alpha)_n$ is the in-plane coordinate vector for the node n. Notation $(\widehat{\cdot})_n$ indicates a nodal quantity for the node n.

The following interpolation schemes for the displacement fields are introduced:

- the in-plane displacement interpolation is bi-linear in the nodal parameters $\widehat{u}_\alpha$, i.e. $u_\alpha = \Psi_n(\widehat{u}_\alpha)_n$;
- the transverse displacement interpolation is bi-linear in the nodal parameters $\widehat{w}_n$ and it is enriched with linked quadratic functions in terms of nodal rotations $(\widehat{\varphi}_\alpha)_n$. In detail, following [1] it results:

$$w = \Psi_n \widehat{w}_n + \Psi_n^{(w\varphi)} \frac{L_n}{16} \left(\widehat{\varphi}_n^o - \widehat{\varphi}_s^o\right) \tag{35}$$

where $\widehat{\varphi}_n^o$ and $\widehat{\varphi}_s^o$ are the components of the n and s node rotations in the direction normal to the $n - s$ side, whose length is L_n. Furthermore, the shape functions $\Psi_n^{(w\varphi)}$ are set as:

$$\begin{cases} \Psi_1^{(w\varphi)} = (1 - \xi_1^2)(1 - \xi_2) \\ \Psi_2^{(w\varphi)} = (1 + \xi_1)(1 - \xi_2^2) \\ \Psi_3^{(w\varphi)} = (1 - \xi_1^2)(1 + \xi_2) \\ \Psi_4^{(w\varphi)} = (1 - \xi_1)(1 - \xi_2^2) \end{cases} ; \tag{36}$$

- the rotational field is bi-linear in the nodal parameters $\widehat{\varphi}_n$ with added internal degrees of freedom $\widehat{\varphi}_n^{(b)}$, i.e. $\varphi_\alpha = \Psi_n(\widehat{\varphi}_\alpha)_n + \Psi_{\alpha n}^{(b)}\widehat{\varphi}_n^{(b)}$, where $\Psi_{\alpha n}^{(b)}$ are bubble functions defined as [1]:

$$\left[\Psi_{\alpha n}^{(b)}\right] = \frac{(1 - \xi_1^2)(1 - \xi_2^2)}{det[\mathbf{J}]} \begin{bmatrix} J_{22}^o & -J_{12}^o & J_{22}^o\xi_2 & -J_{12}^o\xi_1 \\ -J_{21}^o & J_{11}^o & -J_{21}^o\xi_2 & J_{11}^o\xi_1 \end{bmatrix} \tag{37}$$

where $J_{\alpha\beta} = \partial x_\alpha / \partial \xi_\beta$ is the jacobian of the iso-parametric map and $J_{\alpha\beta}^o = J_{\alpha\beta}|_{\xi_1 = \xi_2 = 0}$.

Moreover, the following interpolation schemes are adopted for the remaining fields:

$$\begin{cases} e_{\alpha\beta} = \Psi_n^{(e)}(\widehat{e}_{\alpha\beta})_n \\ \kappa_{\alpha\beta} = \Psi_n^{(\kappa)}(\widehat{\kappa}_{\alpha\beta})_n \\ N_{\alpha\beta} = \Psi_n^{(N)}(\widehat{N}_{\alpha\beta})_n \\ M_{\alpha\beta} = \Psi_n^{(M)}(\widehat{M}_{\alpha\beta})_n \end{cases} \tag{38}$$

where $\Psi_n^{(e)}, \Psi_n^{(\kappa)}, \Psi_n^{(N)}, \Psi_n^{(M)}$ are proper interpolation functions. In this work they are set equal to piece-wise discontinuous linear functions and such that:

$$\Psi_n^{(e)} = \Psi_n^{(\kappa)} = \Psi_n^{(N)} = \Psi_n^{(M)}. \tag{39}$$

It can be observed that the degrees of freedom $(\widehat{\varphi}^{(b)})_n$, $(\widehat{e}_{\alpha\beta})_n$, $(\widehat{\kappa}_{\alpha\beta})_n$, $(\widehat{\mathcal{M}}_{\alpha\beta})_n$ and $(\widehat{\mathcal{N}}_{\alpha\beta})_n$ are local to each element and consequently they can be statically condensed element-by-element.

Introducing the proposed interpolation schemes, the functional (33) may be arranged in a discrete form as a function of the unknown nodal parameters. As an example, the resultant shear force $\mathcal{S}_\alpha$, which has been introduced in (26), may be expressed as:

$$\mathcal{S}_\alpha = -\left[\mathcal{Z}^{(e)}_{\alpha\beta\gamma\delta}\Psi^{(e)}_{n,\beta}(\widehat{e}_{\gamma\delta})_n + \mathcal{Z}^{(\kappa)}_{\alpha\beta\gamma\delta}\Psi^{(\kappa)}_{n,\beta}(\widehat{\kappa}_{\gamma\delta})_n\right]$$
$$= \left(\mathcal{S}^{(e)}_{\alpha\gamma\delta}\right)_n (\widehat{e}_{\gamma\delta})_n + \left(\mathcal{S}^{(\kappa)}_{\alpha\gamma\delta}\right)_n (\widehat{\kappa}_{\gamma\delta})_n \qquad (40)$$

where $(\mathcal{S}^{(e)}_{\alpha\gamma\delta})_n = -\mathcal{Z}^{(e)}_{\alpha\beta\gamma\delta}\Psi^{(e)}_{n,\beta}$ and $(\mathcal{S}^{(\kappa)}_{\alpha\gamma\delta})_n - \mathcal{Z}^{(\kappa)}_{\alpha\beta\gamma\delta}\Psi^{(\kappa)}_{n,\beta}$ can be interpreted as new interpolation functions.

The element balance equation is deduced performing the functional stationary conditions for a single element and statically condensing the internal degrees of freedom.

In this way, although the use of a mixed formulation is associated to an increase of fields to be interpolated, a finite element with only five degrees of freedom per node is obtained, as it would be the case of the classical displacement-based approach.

6   Numerical Examples

The proposed laminate element has been implemented in *FEAP* (Finite Element Analysis Program) [12] and several numerical examples are investigated in order to assess its performances.

In the following, the new element is denoted as EML4S and it is compared with the laminate finite-element EML4, which has been proposed in [2]. In detail, the element EML4 is based on an iterative technique for the correct evaluation of the shear correction factors. The EML4's solution at the beginning of the iterative procedure is indicated as EML4(i) and it corresponds to the classical Reissner–Mindlin theory with direct shear factors set equal to 5/6, whereas the EML4's solution at the end of the iterative procedure is indicated as EML4(f).

Computations are developed considering square cross-ply laminates with side length L and several side-to-thickness ratios $\eta = L/h$. The plates are subjected to a sinusoidal distributed load $q = q_o \sin(\phi x_1)\sin(\phi x_2)$, with $\phi = \pi/L$, and they are assumed to be simply supported along their boundaries:

$$s_2 = 0, \quad s_3 = 0, \quad \varphi_2 = 0 \ \text{ at } \ x_1 = 0, \ \ x_1 = L;$$
$$s_1 = 0, \quad s_3 = 0, \quad \varphi_1 = 0 \ \text{ at } \ x_2 = 0, \ \ x_2 = L.$$

Moreover, the laminae are assumed to be orthotropic with properties corresponding to an high modulus graphite/epoxy composite:

$$\frac{E_L}{E_T} = 25, \quad \nu_{TT} = 0.25, \quad \frac{G_{LT}}{E_T} = 0.5, \quad \frac{G_{TT}}{E_T} = 0.2 \tag{41}$$

where E refers to the elasticity moduli, G to the shear moduli and ν to the Poisson's ratios. Furthermore, the subscript L refers to the along-the-fiber direction, whereas the subscript T refers to any orthogonal-to-fiber direction.

The numerical analyses are performed using regular meshes based on square elements and the relevant results are compared with:

- the classical FSDT analytical solution [9], with direct shear correction factors set equal to 5/6 (denoted as $\text{FSDT}_{5/6}$);
- the classical FSDT analytical solution, with shear correction factors set as suggested by Whitney in [14] (denoted as FSDT_W);
- the analytical solution for the proposed refined FSDT model (denoted as FSDT_R);
- the exact three-dimensional solution proposed by Pagano [8] (denoted as $3D - AS$).

Three different cross-ply lamination sequences are considered: the symmetrical 0/90/0 one, the unsymmetrical 0/90 and 90/0/90/0 ones.

Figure 1 shows the relative error for the transverse displacement field w versus (a) the number of nodes per side (with $\eta = 10$) and (b) versus the side-to-thickness ratio η.

In detail, the relative error for the discrete solution is measured as:

$$E_w^2 = \frac{\sum_N \left[w_\varrho(\hat{x}_{1N}, \hat{x}_{2N}) - w(\hat{x}_{1N}, \hat{x}_{2N}) \right]^2}{\sum_N w(\hat{x}_{1N}, \hat{x}_{2N})^2} \tag{42}$$

where the sum is performed over all the nodal points, $w(\hat{x}_{1N}, \hat{x}_{2N})$ denotes the value of w at the coordinate of node N and it is given by the analytical solution FSDT_R, whereas $w_\varrho(\hat{x}_{1N}, \hat{x}_{2N})$ is the value obtained by numerical computation and corresponding to a mesh parameter ϱ. It is worth to observe that the above-mentioned error measure can be intended as a discrete L^2–type error.

It turns out that element EML4S is characterized by a quadratic convergence rate (in Fig. 1a the theoretical L^2-convergence slope is also indicated) and it does not exhibit locking phenomena.

Table 1 reports the dimensionless transverse displacement $\overline{w}$ at the plate center for the three analysed cross-ply lamination sequences. It is defined as $\overline{w} = 10^2 w E_T / q_o L \eta^3$. Comparison between the EML4S's numerical solutions and the reference ones (numerically and analytically obtained) is performed. The numerical solutions are computed employing regular meshes of 10×10 elements and setting $\eta = 10$.

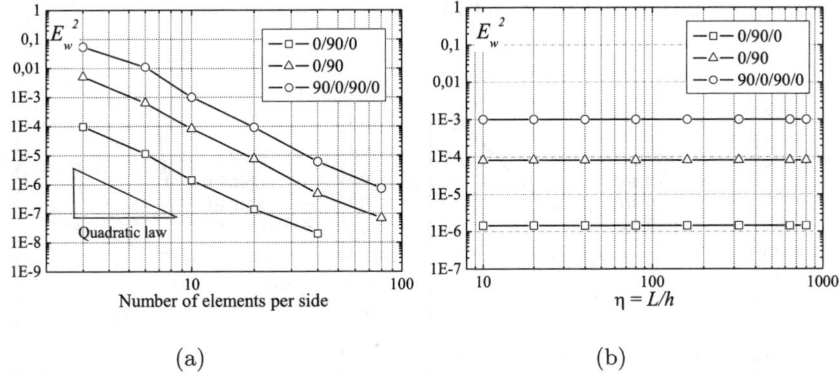

Fig. 1. Relative L^2-norm error for the transversal displacement w vs. (a) mesh refinement ($\eta = 10$) and (b) side-to-thickness ratio

Table 1. Transversal displacement $\overline{w}$ for simply supported laminates subjected to sinusoidal load: comparison between laminate and 3D solutions ($\eta = 10$)

	0/90/0	0/90	90/0/90/0
EML4(i)	0.6694	1.2373	0.6810
EML4(f)	0.7640	1.2522	0.7724
EML4S	0.7647	1.2494	0.7482
$FSDT_{5/6}$	0.6693	1.2373	0.6810
$FSDT_W$	0.8267	1.2751	0.8768
$FSDT_R$	0.7638	1.2382	0.7728
$3D - AS$	0.7514	1.2248	0.7604

Table 2. Horizontal displacement $\overline{u}$ for simply supported laminate 0/90 subjected to sinusoidal load: comparison between laminate and 3D solutions ($\eta = 10$)

mesh	6×6	10×10	20×20
EML4(i)	-0.0811	-0.0793	-0.0790
EML4(f)	-0.0805	-0.0793	-0.0786
EML4S	-0.0804	-0.0786	-0.0786
$FSDT_{5/6}$	$FSDT_W$	$FSDT_R$	$3D - AS$
-0.0793	-0.0766	-0.0786	-0.0784

Furthermore, Table 2 reports the dimensionless horizontal displacement $\overline{u} = 10^2 s_1 E_T / q_o L \eta^3$ at $x_1 = 0$, $x_2 = L/2$ and $z = 0$ for the unsymmetrical 0/90 cross-ply laminate with $\eta = 10$.

Figures from 2 to 5 depict the dimensionless normal and shear stress profiles evaluated at $x_1 = L/10$ and $x_2 = L/2$ for 0/90/0 and 90/0/90/0 laminates ($\eta = 10$). In particular, the stress profiles computed via the element EML4S are compared with the stress profiles provided by the EML4's numerical solution and the exact 3D one.

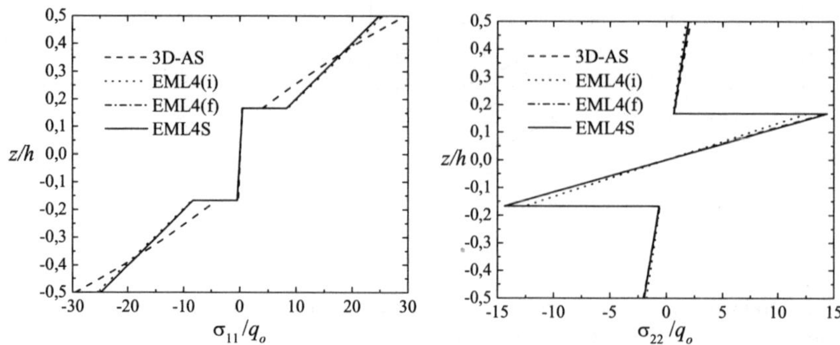

Fig. 2. Normal stress profiles σ_{11}/q_o and σ_{22}/q_o for a simply supported square $0/90/0$ laminate at $x_1 = L/10$ and $x_2 = L/2$: comparison between the element EML4, the EML4S one and the 3D analytical solution

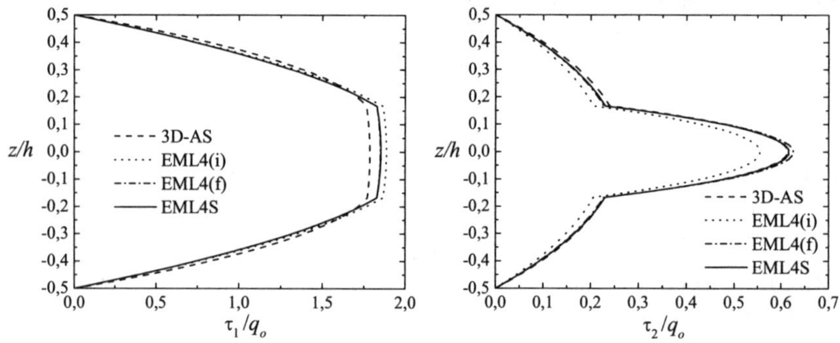

Fig. 3. Shear stress profiles τ_1/q_o and τ_2/q_o for a simply supported square $0/90/0$ laminate at $x_1 = L/10$ and $x_2 = L/2$: comparison between the element EML4, the EML4S one and the 3D analytical solution

7   Concluding Remarks

A partial mixed finite-element formulation for the analysis of composite laminated plates has been developed. It does not require either the introduction

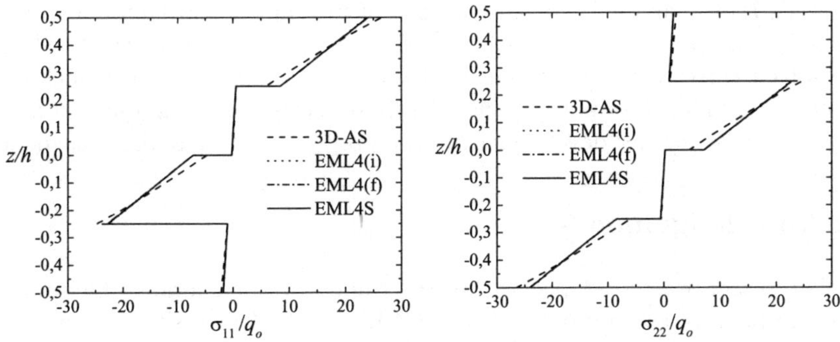

Fig. 4. Normal stress profiles σ_{11}/q_o and σ_{22}/q_o for a simply supported square $90/0/90/0$ laminate at $x_1 = L/10$ and $x_2 = L/2$: comparison between the element EML4, the EML4S one and the 3D analytical solution

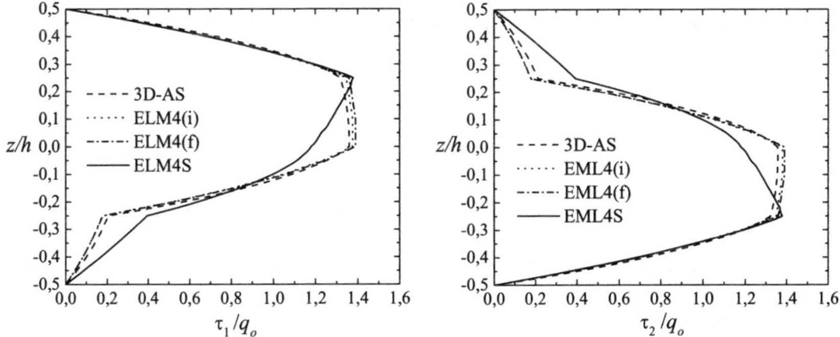

Fig. 5. Shear stress profiles τ_1/q_o and τ_2/q_o for a simply supported square $90/0/90/0$ laminate at $x_1 = L/10$ and $x_2 = L/2$: comparison between the element EML4, the EML4S one and the 3D analytical solution

of shear correction factors or post-processing procedures to obtain transverse shear stress profiles.

The presented numerical results show the effectiveness of the proposed formulation and the satisfactory performances of the new laminate finite-element ELM4S. The element is locking free and it exhibits a quadratic convergence rate. Furthermore, both analytical and numerical laminate solutions are very close to the exact 3D one, not only in terms of transverse and in-plane displacements but also in terms of shear and normal stress profiles, especially for symmetrical laminates. Moreover, it can be emphasized that even coarse mesh discretizations produce accurate results.

The little disagreement of the ELM4S's solutions with respect to the ELM4's and analytical ones in the case of unsymmetrical laminates (cf. figure 5) may be attributed to the choices (38) and (39). However, it should be solved through an enhancement of the interpolation schemes which have been adopted for the auxiliary fields. Therefore, this aspect represents a future guide-line for further development of the proposed laminate finite-element.

Acknowledgements

This research was developed within the framework of Lagrange Laboratory, an European research group between CNRS, CNR, University of Rome "Tor Vergata", University of Montpellier II, ENPC and LCPC.

References

1. Auricchio F., Taylor R. (1994) A shear-deformable plate element with an exact thin limit, Comp. Meth. App. Mech. Engrg. **118**, 393-412.
2. Auricchio F., Sacco E. (1999) A mixed-enhanced finite element for the analysis of laminated composite plates, Int. J. Numer. Methods Engrg. **44**, 1481-1504.
3. Auricchio F., Sacco E. (1999) Partial-mixed formulation and refined models for the analysis of composite laminates within an FSDT, Comp. Struct. **46**, 103-113.
4. Auricchio F., Lovadina C., Sacco E. (2001) Analysis of mixed finite elements for laminated composite plates, Comput. Methods Appl. Mech. Engrg. **190**, 4767-4783.
5. Bisegna P., Sacco E. (1996) A rational deduction of plate theories from the three-dimensional linear elasticity, Zeit. fur angew. math. und mech. **77**, 349-366.
6. Laitinen M., Lahtinen H., Sjölind S. G. (1995) Transverse shear correction factors for laminates in cylindrical bending, Commun. Num. Meth. Eng. **11**, 41-47.
7. Mindlin R. D. (1951) Influence of rotatory inertia and shear on flexural motions of isotropic, elastic plates, J. App. Mech. **38**, 31-38.
8. Pagano N. J. (1979) Exact solutions for rectangular bidirectional composites and sandwich plates, J. Comp. Mater. **4**, 20-34.
9. Reddy J. N. (1984) Energy and variational methods in applied mechanics, New York, Wiley.
10. Reddy J. N. (1997) Mechanics of Laminated Composite Plates - Theory and Analysis, CRC Press, Boca Raton.
11. Reissner E. (1945) The effect of transverse shear deformation on the bending of elastic plates, J. App. Mech. **12**, 69-77.
12. Taylor R. (2000) A finite-element analysis program. Technical report, Univ. of California at Berkeley.
13. Vlachoutsis S. (1992) Shear correction factors for paltes and shell, Int. J. Num. Meth. Eng. **33**, 1537-1552.
14. Whitney J. M. (1973) Shear correction factors for orthotropic laminates under static load, J. Appl. Mech. **40**, 302-304.
15. Zienkiewicz O. C., Taylor R. L. (1997) The Finite Element Method, vol. 1, 4th Ed., McGraw-Hill.

On the use of Continuous Wavelet Analysis for Modal Identification

Pierre Argoul, Silvano Erlicher

Institut Navier - LAMI
ENPC, 6-8 avenue Blaise Pascal,
Cité Descartes, Champs-sur-Marne,
F-77455 Marne la Vallée Cedex 2, France
E-mail: argoul@lami.enpc.fr

Abstract. This paper reviews two different uses of the continuous wavelet transform for modal identification purposes. The properties of the wavelet transform, mainly energetic, allow to emphasize or filter the main information within measured signals and thus facilitate the modal parameter identification especially when mechanical systems exhibit modal coupling and/or relatively strong damping.

1 Introduction

The concept of wavelets in its present theoretical form was first introduced in the 1970s by Jean Morlet and the team of the Marseille Theoretical Physics Center in France working under the supervision of Alex Grossmann. Grossmann and Morlet[1] developed then the geometrical formalism of the continuous wavelet transform. Wavelets analysis has since become a popular tool for engineers. It is today used by electrical engineers engaged for processing and analyzing non-stationary signals and it has also found applications in audio and video compression. Less than ten years ago, some researchers proposed the use of wavelet analysis for modal identification purposes. However, use of wavelet transforms in real world mechanical engineering applications remains limited certainly due to widespread ignorance of their properties.

This paper presents two ways of using the continuous wavelet transform (CWT) for modal parameter identification according to the type of measured responses : free decay or frequency response functions (FRF).

In 1997, Staszewski [2] and Ruzzene [3] started to use wavelet analysis with the Morlet wavelet for the processing of free decay responses, but the identification of the mode shapes was not performed. Wavelet transform allows to reach the time variation of instantaneous amplitude and phase of each component within the signal and makes the identification procedure of modal parameters much easier. Compared to other mother wavelets, Morlet wavelet provides better energy localizing and higher frequency resolution but has a disadvantage: frequency-coordinate window shifts along frequency axis with scaling while other wavelet types only expands the window. To remedy it, Lardies et al. [4] and Slavic et al. [5] preferred the Gabor wavelet. Argoul et al. [6,7] chose the Cauchy wavelet and Le et al. [8] established a complete modal identification procedure with improvements for numerical implementation, especially for a correct choice of the time-frequency localization. Edge effects are seen by all the authors and some attempts to reduce their negative influence on modal parameter identification are presented in [5], [8] and [9]. More recently, Staszewski [10], Bellizzi et al. [11] and Argoul et al. [6] adapted their identification procedure in order to process free responses of nonlinear systems. In [6], the authors proposed four instantaneous indicators, the discrepancy of which from linear case facilitates the detection and characterization of the non-linear behaviour of structures. In [11], an identification procedure of the coupled non-linear modes is proposed and tested on different types of non-linear elastic dynamic systems.

Typically, CWT is applied to time or spatial signals. Processing frequency signals is unusual, but a first application was proposed in [12] and [13]. Argoul showed in [12] that the weighted integral transform (WIT) previously introduced by Jézéquel et al. [14] for modal identification purposes can be expressed by means of a wavelet transform using a complex-valued mother wavelet close to the Cauchy wavelet. The WIT can be applied to either the ratio of the time derivative of the FRF over the FRF or the FRF itself. The representation with CWT provided a better understanding of the amplification effects of the WIT. When the FRF signal is strongly perturbed by noise, its derivative is hard to obtain; thus, Yin et al. [13] then proposed a slightly modified integral transform directly applied to the FRF and called the Singularities Analysis Function (SAF) of order n. It allows the influence of the FRF's poles to be emphasized and direct estimation of the eigen frequencies, eigen modes, and modal damping ratios is performed from the study of the extrema of the WIT.

2 Theoretical background for the continuous wavelet analysis

A wavelet expansion uses translations and dilations of an analyzing function called the mother wavelet $\psi \in L^1(\mathbb{R}) \cap L^2(\mathbb{R})$. For a continuous wavelet transform (CWT), the translation and dilation parameters: b and a respectively, vary continuously. In other words, the CWT uses shifted and scaled copies of $\psi(x) : \psi_{b,a}(x) = \frac{1}{a}\psi(\frac{x-b}{a})$ whose $L^1(\mathbb{R})$ norms ($\|.\|_1$) are independent of a. In the following, $\psi(x)$ is assumed to be a smooth function, whose modulus of Fourier transform is peaked at a particular frequency Ω_0 called the "central" frequency. The variable x may represent either time or frequency; when necessary, the time and circular frequency variables will be referred to t and to ω respectively. The CWT of a function $u(x) \in L^2(\mathbb{R})$ can then be defined by the inner product between $u(x)$ and $\psi_{b,a}(x)$

$$T_\psi[u](b,a) = \langle u, \psi_{b,a} \rangle = \frac{1}{a} \int_{-\infty}^{+\infty} u(x)\,\overline{\psi}\left(\frac{x-b}{a}\right) dx \tag{1}$$

where $\overline{\psi}(.)$ is the complex conjugate of $\psi(.)$. Using Parseval's identity, Eq. (1) becomes

$$T_\psi[u](b,a) = \frac{1}{2\pi} \left\langle \widehat{u}, \widehat{\psi}_{b,a} \right\rangle = \frac{1}{2\pi} \int_{-\infty}^{+\infty} \widehat{u}(\Omega)\, \overline{\widehat{\psi}}(a\Omega)\, e^{i\Omega b}\, d\Omega \qquad (2)$$

where $\widehat{u}(\Omega)$, $\widehat{\psi}_{b,a}(\Omega)$ and $\widehat{\psi}(\Omega)$ are respectively the Fourier transform (FT) of $u(x)$, $\psi_{b,a}(x)$ and $\psi(x)$; for instance for $u(x): \widehat{u}(\Omega) = \int_{-\infty}^{+\infty} u(x)\, e^{-i\Omega x} dx$. Moreover, when ψ and u are continuous and piece-wise differentiable, and $\dot{\psi}$ is square and absolutely integrable, and $\dot{u}$ is of finite energy, the CWT of $\dot{u}$ with ψ is linked to the CWT of u with $\dot{\psi}$:

$$T_\psi[\dot{u}](b,a) = -\frac{1}{a} T_{\dot{\psi}}[u](b,a) \qquad (3)$$

From [15], it can be seen that the CWT at point $(b, \Omega = \frac{\Omega_\psi}{a})$ picks up information about $u(x)$, mostly from the localization domain $D(b, \Omega = \frac{\Omega_\psi}{a})$ of the CWT, defined as

$$D(b, \Omega = \frac{\Omega_\psi}{a}) = [b+ax_\psi-a\Delta x_\psi, b+ax_\psi+a\Delta x_\psi] \times [\frac{\Omega_\psi}{a} - \frac{\Delta\Omega_\psi}{a}, \frac{\Omega_\psi}{a} + \frac{\Delta\Omega_\psi}{a}] (4)$$

where x_ψ and Δx_ψ are called centre and radius of ψ, stated in terms of root mean squares : $x_\psi = \frac{1}{\|\psi\|_2^2} \int_{-\infty}^{+\infty} x\, |\psi(x)|^2\, dx$ and

$\Delta x_\psi = \frac{1}{\|\psi\|_2} \sqrt{\int_{-\infty}^{+\infty} (x - x_\psi)^2 |\psi(x)|^2\, dx}$, and similar definitions hold on for the frequency centre Ω_ψ and the radius $\Delta\Omega_\psi$ of $\widehat{\psi}$. The area of D is constant and equal to four times the uncertainty : $4\Delta x_\psi \Delta\Omega_\psi = 4\mu(\psi)$. The Heisenberg uncertainty principle states that this area has to be greater than 2.

Referring to the conventional frequency analysis of constant-Q filters, the parameter Q defined as the ratio of the frequency centre Ω_ψ to the frequency bandwidth $(2\Delta\Omega_\psi)$:

$$Q = \frac{\frac{\Omega_\psi}{a}}{2\frac{\Delta\Omega_\psi}{a}} = \frac{\Omega_\psi}{2\Delta\Omega_\psi}, \qquad (5)$$

is introduced in [8] to compare different mother wavelets and to characterize the quality of the CWT. Q is independent of a.

The notion of spectral density can be easily extended to the CWT (see [16] for others properties such as linearity, admissibility and signal reconstruction, etc.). From the Parseval theorem applied to Eq. 2, it follows that

$$\int_{-\infty}^{+\infty} |T_\psi[u](b,a)|^2\, db = \frac{1}{2\pi} \int_{-\infty}^{+\infty} |\widehat{u}(k)|^2 \left|\overline{\widehat{\psi}}(ak)\right|^2 dk \qquad (6)$$

and it leads to the "energy conservation" property of the CWT expressing that if $u \in L^2(\mathbb{R})$, then $T_\psi \in L^2(\mathbb{R} \times \mathbb{R}_+^*, db\frac{da}{a})$, and $\|T_\psi[u]\|^2 = C_\psi \|u\|^2 =$

$\frac{C_\psi}{2\pi}\|\widehat{u}\|^2$. Finally, after changing the integration variable: $a = \frac{\Omega_o}{\Omega}$ and because $\widehat{u}(\Omega)$ is hermitian ($u(x)$ being real valued), one gets

$$\int_0^{+\infty}\int_{-\infty}^{+\infty}\left|T_\psi[u](b,\frac{\Omega_o}{\Omega})\right|^2 db\frac{d\Omega}{\Omega} = \frac{C_\psi}{\pi}\int_0^{+\infty}|\widehat{u}(\Omega)|^2 d\Omega \tag{7}$$

where $C_\psi = \int_{-\infty}^{+\infty}\frac{|\widehat{\psi}(\omega)|^2}{\omega}d\omega < \infty$.

This allows to define a local wavelet spectrum: $E_{u,CWT}(\Omega,b) = \frac{1}{2C_\psi\,\Omega}\left|T_\psi[u](b,\frac{\Omega_o}{\Omega})\right|^2$ and a mean wavelet spectrum: $E_{u,CWT}(\Omega) = \int_{-\infty}^{+\infty}E_{u,CWT}(\Omega,b)db$; and Eq. 7 can be rewritten: $\int_0^{+\infty}\left[E_{u,CWT}(\Omega) - \frac{1}{2\pi}|\widehat{u}(\Omega)|^2\right]d\Omega = 0$.

When the mother wavelet ψ is progressive (i.e. it belongs to the complex Hardy space $\widehat{\psi}(\Omega) = 0 :$ for $: \Omega \leq 0$), the CWT of a real-valued signal u is related to the CWT of its analytical signal Z_u (see. [16])

$$T_\psi[u](b,a) = \frac{1}{2}T_\psi[Z_u](b,a) \tag{8}$$

From 4, the CWT has sharp frequency localization at low frequencies, and sharp time localization at high frequencies. Thanks to a set of mother wavelets depending on one(two) parameter(s), the desired time or frequency localization can be yet obtained by modifying its(their) value(s). In [8], two complex valued mother wavelets were analyzed : Gabor and Cauchy. Morlet wavelet is a complex sine wave localized with a Gaussian envelope and the Gabor wavelet function is a modified Morlet wavelet with a parameter controlling its shape. Cauchy wavelet ψ_n of n order for $n \geq 1$, is defined by:

$$\psi_n(x) = \left(\frac{i}{x+i}\right)^{n+1} = \left(\frac{1}{1-ix}\right)^{n+1} = \left[\frac{1}{\sqrt{x^2+1}}\right]^{n+1}e^{i\,(n+1)Arctg(x)}. \tag{9}$$

The main characteristics of $\psi_n(x)$ follow: its FT: $\widehat{\psi}_n(\Omega) = \frac{2\pi\Omega^n e^{-\Omega}}{n!}\Theta(\Omega)$ where $\Theta(\Omega)$ is the Heaviside function (ψ_n is progressive), its central frequency: $\Omega_{0_n} = n$, its L2 norm: $\|\psi_n\|_2^2 = \frac{(2n)!}{2^{2n}(n!)^2}\pi$, its x- and frequency centres: $x_\psi = 0$ and $\Omega_\psi = n + \frac{1}{2}$, its radius: $\Delta x_\psi = \sqrt{\frac{1}{2n-1}}$, its frequency radius: $\Delta\Omega_\psi = \frac{\sqrt{2n+1}}{2}$, its uncertainty: $\mu_\psi = \frac{1}{2}\sqrt{1+\frac{2}{2n-1}}$, its quality factor: $Q = \frac{n+\frac{1}{2}}{\sqrt{2n+1}}$ and its admissibility factor: $C_\psi = 4\pi^2\frac{1}{2^{2n}}\frac{(2n-1)!}{(n!)^2} = \frac{2\pi}{n}\|\psi_n\|_2^2$. The use of the Cauchy wavelet is legitimate when Q is less than $5/\sqrt{2}$; as Q increases, both wavelets give close results, a little better with Morlet wavelet due to its excellent time-frequency localization, and behave similarly in the both time and frequency domains when Q tends toward infinity.

Some numerical and practical aspects for the computation of the CWT are detailed in [8], especially when applied to signals with time bounded support. Let us note $D_u = [0, L] \times [0, 2\pi f_{Nyq}]$ the validity domain of the discretized signal in the (b, Ω) plane where f_{Nyq} is the Nyquist frequency ($f_{Nyq} = \frac{1}{2T}$ with sampling period T). The edge-effect problem due to the finite length (L) and to the discretization of measured data record and to the nature of the CWT (convolution product) is tackled. An extended domain $D_{ext}(b, \Omega) \supset D(b, \Omega)$ where $\Omega = \frac{\Omega_\psi}{a}$), is then proposed to take into account the decreasing properties of ψ and $\widehat{\psi}$ by introducing two real positive coefficients c_x and c_Ω:

$$D_{ext}(b, \Omega = \tfrac{\Omega_\psi}{a}) = [b + ax_\psi - a\, c_x \Delta x_\psi, b + ax_\psi + ac_x \Delta x_\psi] \times [\tfrac{\Omega_\psi}{a} - c_\Omega \tfrac{\Delta \Omega_\psi}{a},$$

$\tfrac{\Omega_\psi}{a} + c_\Omega \tfrac{\Delta \Omega_\psi}{a}]$. Forcing D_{ext} to be included into D_u leads the authors to define a region in the (b, Ω) plane where the edge effect can be neglected; this region is delimited by two hyperbolae whose equations are: $\Omega = \frac{2}{b} c_x Q \mu_\psi$ and $\Omega = \frac{2}{L-b} c_x Q \mu_\psi$ and two horizontal lines whose equations are: $\Omega = 0$ and $\Omega = \frac{2\pi f_{Nyq}}{1 + c_\Omega \frac{1}{2Q}}$. Let us consider a frequency Ω_j for which $\widehat{u}(\Omega)$ exhibits a peak and introduce a frequency discrepancy $d\Omega_j$ (for example, the distance between two successive peaks of $\widehat{u}(\Omega)$). From the intersection of the two hyperbolae at the point $(b = \frac{L}{2}, \Omega = \Omega_j)$ and by imposing the frequency localization along the straight line: $a = \frac{\Omega_0}{\Omega_j}$, to be included into $[\Omega_j - d\Omega_j, \Omega_j + d\Omega_j]$, some upper and lower bounds are found for Q: $c_\Omega \frac{\Omega_j}{2d\Omega_j} \leq Q \leq \frac{L\,\Omega_j}{2c_x}$ and finally, the authors proposed $c_x = c_\Omega \simeq 5$.

3   Modal analysis and modal identification with CWT

Experimental identification of structural dynamics models is usually based on the modal analysis approach. One basic assumption underlying modal analysis is that the behaviour of the structure is linear and time invariant during the test. Modal analysis and identification involve the theory of linear time-invariant conservative and non-conservative dynamical systems. In this theory, the normal modes are of fundamental importance because they allow to uncouple the governing equations of motion. Also, they can be used to evaluate the free or forced dynamic responses for arbitrary sets of initial conditions. Modal analysis of a structure is performed by making use of the principle of linear superposition that expresses the system response as a sum of modal responses.

For linear MDOF systems with N degrees of freedom, the transfer function $\mathbb{H}_{ij}(p)$ of receptance type is defined as the Laplace transform of the displacement at point j to the impulse unit applied at point i and it can be expressed as a sum of simple rational fractions

$$\mathbb{H}_{ij}(p) = \sum_{r=1}^{N} \left(\frac{(A_r)_{ij}}{p - p_r} + \frac{\overline{(A_r)}_{ij}}{p - \overline{p_r}} \right) \tag{10}$$

where $\overline{(A_r)}_{ij}$ and $\overline{p_r}$ are respectively the conjugate of the residues $(A_r)_{ij}$ and of the poles p_r of $\mathbb{H}_{ij}(p)$. When the system is stable, $p_r = -\xi_r\, \omega_r + i\widetilde{\omega}_r$ where ω_r, $\widetilde{\omega}_r$ are the undamped and the damped vibration angular frequencies and

ξ_r is the damping ratio for the mode r. $(A_r)_{ij}$ can also be expressed from the complex modes χ_r by $(A_r)_{ij} = \frac{(\chi_r)_i \, (\chi_r)_j}{\gamma_r}$ where $\gamma_r = \chi_r^T \underline{\underline{C}} \chi_r + 2 p_r \chi_r^T \underline{\underline{M}} \chi_r$, $\underline{\underline{M}}$ and $\underline{\underline{C}}$ being respectively the mass and viscous damping matrices. Moreover, the $\overline{\text{FRF}}$ of receptance type $H_{ij}(\omega)$ can be usually related to $\mathbb{H}_{ij}(p)$ by: $H_{ij}(\omega) = \mathbb{H}_{ij}(p = i\omega)$, leading to

$$H_{ij}(\omega) = \sum_{r=1}^{N} \left(\frac{-i \, (A_r)_{ij}}{(\omega - \widetilde{\omega}_r) - i\xi_r \, \omega_r} + \frac{-i \, (\overline{A_r})_{ij}}{(\omega + \widetilde{\omega}_r) - i\xi_r \, \omega_r} \right) \tag{11}$$

The free responses at point j in terms of displacements $u_j^{(free)}(t)$ can be expressed according to the modal basis of complex modes: $\chi_{jr} = \eta_{jr} + i\,\kappa_{jr}$ (r being the mode number, $1 \leq r \leq N$)

$$u_j^{(free)}(t) = \sum_{r=1}^{N} A_{rj}(t) \cos\left(\Phi_{rj}(t)\right) \tag{12}$$

where $A_{rj}(t) = |\chi_{jr}| \, \rho_r \, e^{-\xi_r \omega_r t}$ and $\Phi_{rj}(t) = \widetilde{\omega}_r t - \varphi_r + \arctan\left(\frac{\kappa_{jr}}{\eta_{jr}}\right)$. ρ_r and φ_r are defined from initial displacement and velocity of mode r (cf. [17]). In the case of proportional viscous damping (Basile assumption), introducing the real eigenvectors Ψ_r of mode r for the associated conservative system, and replacing the residues $(A_r)_{ij}$ with $\frac{(\Psi_r)_i \, (\Psi_r)_j}{2i\,\widetilde{\omega}_r m_r}$ in Eq. 11 lead to

$$H_{ij}(\omega) = \sum_{r=1}^{N} \frac{(\Psi_r)_i \, (\Psi_r)_j}{m_r \left(\omega_r^2 - \omega^2 + 2i\,\xi_r \, \omega_r \omega\right)} \tag{13}$$

Moreover, in Eq. 12, $\kappa_{jr} = 0$; thus $|\chi_{jr}|$ can be replaced by $\left|(\Psi_r)_j\right|$ and the term $\arctan\left(\frac{\kappa_{jr}}{\eta_{jr}}\right)$ can be equal to 0 or π according to the sign of the real part η_{jr}.

3.1   Modal identification using free decay responses

The processed signals $u_j^{(free)}(t)$, whose expression is given in Eq. 12, can be considered under the assumption of weak damping ($\xi_r \ll \frac{1}{\sqrt{2}}$), as a sum of N modal components: $A_{rj}(t) \cos(\Phi_{rj}(t))$ consisting in asymptotic signals. The approximation of asymptotic signal means that the oscillations resulting from the phase term: $\Phi_{rj}(t)$ are much faster than the variation coming from the amplitude term: $A_{rj}(t)$; it entails that the analytical signal associated with $A_{rj}(t) \, e^{i\Phi_{rj}(t)}$ can be approximated by: $A_{rj}(t) \, e^{i\Phi_{rj}(t)}$. Therefore, Eq. (8) and the linearity of the CWT entail that: $T_\psi[u_j^{(free)}](b,a) = \frac{1}{2} \sum_{r=1}^{N} T_\psi[A_{rj}(t) \, e^{i\Phi_{rj}(t)}](b,a)$. Taking the Taylor expansion of each amplitude term of this sum, it follows that

$$T_\psi[u_j^{(free)}](b,a) = \frac{1}{2} \sum_{r=1}^{N} \left(A_{rj}(b) \, e^{i\Phi_{rj}(b)} \, \overline{\widehat{\psi}(a\widetilde{\omega}_r)} + R_{rj}(b,a) \right). \tag{14}$$

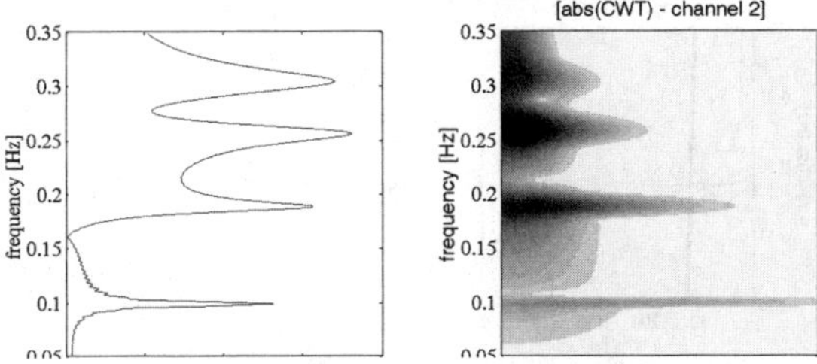

Fig. 1. Modulus of FT and of the CWT for the 4DoF system ($Q = 18$)

Assuming that each $R_{rj}(b, a)$ is small enough to be neglected, $\left|T_\psi[u_j^{(free)}](b, a)\right|$ is peaked in the (b, a) plane near straight lines of equation: $a = \frac{\Omega_0}{\Phi'_{rj}(b)} = \frac{\Omega_0}{\widetilde{\omega}_r}$. When each modal pseudo-pulsation $\widetilde{\omega}_r$ is far enough from the others, this straight line's equation easily provides an approximation of $\widetilde{\omega}_r$. Eq. 14 implies that the CWT of asymptotic signals has a tendency to concentrate near a series of curves in the time-frequency plane called ridges, which are directly linked to the amplitude and phase of each component of the measured signal. In [8], a complete modal identification procedure for natural frequencies, viscous damping ratios and mode shapes, is given in the case of viscous proportional damping and applied to a numerical case consisting in the free decay responses of a mass-spring-damper system with four degrees of freedom (4-DoF). The procedure is applied here to the free responses of a 4-DoF system whose natural frequencies are: $f_1 = 0.0984$ Hz, $f_2 = 0.1871$ Hz, $f_3 = 0.2575$ Hz and $f_4 = 0.3027$ Hz and modal damping ratios: $\xi_1 = 0.0124$, $\xi_2 = 0.0235$, $\xi_3 = 0.0324$, $\xi_4 = 0.0380$. Fig. 1 shows the FT and the CWT of the displacement of the second mass for which the four eigenfrequencies are clearly visible. Fig. 2 presents for the third mode and for the four masses, the modulus of the CWT and its logarithm that will allow to estimate the value of the damping ratio ξ_3 (cf. [8]). The edge effect is delimited by two hyperbolae in the time frequency plane; Q being chosen to 20 ($15.3 \leq Q \leq 43.1$). Identified values are very close to the exact ones and identification errors are negligible inside the domain D_{ext}. This procedure was also applied in [7] to a set of accelerometric responses of modern buildings submitted to non destructive shocks. From the CWT of measured responses, the ridge and the corresponding amplitude and phase terms for the two first modes were extracted. For the first instantaneous frequency $\widetilde{\omega}_1$, the processing revealed a slight increase of $\widetilde{\omega}_1$ just after the shock. The origin of this non linear effect was attributed to the non-linear behaviour of the soil-structure interaction. Other preliminary results can be found in [6] where instantaneous indicators based on CWT are computed from the accelerometric responses of a non-linear beam to an impact force. A Duffing non-linearity effect was then identified thanks to the first and the super-harmonic components.

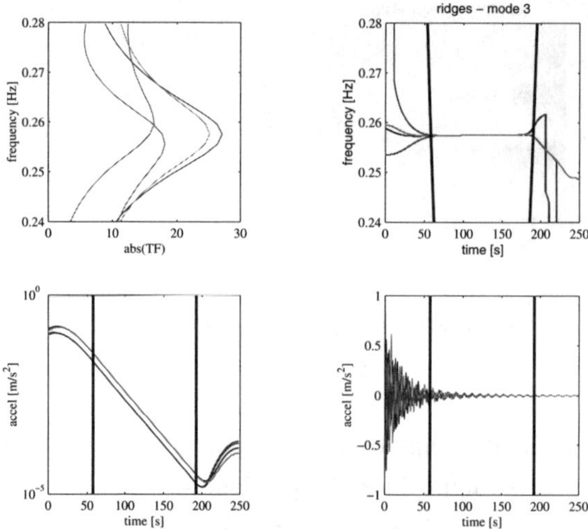

Fig. 2. Modulus of FT and of CWT around the 3rd mode for all channels. Logarithm of the modulus of the CWT ($L = 250$s, $T = 0.175$s, $c_t = c_\Omega = 4.7$, $Q = 20$).

3.2   Modal identification using FRF functions

The first definition of the SAF of order n ($n \in \mathbb{N}$) uses the derivative of a transfer function $\mathbb{H}(p)$ of the linear mechanical system, where $p = a + ib$:

$$\mathbf{H}_n(b, a) = (-1)^n 2\pi a^{\frac{n+1}{2}} \frac{d^n}{dp^n} \{\mathbb{H}(p)\} \tag{15}$$

$\mathbf{H}_n(b, a)$ can also be expressed as the FT of the impulse response function $h(t)$ filtered by: $2\pi a^{\frac{n+1}{2}} t^n e^{-at} \Theta(t)$ where $\Theta(t)$ is the Heaviside function :

$$\mathbf{H}_n(b, a) = 2\pi a^{\frac{n+1}{2}} \int_{\mathbb{R}^+} t^n e^{-at} h(t) e^{-ibt} dt \tag{16}$$

Finally, $\mathbf{H}_n(b, a)$ is equal to the CWT of the FRF $H(\omega)$ with a mother wavelet being the conjugate of the Cauchy wavelet of order n multiplied by $\frac{n!}{a^{\frac{n-1}{2}}}$:

$$\mathbf{H}_n(b, a) = \frac{n!}{a^{\frac{n+1}{2}}} T_{\overline{\psi}_n}[H](b, a) = \frac{n!}{a^{\frac{n+1}{2}}} \int_{\mathbb{R}} H(\omega) \psi_n\left(\frac{\omega - b}{a}\right) d\omega \tag{17}$$

Let us now consider a transfer function restricted to only one term $F_{\lambda,\omega_0}(\omega)$ of the sum given in (11) and perturbed by a Gaussian white noise $n(\omega)$ so that : $H(\omega) = F_{\lambda,\omega_0}(\omega) + n(\omega) = \frac{-B\,i}{(\omega-\omega_0)-i\lambda} + n(\omega)$. Let us introduce the coefficient ρ^2 defined as the ratio of the square absolute value of the CWT of the pure signal upon the common variance of the CWT of the noise itself:

$\rho^2 = \dfrac{|T_\psi[H](b,a)|^2}{E\{|T_\psi[n](b,a)|^2\}} = \dfrac{|\langle F_{\lambda,\omega_0}, \psi_{b,a}\rangle|^2}{E\{|\langle n, \psi_{b,a}\rangle|^2\}}$, it looks like a signal to noise ratio.

The use of the Cauchy-Schwarz inequality gives that : $\rho^2 \leq \|F_{\lambda,\omega_0}\|^2$ and the maximum of ρ^2 is reached for $(b, a) = (\omega_0, \lambda)$ when $\psi_{b,a}(\omega) = K\,F_{a,b}(\omega) = K\frac{-B\,i}{(\omega-b)-ia}$ where K is a constant. In conclusion, $\psi(\omega) = K\,B\frac{-i}{\omega-i} = K\,B\overline{\psi}_0(\omega)$; the analyzing functions $\psi(\omega)$ which allow the minimization of the noise effect on the peaks of $|T_\psi[H](b,a)|$ are proportional to $\overline{\psi}_0(s)$ defined in Eq. 9.

The above definitions of the SAF of order n ($n \in \mathbb{N}^*$) allow to better understand its effect when applied to discrete causal linear mechanical systems. In Eq. 15, the successive derivatives of a rational fraction make the degree of the denominator increase and cause an amplification of the pole's effect. So, when Eq. 10 is introduced in Eq. 15, the SAF of order n becomes

$$\mathbf{H}_n(b, a) = 2\pi\, n!\, a^{\frac{n+1}{2}} \sum_{r=1}^{N} \left(\frac{A_r}{(a + ib - p_r)^{n+1}} + \frac{\overline{A_r}}{(a + ib - \overline{p_r})^{n+1}} \right) \qquad (18)$$

The absolute value of $\mathbf{H}_n(b, a)$ exhibits $2N$ maxima located symmetrically from the vertical axis in the phase plane. The effect of parameter n is to facilitate the modal identification, especially when the structure has neighboring poles. As soon as n is large enough, the coordinates (b_{max}, a_{max}) of these $2N$ maxima allow the estimation of the real and imaginary parts, respectively $\Re\{p_r\}$ and $\Im\{p_r\}$, of each of the $2N$ conjugate poles of the system :

$$\Re\{p_r\} = -\xi_r\,\omega_r \approx -a_{\max_r} \qquad \text{and} \qquad \Im\{p_r\} = \widetilde{\omega}_r \approx b_{\max_r} \qquad (19)$$

The residue A_r can be then estimated from the complex valued amplitude of the extremum called $\mathbf{H}_n^{\max_r} = \mathbf{H}_n^{\max}(a_{\max_r}, b_{\max_r})$

$$A_r \approx \frac{2^n\,(a_{\max_r})^{\frac{n+1}{2}}}{\pi\,n!}\,\mathbf{H}_n^{\max_r} \qquad (20)$$

As n increases, the estimation of the extrema is better for signals without noise; but for noisy signals, noise effects are also increased. Recently, Yin et al. [18] proposed the use of analyzing wavelets which are the conjugate of Cauchy wavelets in which n is replaced by a positive real number. Applications both to numerical and experimental data showed the efficiency of this technique using SAF [19]. It was applied to the FRFs obtained with H1 estimators, of a test structure designed and build by ONERA for the GARTEUR-SM-AG19 Group. This structure was made of two aluminum sub-structures simulating wings/drum and fuselage/tail. Finally, the results obtained with SAF were very similar to those obtained with the broadband MIMO modal identification techniques provided by the MATLAB toolbox IDRC.

In conclusion, the pre-processing of measured signals (free decay responses or FRFs) with CWT proved to be an efficient tool for modal identification.

References

1. Grossman A., Morlet J. (1985) Decompositions of functions into wavelets of constant shape and related transforms, L. Streit, ed., World Scientific, Singapore, 135-165

2. Staszewski W.J. (1997) Identification of damping in MDOF systems using time-scale decomposition, Journal of Sound and Vibration, **203**, 283-305
3. Ruzzene M., Fasana A., Garibaldi L., Piombo B. (1997) Natural frequencies and dampings identification using wavelet transform: application to real data, Mechanical Systems and Signal Processing, **11**, 207-218
4. Lardies J., Ta M.N., Berthillier M. (2004) Modal parameter estimation based on the wavelet transform of output data, Archive of Applied Mechanics, **73**, 718-733
5. Slavic J., Simonovski I., Boltezar M. (2003) Damping identification using a continuous wavelet transform: application to real data, Journal of Sound and Vibration, **262**, 291-307
6. Argoul P., Le T-P. (2003) Instantaneous indicators of structural behaviour based on continuous Cauchy wavelet transform. Mechanical Systems and Signal Processing **17**, 243-250
7. Argoul P., Le T-P. (2004) Wavelet analysis of transient signals in civil engineering. in M. Frémond & F. Maceri (Eds) Novel Approaches in Civil Engineering, Lecture Notes in Applied and Computational Mechanics, Springer-Verlag, **14**, 311-318
8. Le T-P., Argoul P. (2004) Continuous wavelet transform for modal identification using free decay response, Journal of Sound and Vibration, **277**, 73-100
9. Boltezar M., Slavic J. (2004) Enhancements to the continuous wavelet transform for damping identifications on short signals. Mechanical Systems and Signal Processing **18**, 1065-1076
10. Staszewski W.J. (1998) Identification of non-linear systems using multi-scale ridges and skeletons of the wavelet transform, Journal of Sound and Vibration, **214**, 639-658
11. Bellizzi S., Guillemain P., Kronland-Martinet R. (2001) Identification of coupled non-linear modes from free vibration using time-frequency representations, Journal of Sound and Vibration, **243**2, 191-213
12. Argoul P. (1997) Linear dynamical identification : an integral transform seen as a complex wavelet transform. Meccanica, **32**(3), 215-222
13. Yin H.P., Argoul P. (1999) Transformations intégrales et identification modale. C. R. Acad. Sci. Paris, Elsevier Paris, t. 327, Série II b, 777-783
14. Jézéquel L., Argoul P. (1986) A new integral transform for linear systems identification. Journal of Sound and Vibration, **111**(2), 261-278
15. Chui C. K.(1991) An introduction to Wavelets, (Charles K. Chui, Series Editor); San Diego: Academic Press
16. Carmona R., Hwang W-L., Torrésani B. (1998) Practical Time-Frequency Analysis. (Charles K. Chui, Series Editor); San Diego: Academic Press
17. Geradin M., Rixen D. (1996) Théorie des vibrations Application à la dynamique des structures, 2nd edition, Masson Paris (in french)
18. Yin H.P., Duhamel D., Argoul P. (2004) Natural frequencies and damping estimation using wavelet transform of a frequency response function, Journal of Sound and Vibration, **271**, 999-1014
19. Argoul P., Guillermin B., Horchler A., Yin H-P. (1999) Integral transforms and modal identification in M.I. Friswell, J.E. Mottershead and A.W. Lees (Eds) Identification in Engineering Systems : Proceedings of the 2nd International Conference, Swansea, pp. 609-617

Propagation of Phase Change Front in Monocrystalline SMA

André Chrysochoos, Christian Licht, Robert Peyroux

Laboratoire de Mécanique et Génie Civil,
cc 048, Université Montpellier II,
34095 Montpellier Cedex 5, France

Abstract. Calorimetric effects related to the propagation of phase change front in a monocrystalline sample of CuZnAl shape memory alloy were derived from thermographic data analysis. During a load-controlled test, the displacement of the front induces a creep of the sample strongly depending on thermal exchanges with the surroundings. The main role played by the thermomechanical couplings can be pointed out by reversing the heat flux at the boundary of the sample: this leads to an inversion of the front propagation way associated with a recovery of the creep strain. We propose a behavioral modelling that takes into account the thermomechanical couplings accompanying the phase transition in single–crystal CuZnAl samples. The goal of this model is to put forward the significant role played by the heat diffusion in the propagation mode of the phase change fronts. Numerical simulations show the existence of phase change fronts such as the observed ones, and give good predictions of the calorimetric and kinematic effects accompanying the propagation.

1   Introduction

Strain induced by the propagation of phase change fronts during tensile tests on monocrystalline shape memory alloys (SMA) has already been evidenced in [1–3]. In this work, we propose to analyse the calorimetric effects linked to the propagation of these fronts using infrared image processing, and to employ a model, already presented in [4,5], to show that localization and propagation of phase change can be attributed without any ambiguity to coupling effects and heat diffusion. This model was already used to simulate the anisothermal behavior of polycrystalline sample of SMA, and allowed a correct prediction of the kinetics of phase change, the temperature variations and the associated strain (figures 1 and 2). The specificity of such o model is the way of taking into account the phase change mechanisms through strong thermomechanical couplings, the heat source related to phase change being the latent heat rate. Regarding the propagation of a phase change front, these notions are crucial, as they make it possible to determine the fields of strain, stress, temperature and volume fraction of phase change, and to track the propagation of the front.

In the following, we briefly recall the thermomechanical frame in which all the work will be developed, present the experiments realized on monocrystalline sample, and propose a simple modelling of phase change front nucleation and propagation.

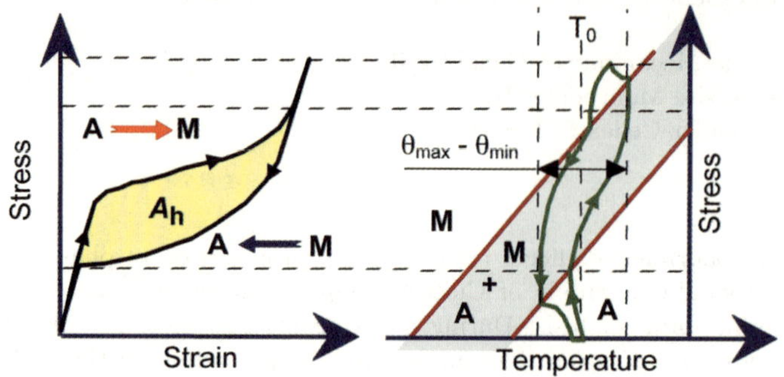

Fig. 1. Pseudoelastic behaviour and width of the transition domain

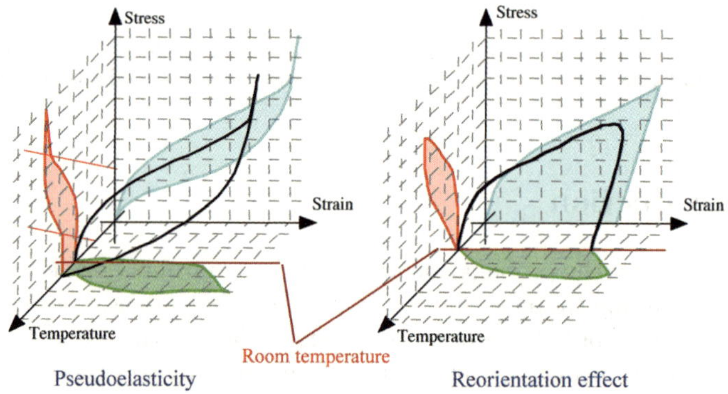

Fig. 2. Pseudoelastic and reorientation effects in a stess-strain-temperature space

2   A convenient thermomechanical framework

A convenient thermodynamic framework is the Continuum Thermodynamics [9], that postulates the local state axiom and considers the material both as a continuum and as a thermodynamic system described by a set of state variables. More precisely, we use the formalism of Generalised Standard Materials (GSM) [10] for which the constitutive equations can be derived from a thermodynamic potential and a dissipation potential. The chosen set of variables is $\{T, \varepsilon, x\}$, T being the temperature, ε the strain tensor and x referring to the volume fractions of martensite.

Using the formalism of GSM, a thermodynamic potential $\psi(T, \varepsilon, x)$ (specific Helmholtz free energy) and a dissipation potential $\varphi(grad T, \dot{\varepsilon}, \dot{x})$ are introduced. Their derivatives with respect to the state variables and their time derivatives give the state equations (1a) and the evolution equations (1b).

$$a) \begin{cases} s = -\Psi,_T \\ \sigma^r = \varrho\Psi,_\varepsilon \\ A = \varrho\Psi,_x \end{cases} \qquad b) \begin{cases} q = \varphi,_{grad\ T} \\ \sigma^{ir} = \varphi,_{\dot{\varepsilon}} \\ A = -\varphi,_{\dot{x}} \end{cases} \qquad with\ \sigma = \sigma^r + \sigma^{ir} \qquad (1)$$

where σ, q, s, ϱ and A stand for the stress tensor, the heat influx vector, the specific entropy, the mass density and the thermodynamic force associated with x, respectively. Note that A is defined such as the product $-A\dot{x}$ is the part of dissipated energy corresponding to the phase transformation.

The local heat conduction equation can be derived from the two principles of thermodynamics in the following form,

$$\varrho C\dot{T} - k\Delta T = d_1 + \varrho T\psi,_{T\varepsilon} : \dot{\varepsilon} + \varrho T\psi,_{Tx} .\dot{x} = w_h \qquad (2)$$

where C denotes the specific heat capacity, and k the coefficient of thermal conduction. The quantity d_1 represents the intrinsic dissipation and w_h stands for the volume heat source generated by all the terms in the right hand of equation 2. This last term can be experimentally evaluated using temperature fields given by an infrared camera (see [6] for more details).

3   Experimental analysis.

3.1   Material data

Quasi-static tension tests were performed at constant room temperature on a cylindric sample (diameter 3.85mm, length 24mm). The material is a monocrystalline shape memory alloy of type $Cu_{70}Zn_{25}Al_4$. The main thermomechanical data of this alloy are gathered in the following table. The room temperature is chosen such as the sample is austenitic at the beginning of the test.

specific mass	specific heat capacity	thermal conduction	expansion coefficient	latent heat
ϱ (kg.m^{-3})	C (J.kg^{-1}.K^{-1})	k (W.m^{-1}.K^{-1})	α (K^{-1})	C (J.kg^{-1}.K)
7800	390	120	1.9 10^{-7}	7000

elastic modulus	transformation strain	expansion coefficient	austenite temperatures		martensite temperatures	
E (GPa)	β	α (K^{-1})	As °C	Af	Ms °C	Mf
7800	8. 10^{-2}	19. 10^{-5}	22	29	23	12

3.2 Experiments

The mechanical loading imposed during this test consists of three stages (see figure 3a):
- stage a–b: force loading of the sample ($\dot{F} = 93$ N.s^{-1}).
- stage b–c: force hold at 885 N during 2 minutes.
- stage c–d: return to zero force ($\dot{F} = -93$ N.s^{-1}).
The evolution of a temperature profile along one generating line of the sample is presented versus time in figure 3b. In the same way, the corresponding heat sources are plotted in figure 3c.
The very first part of the loading (a–a') is associated with the elastic behavior of the material, involving small temperature variations and slight deformation of the sample. At time a', strong and positive heat sources concentrate at the bottom of the sample, in a zone moving during the loading (a'–b). The displacement of this zone is linked to the propagation of a phase change front (austenite/martensite exothermal transformation). When the stress is held constant (b–c), sources of lower intensity are observed, and the zone where they concentrate moves slower. The associated phase transformation involves also a localized strain and consequently a displacement of the cross–head (also plotted in figure 3a). This displacement is comparable to the one observed in creep when dealing with viscoelastic materials, but here the time–dependency is not linked to viscosity but to heat diffusion. Finally, we observe during the unloading stage (c–d) the propagation of a zone where negative heat sources are concentrated (martensite/austenite endothermal transformation).
For all the different stages, and due to heat diffusion, phase change localization is easier to track from heat sources analysis than from temperature fields. Indeed, the evolution in time of temperature in one characteristic point results from all heat transfer modes: conduction of the latent heat within the sample and convection, radiation and conduction of heat between the sample and the surroundings. Nevertheless, analysis of heterogeneous temperature profiles is consistent with the existence of wave–like front moving along the sample, the maximum amplitude of temperature variations being roughly associated with the current position of the front.

The propagation of the phase change front is tightly related to thermomechanical couplings. Calorimetric effects associated to phase change result in temperature variations, these variations playing a part in the phase change kinetics. Remind that the amplitude of temperature evolution during the test (around 5°C) is of the same order of magnitude that the width of the transition domain (around 11°C) (Fig. 1). The local mechanical response is consequently depending on temperature, the temperature resulting from the rate of latent heat and from the thermal exchange in the sample and with the surroundings. During stage b–c, when the force is kept constant, the front propagation speed is directly linked to the heat exchanges with the exterior. The creep–like displacement is governed by thermal diffusion. As a matter of

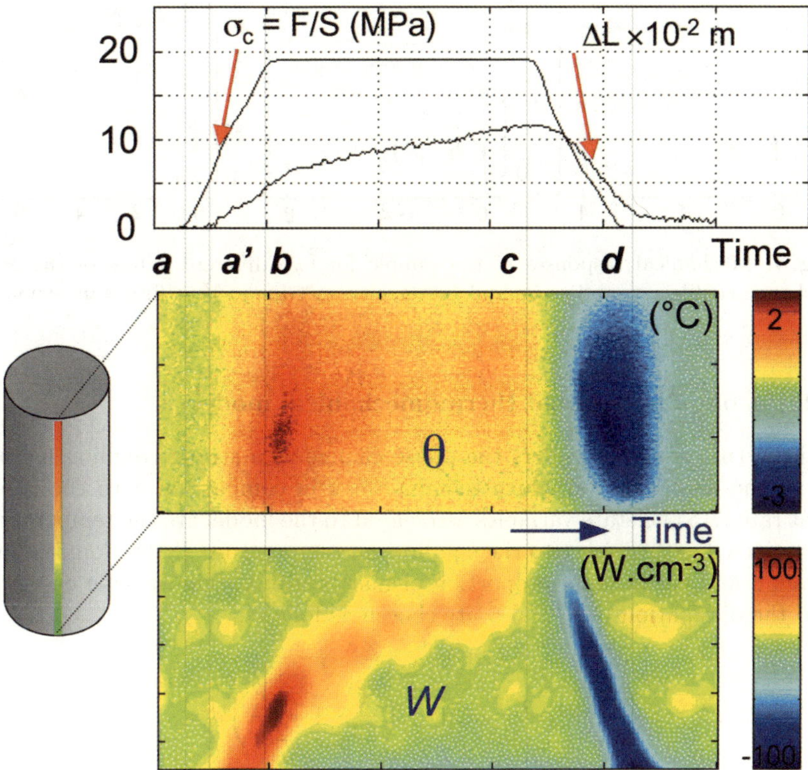

Fig. 3. Time evolutions of stress and cross-head displacement (a), of temperature variations θ (b), and of heat sources W_h (c) along the sample

proof, we increased the convection exchange coefficients using a fan, and we observed higher front propagation speeds (figure 4a and b). One can obtain up to an inversion of phase change way (figure 4c) by reversing the exchanged heat flux (this is done by a rapid increase of the surrounding temperature above the sample maximum temperature).

4   Modelling and numerical simulations

In this part, a one-dimensional version of the model developed in [4] is presented. The aim of this modeling is to show that localization and propagation of phase change, as experimentally observed, may be attributed without ambiguity to coupling effects and to heat diffusion.

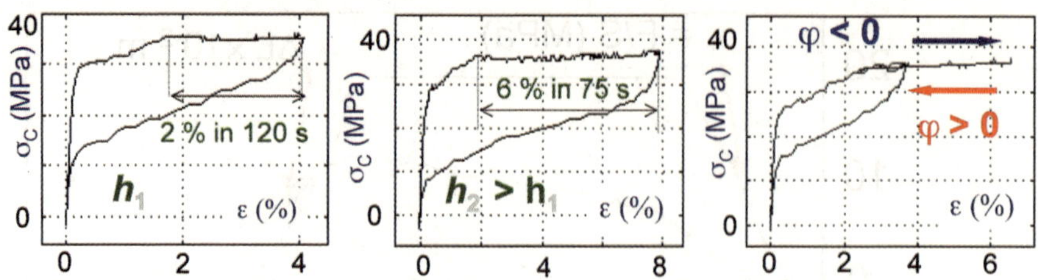

Fig. 4. Mechanical responses of the sample for two different values of the heat exchange coefficient, and creep and "reverse creep" due to heat flux ϕ inversion

4.1   A one-dimensional thermomechanical model

Indeed, this model considers the phase change as a strong coupling mechanism, the dissipation of which remains weak in comparison with the latent heat rate [7]. The state variables associated to the model are the temperature variations θ with respect to the equilibrium temperature T_0, the longitudinal strain ε and the volume proportion of martensite X. The specific energy ψ and the dissipation potential ϕ are chosen as follows:

$$\psi(\theta, \varepsilon, X) = \frac{E}{2\varrho}(\varepsilon - \alpha\theta - X\beta)^2 - \left(\frac{C}{2T_0} + \frac{E\alpha^2}{2\varrho}\right)\theta^2$$

$$+ \frac{K\beta}{\varrho}\left(\frac{A-M}{2}X^2 + (T-A)X\right) + I_{[0,1]}(X) \tag{3}$$

$$\varphi\left(\frac{\partial\varepsilon}{\partial t}, \frac{\partial X}{\partial t}, \frac{\partial\theta}{\partial x}; T\right) = \frac{k}{2T}\left(\frac{\partial\theta}{\partial x}\right)^2 \tag{4}$$

where K, A and M are three characteristic constants defining the transition domain in the plane (θ, σ) (Fig. 5). According to this choice of ψ and φ, the strain can be split into an elastic, a thermal and a transformation part ($\varepsilon = \sigma/E + \alpha\beta + X\beta$), and the kinetics of phase change is found to be a linear evolution of X between the borders of the domain (Fig. 5). The heat equation (5) neglects the convective terms of the particular time derivative, takes account of lateral heat losses (characteristic time τ) and considers the two coupling heat sources induced by thermal dilatation and phase change. Finally, heat exchanges at the ends of the sample are characterised by Fourier conditions (Fig. 5).

$$\varrho C\left(\frac{\partial\theta}{\partial t} + \frac{\theta}{\tau}\right) - k\frac{\partial^2\theta}{\partial x^2} = -E\alpha T_0\frac{\partial\varepsilon}{\partial t} + \varrho L_0\frac{\partial X}{\partial t} \tag{5}$$

Note that the latent heat L_0 can be written $L_0 = (E\alpha + K)\beta T_0/\varrho$.

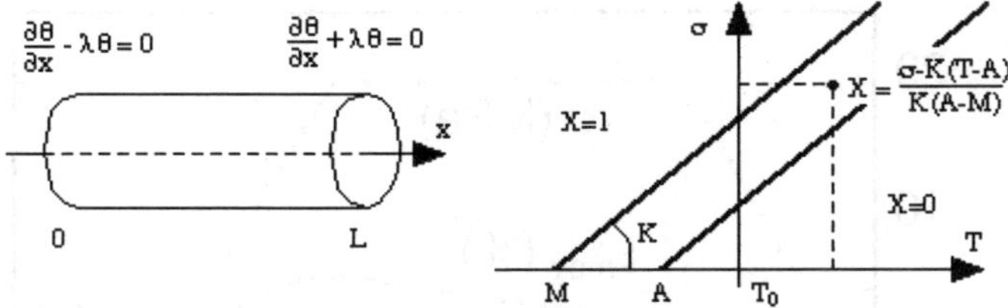

Fig. 5. Gauge part of the sample with thermal boundary conditions and transition domain

4.2 Numerical results

The numerical simulations were performed using a sample of 24 mm length and 3.855 mm diameter. The value of the longitudinal heat exchange coefficient λ is 1872 $W.m^2.K^{-1}$, and the characteristic time τ is 586 s. Owing to the symmetry of the problem, only half a sample is considered.

On the one hand, diffusion effects, amplified by couplings mechanisms, yield to uniform temperature fields. On the other hand, stress remains uniform within the sample during uniaxial quasi-static loading. One can thus expect that phase transition occurs within as narrow as the width $A - M$ of the transition domain is small. The limit case corresponds with a unique transition line, traditionally introduced in thermodynamics of phase transition with change of state [8]. Consequently, a value of 0.01 °C has been chosen for $A - M$. The numerical results in term of temperature and heat sources evolutions are found to be consistent with the experimental ones (Fig. 6), and the "creep" and "anti-creep" phenomena are qualitatively satisfying. Computations also show a hysteresis loop in good agreement with experimental observations even though the model supposes that the intrinsic dissipation equals to zero. In this case, hysteresis arises from two conjugate effects: thermomechanical couplings (material effect) and heat diffusion (material and structure effect).

4.3 On transition temperatures

To characterize a SMA monocrystal, the present model uses only two transition temperatures named A and M. The choice of a very small value for the difference $A - M$, corresponding to the width of the transition domain, is not incompatible with the classical way of defining the four transitions temperature A_s, A_f, M_s, M_f. Figure 7 presents a virtual dilatometry test where these four temperatures correspond to slope changes in the stress–strain response. The difference between M_s et M_f can reach 10°C for a transition domain of the monocrystal of 0.01 °C. The hysteresis area is then induced by the heat losses during the phase transition. Consequently, the four above-mentioned temperatures are depending on the sample geometry and on the heat exchange conditions. They are no longer intrinsic parameters of the material.

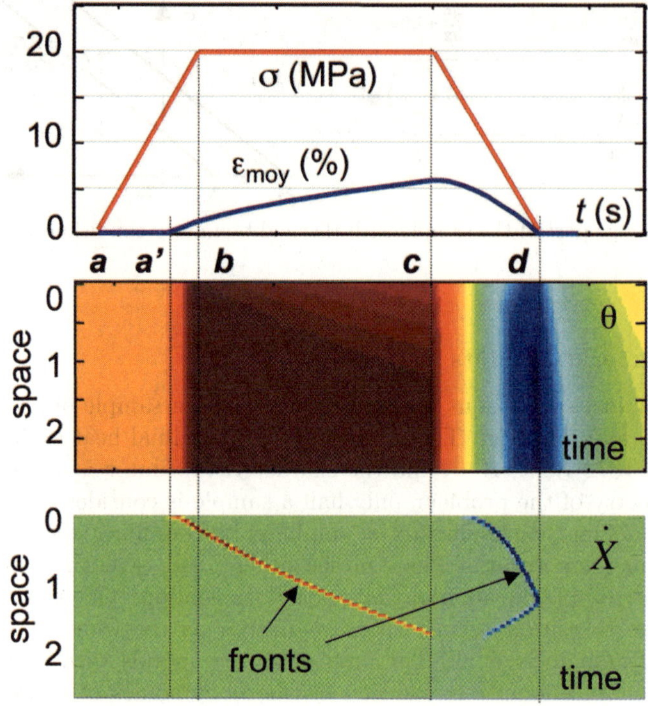

Fig. 6. Mechanical loading and sample stretching, temperature field θ, and latent heat rate pattern $\dot{X}$

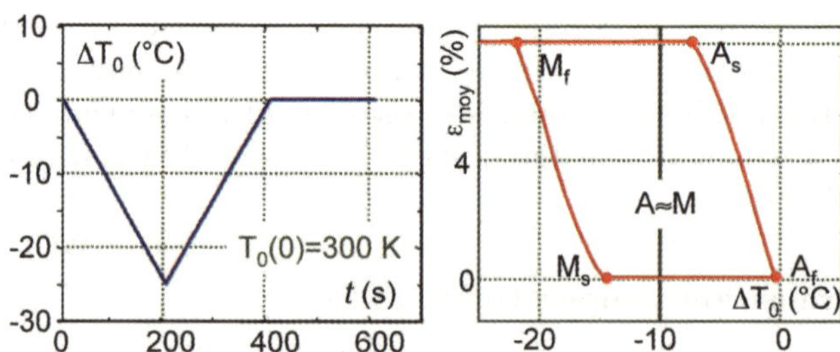

Fig. 7. Virtual dilatometry test showing the four transitions temperatures of the sample

5 Discussion

The main purpose of this paper is to point out the importance of the thermo-mechanical couplings on the material behavior. In particular, it underlines the dependance of the mechanical response on the temperature and the structure effects induced by the heat diffusion in the volume of the sample. The amazing mechanical effects observed and modelled in this work (for instance "creep" and "anti-creep" phenomema) are the consequences of the time–dependency through heat diffusion. When the phase change is localized in a front, the heterogeneous distribution of heat sources increases the influence of heat losses and amplifies the interaction between material and structure.

References

1. Shaw J.A., Kyriakides S., Thermomechanical aspects of NiTi, J. Mech. Phys. Solids 43 (1995) 1243-81.
2. Orgeas L., Favier D., Stress-induced martensitic transformation of a NiTi alloy in isothermal shear, tension and compression, Acta mater. 46 (1998) 5579-91.
3. Liu Yi., Liu Yo., Van Humbeeck J., Luders-like deformation associated with martensite reorientation in NiTi, Scripta Materialia 39 (1998) 1047-55.
4. Peyroux, R., Chrysochoos, A., Licht, C., Löbel, M., 1998; Thermomechanical couplings and pseudoelasticity of shape memory alloys, Int. J. of Engng. Sci., Vol. 36, n° 4, pp. 489-509.
5. Balandraud X., Chrysochoos A., Leclercq S., Peyroux R., Effet du couplage thermomcanique sur la propagation dun front de changement de phase, C.R. Acad. Sci. Paris, t.329, Srie II b, (2001), 621-626.
6. Chrysochoos, A., 1995; Analyse du comportement thermomé canique des matériaux par thermographie infrarouge, Photomécanique 95, Ed. Eyrolles, pp.203-211.
7. Chrysochoos, A., Löbel, M., 1996; A thermomechanical approach to martensite phase transition during pseudoelastic transformation of SMAs, in K.Z. Markov (Ed.), Proc. of the 8th Int. Symp. n Continuum Models and Discrete Systems, World Scientific, pp. 21-29.
8. Callen, H.B., 1985; Thermodynamics and an Introduction to Thermostatics, Wiley, pp. 215-253.
9. Germain, P., 1973; Cours de Mécanique des Milieux Continus, Masson et Cie Ed., p. 417
10. Germain,P., Nguyen Q.S., Suquet, P., 1983; Continuum Thermodynamics, J. of Appl. Mech., Transactions of the ASME, 50, pp. 1010-1020.
11. Halphen, B., Nguyen Q.S., 1975; Sur les Matériaux Standards Généralisés, Journal de Mécanique, Vol. 14, pp 39-63.

Entropy balance versus energy balance Application to the heat equation and to phase transitions

Pierluigi Colli, Elena Bonetti[1], Michel Frémond[2]

[1] Dipartimento Matematica "F.Casorati"
   Università degli Studi di Pavia
   via Ferrata, 1 - 27100 Pavia, Italy
[2] Laboratoire Central des Ponts et Chaussées
   58, boulevard Lefbvre,
   75732 Paris Cedex, France

Abstract. In these notes we are concerned with heat conduction described by using the entropy balance. The main advantage of this approach consists in recovering the positivity of the absolute temperature, necessary to prove thermodynamical consistency, directly by solving the equation. We introduce the model and discuss its thermomechanical consistency. Then, we investigate from the analytical and mechanical point of view, the Stefan problem written in terms of the entropy balance and using a generalized version of the principle of virual power including the effects of microscopic forces, responsible for the phase transition process. We prove existence of a solution in a fairly general physical framework, accounting for possible thermal memory effects and local interactions between the phases. Uniqueness is proved in the case no thermal memory nor local interactions are considered.

1 Introduction

In continuum mechanics the first and second laws of thermodynamics result in two partial differential equations. They may be combined in only one, the entropy balance,

$$\frac{ds}{dt} + \operatorname{div} \boldsymbol{Q} = R + H, \tag{1}$$

where s is the specific entropy, $\boldsymbol{Q}$ an entropy flux vector, R and H, with $H \geq 0$, are exterior and interior entropy sources, [9]. The existence of a solution for the initial and boundary values problem associated to the above system is proved under physically meaningful assumptions on R, but without any prescription on its sign. In this setting we can prove that the absolute temperature θ is always positive, which is essential for the thermodynamical consistence.

We use the entropy balance to describe the heat conduction, considering both simple and sophisticated phase change problems. In the past years several models have been proposed to describe phase change phenomena, see, e.g., [9], [15], and references therein. These models, mostly obtained by generalizing and regularizing the Stefan problem, have been widely investigated and several results have been obtained from the point of view of existence, uniqueness, and regularity of solutions, at least for some weak formulations, see, e.g., [6], [8], [15]. Moreover, we recall that a dissipative model describing phase transitions (possibly irreversible) and including microscopic movements in the virtual power of interior forces has been recently introduced in [5], [9] (cf. also [1] and [14]).

We follow the theory of microscopic movements in phase transitions, but we rewrite the first law of thermodynamics in terms of the entropy balance. However, we point out that in our approach, under the small perturbations assumption, it is possible to recover the classical phase change models.

The main advantages which are illustrated in the sequel are the versatility of the entropy balance and the insurance that the absolute temperature is positive.

2   The two laws of thermodynamics

We recall some basic results of thermo-mechanics, [10], [9]. Consider a continuous medium located in a smooth bounded domain $\Omega \subset \mathbb{R}^3$ with temperature $\theta(x,t)$ $((x,t) \in \Omega \times (0,T)$, T final time) and, for the sake of simplicity, small deformation ε. We denote by $\Psi(\varepsilon, \theta)$, e, s the specific free energy, the internal energy, the entropy, respectively. Then, we have

$$s = -\frac{\partial \Psi}{\partial \theta}, \quad e = \Psi + s\theta. \tag{2}$$

The stress σ is split into a dissipative component σ^d, and a non-dissipative one σ^{nd}, i.e.

$$\sigma = \sigma^d + \sigma^{nd}, \quad \sigma^{nd} = \frac{\partial \Psi}{\partial \varepsilon}. \tag{3}$$

2.1   The first law in a domain $\mathcal{D}$

The first law of thermodynamics, in any domain $\mathcal{D} \subset \Omega$, is

$$\begin{aligned}
\frac{de}{dt} + \operatorname{div} \boldsymbol{q} &= r + \sigma : \frac{d\varepsilon}{dt} \quad \text{in } \mathcal{D}, \\
-\boldsymbol{q} \cdot \boldsymbol{N} &= \varpi \quad \text{in } \partial\mathcal{D},
\end{aligned} \tag{4}$$

where $\boldsymbol{q}$ is the heat flux vector, r is the rate of exterior heat production and ϖ the heat provided to the domain $\mathcal{D}$ by contact actions ($\boldsymbol{N}$ is the outward normal vector to $\mathcal{D}$). The power of the interior forces is $-\sigma : d\varepsilon/dt$.

According to thermomechanical rules (cf. [10], [9]), we aim to emphasize the contribution of the temperature in the basic thermodynamical quantities Thus, we set

$$\boldsymbol{q} = \theta \boldsymbol{Q}, \quad r = \theta R, \quad \varpi = \theta \Pi, \tag{5}$$

and recall that Q is the entropy flux vector, R is the exterior entropy production, and Π the contact entropy production. Owing to these choices, the first law becomes

$$\frac{de}{dt} + \theta \operatorname{div} Q = \theta R + \sigma : \frac{d\varepsilon}{dt} - Q \cdot \operatorname{grad} \theta,$$
$$-\theta Q \cdot N = \theta \Pi. \tag{6}$$

By using relations (2) we get

$$\theta \frac{ds}{dt} + \theta \operatorname{div} Q = \theta R + \sigma^d : \frac{d\varepsilon}{dt} - Q \cdot \operatorname{grad} \theta. \tag{7}$$

The quantity

$$\sigma^d : \frac{d\varepsilon}{dt} - Q \cdot \operatorname{grad} \theta = \sigma^d : \frac{d\varepsilon}{dt} - q \cdot \frac{\operatorname{grad} \theta}{\theta} \tag{8}$$

represents the dissipation, given by mechanical and thermal contributions. We denote this dissipation by

$$\theta H = \sigma^d : \frac{d\varepsilon}{dt} - Q \cdot \operatorname{grad} \theta, \tag{9}$$

after observing that it is low when temperature is low. Assuming $\theta \neq 0$, we get

$$H = \frac{\sigma^d}{\theta} : \frac{d\varepsilon}{dt} - Q \cdot \frac{\operatorname{grad} \theta}{\theta}. \tag{10}$$

Equations (6) and (6) give

$$\frac{ds}{dt} + \operatorname{div} Q = R + H,$$
$$-Q \cdot N = \Pi. \tag{11}$$

which is equivalent to the first law (4), (4) provided $\theta \neq 0$. Hence, the entropy system (11)-(11) yields that H and R are interior and exterior entropy sources, respectively.

2.2 The first law on a discontinuity surface Γ

We denote by

$$[A] = A_2 - A_1, \quad \underline{A} = \frac{A_1 + A_2}{2}, \tag{12}$$

the discontinuity and the average of a quantity A on a discontinuity surface with normal vector N directed from side 1 toward side 2, respectively. The first law on the discontinuity surface is, [9]

$$m([e] - \underline{T}_N[\rho^{-1}]) - \underline{T}_T \cdot [U_T] - \underline{\theta}[Q] - \underline{Q}[\theta] = 0, \tag{13}$$

where m is the mass flow on the discontinuity surface, $T = \sigma N$ the stress ($\underline{T}_N$ is the average normal stress, $\underline{T}_T$ the average tangential stress), $Q = Q \cdot N$ the entropy flux on the discontinuity surface, $\underline{\theta}$ the average temperature, $\underline{Q}$ the average entropy flux, $\boldsymbol{U}_T$ the tangential velocity and ρ the density. Relations (2) and (13) give

$$\underline{\theta}\{m[s] - [Q]\} = \underline{\theta}h, \tag{14}$$

where

$$\underline{\theta}h = (-([\Psi] + \underline{s}[\theta] - \underline{T}_N[\rho^{-1}]))m + \underline{T}_T \cdot [\boldsymbol{U}_T] + \underline{Q}[\theta] \tag{15}$$

is the dissipation on the discontinuity surface, [9]. Thus the first law is

$$\begin{aligned} &m[s] - [Q] = h, \\ &h = \left(\frac{-([\Psi] + \underline{s}[\theta] - \underline{T}_N[\rho^{-1}])}{\underline{\theta}}\right) m + \frac{\underline{T}_T}{\underline{\theta}} \cdot [\boldsymbol{U}_T] + \frac{\underline{Q}}{\underline{\theta}}[\theta], \end{aligned} \tag{16}$$

if $\underline{\theta} \neq 0$ (which is equivalent to $\theta \neq 0$).

2.3   The second law in a domain $\mathcal{D}$ and on a discontinuity surface Γ

Thermodynamics (cf., e.g., [10]) requires that the temperature is positive

$$\theta > 0, \tag{17}$$

and that in any domain $\mathcal{D}$

$$\frac{ds}{dt} + \operatorname{div}\left(\frac{\boldsymbol{q}}{\theta}\right) \geq \frac{r}{\theta} \quad \text{in } \mathcal{D}. \tag{18}$$

By using the notation introduced in the previous sections, we have that (18) can be rewritten as follows

$$\frac{ds}{dt} + \operatorname{div}\boldsymbol{Q} \geq R \quad \text{in } \mathcal{D}. \tag{19}$$

Analogously, on every discontinuity surface Γ, the second law is

$$m[s] - [Q] \geq 0. \tag{20}$$

3   An equivalent formulation. The entropy balance

Relations (4), (4), (16) and (17), (18), (20) give

$$\begin{aligned} &\frac{ds}{dt} + \operatorname{div}\boldsymbol{Q} = R + H \quad \text{in } \mathcal{D}, \\ &-\boldsymbol{Q} \cdot \boldsymbol{N} = \Pi \quad \text{in } \partial\mathcal{D}, \\ &m[s] - [Q] = h \quad \text{in } \Gamma \cap \mathcal{D}, \end{aligned} \tag{21}$$

with the thermodynamical constraints

$$\begin{aligned} &\theta > 0, \\ &H \geq 0, \ h \geq 0. \end{aligned} \tag{22}$$

We observe that the interior entropy source H, required to be positive, tends to increase the entropy s, while the entropy flux $\boldsymbol{Q}$ tends to homogenize it. Conversely, relations (21) and (22) give the classical laws of Thermodynamics.

Theorem 1. *Classical laws (4), (4), (16) and (17), (18), (20) are equivalent to relations (21) and (22).*

As already said, it is wise to consider the relations specified by (21) as a conservation law

$$\forall \mathcal{D} \quad \frac{d}{dt}\int_{\mathcal{D}} s\,d\Omega = \int_{\mathcal{D}}(H+R)\,d\Omega + \int_{\Gamma\cap\mathcal{D}} h\,d\Gamma + \int_{\partial\mathcal{D}} \Pi(\boldsymbol{N})\,d\Gamma. \qquad (23)$$

By this point of view, the second law (22) corresponds to the positivity of the temperature and the positivity of the internal entropy sources, H and h.

4   An example. Heat conduction with the entropy balance

We recall that our state quantities are the absolute temperature θ and the small deformation ε. The free energy and the entropy of the material are (recall that $s = -\partial\Psi/\partial\theta$)

$$\Psi = -C\theta\log\theta + \tilde{\Psi}(\varepsilon), \quad s = C(1+\log\theta), \qquad (24)$$

where C is the heat capacity of the material and $\tilde{\Psi}$ the free energy of deformation. The constitutive law for the entropy flux is

$$\boldsymbol{Q} = -\lambda\frac{\mathrm{grad}\,\theta}{\theta}. \qquad (25)$$

We observe that letting $\boldsymbol{q} = \theta\boldsymbol{Q}$ corresponds to the classical Fourier law for the heat flux

$$\boldsymbol{q} = -\lambda\mathrm{grad}\,\theta. \qquad (26)$$

Due to the previous constitutive relations, the entropy balance (21) gives

$$\begin{aligned}
C\frac{d\log\theta}{dt} - \lambda\Delta\log\theta &= R+H \quad \text{in } \Omega,\\
\frac{\lambda\partial\log\theta}{\partial N} &= \Pi \quad \text{in } \partial\Omega,\\
\theta(0) = \theta_{initial} &\geq 0 \quad \text{in } \Omega,
\end{aligned} \qquad (27)$$

where $\frac{\partial}{\partial N}$ stands for the usual normal derivative on the boundary. Within the small perturbation assumption (cf. [10]), the interior entropy source H is neglected (for the moment) and material derivative $d\log\theta/dt$ becomes partial derivative $\partial\log\theta/\partial t$. Equation (27) becomes

$$C\frac{\partial\log\theta}{\partial t} - \lambda\Delta\log\theta = R. \qquad (28)$$

With standard assumptions on the data, the partial differential system (28), (27), (27) has a unique solution $\log\theta_1$. On account of the positivity of H, denoting by θ a possible solution of (27)-(27), the maximum principle allows us to deduce

$$\log \theta \geq \log \theta_1, \tag{29}$$

from which we have

$$\theta \geq \theta_1 > 0. \tag{30}$$

Notice that the positivity of the temperature θ follows as a consequence of the theory. This is an important advantage of this modelling approach.

Within the small perturbation assumption, let us now discuss a relation with the standard heat equation. If we consider θ_1 in a neighborhood of a reference temperature θ_0, assumed to be constant, we can provide a first order approximation of the logarithm, i.e. we can write

$$\log \theta_1 = \log(\theta_0 + (\theta_1 - \theta_0)) = \log(\theta_0 + \tau) = \log \theta_0 + \frac{\tau}{\theta_0} + \ldots$$

Under this approximation, the entropy balance is rewritten as the classical heat equation

$$\begin{aligned} C\frac{\partial \tau}{\partial t} - \lambda \Delta \tau &= \theta_0 R = r, \\ \frac{\lambda \partial \tau}{\partial N} &= \theta_0 \Pi = \varpi, \\ \tau(0) &= \theta_{initial} - \theta_0. \end{aligned} \tag{31}$$

However, this last equation does not allow to deduce the positivity of θ directly.

5   The Stefan problem with the entropy balance

Let us investigate a more complex physical situation. We deal with the solid-liquid phase change, described by two state variables: the absolute temperature θ and the liquid volume fraction β, which is sometimes called "phase parameter". Concerning the admissible values for β, we require that

$$0 \leq \beta \leq 1, \tag{32}$$

and let $(1-\beta)$ be the solid volume fraction, which corresponds to assume that neither void nor overlapping can occur between the two phases. Moreover, as it is an internal property of the thermo-mechanical system, the constraint (32) is included in the free energy functional. We have

$$\Psi = -C\theta \log \theta - \beta \frac{l}{\theta_0}(\theta - \theta_0) + I(\beta) + \tilde{\Psi}(\varepsilon), \tag{33}$$

where l is the latent heat, θ_0 the phase change temperature, and I denotes the indicator function of the interval $[0,1]$, namely $I(\beta) = 0$ if $\beta \in [0,1]$ and

$I(\beta) = +\infty$ otherwise (see, e.g., [12]). In this case, the entropy turns out to be

$$s = C(1 + \log\theta) + \beta\frac{l}{\theta_0}. \tag{34}$$

The interior source of entropy is

$$H = \frac{\sigma^d}{\theta} : \frac{d\varepsilon}{dt} + \frac{B^d}{\theta}\frac{d\beta}{dt} - \boldsymbol{Q} \cdot \frac{\mathrm{grad}\,\theta}{\theta}, \tag{35}$$

where B^d is a dissipative interior forces resulting from microscopic motion, responsible for phase transitions. The constitutive laws for the above new interior forces are

$$0 \in B^d + \frac{\partial\Psi}{\partial\beta} = B^d - \frac{l}{\theta_0}(\theta - \theta_0) + \partial I(\beta), \tag{36}$$

where

$$B^d = c\frac{d\beta}{dt} \tag{37}$$

for some dissipative parameter $c > 0$. Let us recall that $\partial I(\beta)$ is defined by $\partial I(\beta) = 0$ if $\beta \in (0,1)$, $\partial I(1) = [0,+\infty)$, $\partial I(0) = (-\infty,0]$, and $\partial I(\beta) = \emptyset$ if $\beta \notin [0,1]$. Note that the internal constraint (32) turns out to be guaranteed by constitutive law (36). The equations for θ and β are the entropy balance, in which we neglect the contribution by H owing to the small perturbations assumption, and (36) (see also (37)), combined with initial and boundary conditions. It results

$$\begin{aligned}
&C\frac{\partial\log\theta}{\partial t} + \frac{l}{\theta_0}\frac{\partial\beta}{\partial t} - \lambda\Delta\log\theta = R \quad \text{in } \Omega \times (0,T),\\
&c\frac{\partial\beta}{\partial t} - \frac{l}{\theta_0}(\theta - \theta_0) + \partial I(\beta) \ni 0 \quad \text{in } \Omega \times (0,T),\\
&\lambda\frac{\partial\log\theta}{\partial N} = \Pi \quad \text{in } \partial\Omega \times (0,T),\\
&\theta(0) = \theta_{initial}, \ \beta(0) = \beta_0 \quad \text{in } \Omega.
\end{aligned} \tag{38}$$

We take (see [13] for the definition of the Lebesgue and Sobolev spaces L^p and H^1)

$$R \in L^2(0,T;L^2(\Omega)), \quad \Pi \in L^2(0,T;L^2(\partial\Omega)), \tag{39}$$

such that there exist some upper bounds $\widetilde{R}$ and $\widetilde{\Pi}$ (assumed to be constant for simplicity) with

$$R \le \widetilde{R} \quad \text{a.e. in } \Omega \times (0,T), \quad \Pi \le \widetilde{\Pi} \quad \text{a.e. in } \partial\Omega \times (0,T). \tag{40}$$

Finally, we fix the initial values

$$\theta_{initial}, \beta_0 \in L^\infty(\Omega), \quad \beta_0 \in [0,1] \quad \text{a.e. in } \Omega. \tag{41}$$

We write the problem in the precise framework of

$$V = H^1(\Omega) \subset L^2(\Omega) \subset \left(H^1(\Omega)\right)' = V', \tag{42}$$

where $L^2(\Omega)$ is identified as usual with its dual space and V' is the dual space of V. Under the above assumptions the following theorem can be proved, [4]:

Theorem 2. *Let assumptions (39)-(41) hold. Then, there exists a unique (θ, β) fulfilling*

$$\theta \in H^1(0,T;V') \cap L^2(0,T;V) \cap L^\infty(Q),$$
$$\beta, \frac{\partial \beta}{\partial t} \in L^\infty(Q)$$

and solving (38) in a suitable sense.

The proof of the above existence result exploits a truncation technique, a priori estimates, and a maximum principle. Hence, we deduce a comparison principle of the solutions from which uniqueness of the solution follows. Let us point out that by the above prescriptions on the data, it is clear that our results can be applied to a fairly general realistic situation, as we do not require any restrictions on the sign of R or Π, but only some boundedness and regularity properties ((39) and (40)). We note that as a particular case, in (40) we could take $\widetilde{R} = 0$ and $\widetilde{\Pi} = 0$, yielding that R and Π are non positive (a.e.). In this case, R can describe a volume cooling and Π a heat loss through the boundary. This kind of conditions can lead to the appearance of freezing phenomena. We also point out that, in agreement with mechanics, we are actually requiring some boundedness on the heating sources and not on the cooling ones. Finally, also adiabatic systems are included in our set of assumptions as they correspond to the choice $\Pi = 0$. Concerning the comparison principle of solutions, it shows that the model is in agreement with everyday practical properties. Indeed, let two families of data for problem (38)

$$\{\theta^1_{initial}, \beta^1_0, R^1, \Pi^1\} \quad \text{and} \quad \{\theta^2_{initial}, \beta^2_0, R^2, \Pi^2\}, \tag{43}$$

such that there holds

$$\theta^1_{initial} \leq \theta^2_{initial}, \ \beta^1_0 \leq \beta^2_0 \quad \text{a.e. in } \Omega,$$
$$R^1 \leq R^2 \quad \text{a.e. in } \Omega \times (0,T),$$
$$\Pi^1 \leq \Pi^2 \quad \text{a.e. in } \partial\Omega \times (0,T). \tag{44}$$

Then, in [4] we prove that the corresponding solutions (θ^1, β^1) and (θ^2, β^2) to problem (38) evolve in such a way that

$$\theta^1 \leq \theta^2 \text{ and } \beta^1 \leq \beta^2 \quad \text{a.e. in } \Omega \times (0,T). \tag{45}$$

The above property is interesting from a practical point of view. For example, consider two systems that are identical except for their initial data, such that (44) is satisfied. It results that their internal energies e^1 and e^2 satisfy $e^1 \leq e^2$ a.e. in Ω at time $t = 0$. Then, (45) yields that the internal energies evolve during the phase transition in such a way that $e^1 \leq e^2$ a.e. in $\Omega \times (0,T)$.

6   A sophisticated phase change with thermal memory

The entropy balance equation turns out to be elegant and productive in more general and sophisticated situations. Now, we refer to a phase change model in which the microscopic motions that occur are taken into account and where there exists a thermal memory (see [11] and [3] for a detailed derivation of the model, the analytical investigation, and a thermo-mechanical discussion). In this case, the equations, the first being the entropy balance, are

$$
\begin{aligned}
&C\frac{\partial \log \theta}{\partial t} + \frac{l}{\theta_0}\frac{\partial \beta}{\partial t} - \lambda\Delta(\log\theta) - k*\Delta\theta = R \quad \text{in } \Omega \times (0,T),\\
&c\frac{\partial \beta}{\partial t} - \nu\Delta\beta + \partial I(\beta) \ni \frac{l}{\theta_0}(\theta - \theta_0) \quad \text{in } \Omega \times (0,T),\\
&\frac{\partial}{\partial N}\left(\lambda \log\theta(x,t) + k*\theta\right) = \Pi, \quad \frac{\partial \beta}{\partial N} = 0 \quad \text{in } \partial\Omega \times (0,T),\\
&\theta(0) = \theta_{initial}, \quad \beta(0) = \beta_0 \quad \text{in } \Omega,
\end{aligned}
\tag{46}
$$

where $*$ denotes the usual time convolution product over the interval $(0,t)$,

$$
(a*k)(t) := \int_0^t a(t-s)k(s)\,ds, \ t \in [0,T],
\tag{47}
$$

and k is the thermal memory kernel. The data R and Π collect known history terms. Such sets of partial differential equations have nice physical and mathematical properties [3]. In particular, we can prove an existence theorem for (46) (see [3], Theorem 3.1) by applying a fixed point argument to a Lipschitz regularization of the system and then passing to the limit. However, the system is highly nonlinear and the regularity of found solutions does not allow us to prove uniqueness. The anisotropy case is investigated in [2] by exploiting a different mechanical approximation, which could present some interests in itself.

7   Conclusion

The laws of thermodynamics we use involve only one partial differential equation and the positivity of the entropy source. They have two advantages. First, it is possible to treat the nonlinear equations of the thermo-mechanical models, by neglecting only dissipative contributions (cf. (38)), which is a procedure justified by the small perturbations assumption. Hence, and this seems more relevant, the models directly imply the positivity of the temperature, i.e. any a posteriori use of a maximum principle is not required. As a consequence, the thermo-mechanical consistence of the model, which is based on the positivity of the temperature, results directly from the mathematical structure of the equations, [7], [9].

References

1. E. Bonetti, 2002, Global solution to a nonlinear phase transition model with dissipation, Adv. Math. Sci. Appl., **12**, 355-376.

2. E. Bonetti, 2003, A new approach to phase transitions with thermal memory via the entropy balance, submitted.

3. E. Bonetti, P. Colli, M. Frémond, 2003, A phase field model with thermal memory governed by the entropy balance, Math. Models Methods Appl. Sci., **13**, 1565-1588.

4. E. Bonetti, M. Frémond, 2003, A phase transition model with the entropy balance, Math. Methods Appl. Sci., **26**, 539-556.

5. G. Bonfanti, M. Frémond, F. Luterotti, 2000, Global solution to a nonlinear system for irreversible phase changes, Adv. Math. Sci. Appl. **10**, 1-24.

6. P. Colli, M. Frémond, O. Klein, 2001, Global existence of a solution to a phase field model for supercooling, Nonlinear Anal. Real World Appl. **2**, 523-539.

7. P. Colli, G. Gentili, C. Giorgi, 1999, Non linear systems describing phase transition models compatible with thermodynamics, Math. Models Methods Appl. Sci. **9**, 1015-1037.

8. M. Frémond, A. Visintin, 1985, Dissipation dans le changement de phase. Surfusion. Changement de phase irréversible, C. R. Acad. Sci. Paris Sér. I Math. **301**, 1265-1268.

9. M. Frémond, 2001, Non-smooth Thermomechanics, Springer-Verlag, Heidelberg.

10. P. Germain, 1973, Mécanique des milieux continus, Masson, Paris.

11. M.E. Gurtin, A.C. Pikin, 1968, A general theory of heat conduction with finite wave speeds, Arch. Rational Mech. Anal., **31**, 113-126.

12. J. J. Moreau, 1966, Fonctionnelles convexes, Séminaire sur les équations aux dérivées partielles, Collège de France, and 2003, Dipartimento di Ingegneria Civile, Tor Vergata University, Roma.

13. J.L. Lions, E. Magenes, 1972, Non-homogeneous boundary value problems and applications, Vol. I, Springer-Verlag, Berlin.

14. F. Luterotti, G. Schimperna, U. Stefanelli, 2001, Existence result for a nonlinear model related to irreversible phase changes, Math. Models Methods Appl. Sci. **11**, 1-17.

15. A.Visintin, 1996, Models of phase transition, Birkhäuser, Basel.

On the Choice of the Shunt Circuit for Single-mode Vibration Damping of Piezoactuated Structures

Paolo Bisegna[1], Giovanni Caruso[2], Franco Maceri[1]

[1] Dipartimento di Ingegneria Civile
   Università di Roma "Tor Vergata",
   via del Politecnico, 1
   00133 Roma, Italia
[2] Istituto per le Tecnologie della Costruzione
   Consiglio Nazionale delle Ricerche
   viale Marx, 15
   00137 Roma, Italy

Abstract. This paper deals with single-mode passive damping of piezoactuated structures. The problem of shunting in the best way a piezoelectric actuator is discussed, and a new passive shunt circuit, given by the parallel of an inductance and a capacitance in series to a resistance, is here proposed. For a sufficiently high piezoelectric coupling coefficient, it is analytically shown to be more performant than the classical resistive-inductive shunt circuit, in the sense that it guarantees an higher exponential time decay rate of the free vibrations.

1   Introduction

Piezoelectric materials, due to their lightness and easy integrability to the host structure, have been intensely applied for vibration control in special applications, like aeronautic and aerospace structures. Different strategies to vibration control are available: passive, active or hybrid [2,4]. Passive vibration damping, which is dealt with in this paper, is intrinsically stable, cheap and easy to be implemented by using piezoelectric actuators. The actuators are bonded to the mechanical structure and shunted to suitable external electric circuits containing at least resistive components, able to dissipate electric energy [7,5,6]. The basic idea can be easily understood looking on a simple, linearly elastic strut equipped with a piezoelectric patch and vibrating along an eigenmode. Due to the piezoelectric effect, at the electrodes of the piezo-electric device, acting as a capacitor, an alternating voltage arises. It is quite a natural idea to connect the electrodes on a passive circuit, in order to obtain an alternating current and an energy dissipation. The simplest shunt is the resonant one, tuned on the same frequency of the motion to be damped, in order to obtain, at that frequency, the lowest, purely resistive impedance and the maximum current for the induced voltage. It appears that the basic requirement is to dissipate as much energy as possible in the lowest time, with a circuit easy to be implemented, stable and cheap. Another requirement may concern the performance of the shunt in a frequency band large enough to include different structural eigenmodes.

In [7] both a resistive (R) and a resistive-inductive (RL) shunt circuit applied to the vibration damping of a single-mode piezoactuated structure were thoroughly examined. A slight variation of the RL series circuit, i.e. the RL parallel circuit, was presented in [8]. Many authors were involved in the analysis of those circuits [9,10] and some modifications were proposed in order to improve their performances [11–13].

A major problem arises from the values of the electrical parameters of the piezoelectric components commercially available, because the inductors needed to tune the resonant circuit at the mechanical frequencies of interest are usually very high. Consequently, some authors [3] were considering the use of synthetic inductors. Later, others synthetic components of special shunt circuits were proposed, including negative capacitances [14].

At the present time, the RL series circuit is generally accepted as the most performant passive shunt in vibration damping on a single eigenmode. Aim of this paper is to show that different schemes can be proposed, which are more performant than the RL circuit, for a sufficiently high piezoelectric coupling coefficient. This result is obtained by choosing as performance index, and therefore optimization criterium, the exponential time-decay rate (ETDR) of the free vibrations. The maximization of such an objective function can be obtained, due to the linearity of the differential model of the electromechanical system, by means of an application of the pole placement technique to the system transfer function.

The new passive shunt circuit presented here in order to outperform over the classical RL shunt circuit, is indicated by $R(L||C)$ and is shown in Fig. 1. It is composed by the parallel of an inductance L and a capacitance C, in series to a resistance R.

The analysis is analytically carried out, and closed-form expressions for the optimal values of the electrical components and of the achieved ETDR are determined. This analysis turns out to be very useful for design purposes and effective comparisons with the classical RL series shunt circuit, in terms of optimal values of circuit parameters and achieved ETDR.

2   The electromechanical model

A single-degree-of-freedom mechanical system equipped with a piezoelectric device shunted on a passive circuit is here considered. The governing equations, in the Laplace domain, are [1,13]:

$$(s^2 m + k_{mm})x + k_{me}V = sx_o + \dot{x}_o \tag{1a}$$

$$q = -k_{me}x + C_p V \tag{1b}$$

$$V + sZq = 0 \tag{1c}$$

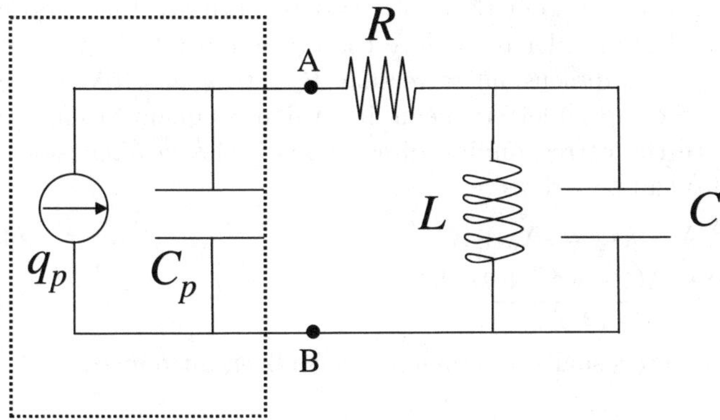

Piezoelectric element

Fig. 1. $R(L\|C)$ shunt circuit

In the domain of the Laplace variable s, x is the mechanical variable, V is the electric potential and q is the electric charge on the piezoelectric electrode. Moreover, m is the mass, k_{mm} is the mechanical stiffness, k_{me} is the electro-mechanical coupling stiffness and x_o, $\dot{x}_o$ are the initial displacement and velocity. The electric initial conditions are assumed to vanish, for the sake of simplicity. Equation (1a) is the dynamic equilibrium of the mass m, equation (1b) is the free electric charge balance equation on the piezoelectric surfaces and equation (1c) is the Kirchhoff equation of the shunt electric circuit connected to the piezoelectric electrodes, whose impedance is denoted by Z.

In the case of a complex structure, equations (1a)–(1b) can be regarded as modal equations describing the evolution along a chosen eigenmode. Accordingly, the involved parameters can be evaluated by means of a suitable finite element formulation (e.g., [15]).

The shunt circuit described in the Introduction and reported in Fig. 1 is considered here. Its impedance is:

$$Z = R + \frac{1}{(sL)^{-1} + sC} \tag{2}$$

By substituting V from equation (1b) into equations (1a) and (1c), and taking into account the above expression of Z, the governing equations are rewritten as follows:

$$\left(s^2 m + k_{mm} + \frac{k_{me}^2}{C_p}\right) x + \frac{k_{me}}{C_p} q = s x_o + \dot{x}_o \tag{3a}$$

$$\frac{k_{me}}{C_p} x + \left(\frac{1}{C_p} + \frac{sR + s^3 LRC + s^2 L}{1 + s^2 LC}\right) q = 0 \tag{3b}$$

In order to study the system (3a)–(3b) and to optimize the electrical parameters R, L and C in order to achieve the maximum ETDR, it is convenient to rewrite it in a dimensionless version. By letting $x = \bar{x}X$, $q = \bar{q}Q$ and $s = \bar{s}S$, where the capital letters mean dimensionless quantities and the overlined letters are the corresponding dimensional scales, a dimensionless version of (3a)–(3b) can be written as:

$$(S^2 + 1 + \kappa^2)X + \kappa Q = sX_o + \dot{X}_o \tag{4a}$$

$$\kappa X + \frac{1 + \rho S + \lambda(1 + c)S^2 + \lambda c\rho S^3}{1 + c\lambda S^2} Q = 0 \tag{4b}$$

where the following dimensionless parameters have been introduced:

$$\kappa = \frac{k_{me}}{\sqrt{k_{mm}C_p}}, \quad c = \frac{C}{C_p}, \qquad \rho = R\omega_m C_p, \quad \lambda = \omega_m^2 C_p L, \tag{5}$$

$$\omega_m = \sqrt{\frac{k_{mm}}{m}}, \quad \frac{\bar{x}}{\bar{q}} = \sqrt{\frac{1}{k_{mm}C_p}}, \quad \bar{s} = \omega_m$$

and X_o, $\dot{X}_o$ are the dimensionless counterparts of x_o, $\dot{x}_o$, respectively. In equation (5), κ is the piezoelectric coupling coefficient and depends only on the piezoactuated structure, c is the ratio between the capacity C of the shunt circuit and the piezoelectric capacity C_p, λ is the dimensionless inductance and ρ is the dimensionless resistance.

3　Optimization of the shunt circuit

In this section an optimization of the dimensionless quantities c, λ and ρ is performed to maximize the exponential time decay rate of the solution. The characteristic polynomial $p(S)$ of the dimensionless system (4a)–(4b) reads explicitly as:

$$p(S) = S^5 + \frac{1+c}{c\rho}S^4 + \left(\frac{1}{c\lambda} + 1 + \kappa^2\right)S^3$$

$$+ \frac{\lambda(1+\kappa^2) + \lambda c + 1}{c\lambda\rho}S^2 + \frac{1+\kappa^2}{c\lambda}S + \frac{1}{c\lambda\rho} \tag{6}$$

The system poles S_i are the roots of the above polynomial $p(S)$. They have a nonpositive real part, due to the inherent passivity of the system. The exponential time decay rate of the free vibrations is given by:

$$\text{ETDR} = -\max_i\{\text{Re}(S_i)\} \tag{7}$$

where Re denotes the real part of a complex number. Consequently, the optimization problem amounts to searching for the values of λ, ρ and c which maximize the ETDR for each fixed value of κ. It is here remarked that the poles of the dimensionless system (4a)–(4b) are proportional to the poles of (3a)–(3b) trough the time-scale coefficient ω_m^{-1}.

3.1   Analysis of the asymptotic case $c \to 0$

In this section it is proved that the addition of a small capacitance in parallel to the inductance L of a RL series shunt circuit increases the ETDR, provided that the piezoelectric coupling coefficient is sufficiently high. To this end, an asymptotic analysis near $c = 0$ is developed.

In the case $c = 0$, the impedance Z in equation (2) reduces $R + sL$. Consequently, the dimensionless system (4a)–(4b) reduces to the case of a RL shunt circuit and the characteristic polynomial becomes [7,13]:

$$\bar{p}(S) = S^4 + \frac{\rho}{\lambda}S^3 + (1 + \kappa^2 + \frac{1}{\lambda})S^2 + \frac{\rho}{\lambda}(1 + \kappa^2)S + \frac{1}{\lambda} \tag{8}$$

which can be also obtained by taking the limit of $c\rho p(S)$ for $c \to 0$. Accordingly, the optimal values of λ and ρ can be computed, in the practical situation $0 < \kappa < 2$, by enforcing that $\bar{p}(S)$ has two coincident couples of complex conjugate roots:

$$\bar{p}(S) = (S^2 - 2x_1 S + x_1^2 + x_2^2)^2 \tag{9}$$

Here x_1 and x_2 are, respectively, the real and the imaginary part of the complex conjugate roots, and the ETDR coincides with $-x_1$. The optimal values of the parameters ρ and λ and the corresponding ETDR are [1,13]

$$\lambda = \frac{1}{(1 + \kappa^2)^2}, \qquad \rho = \frac{2\kappa}{(1 + \kappa^2)^{\frac{3}{2}}}, \qquad \text{ETDR} = \frac{\kappa\sqrt{1 + \kappa^2}}{2} \tag{10}$$

Returning now to the case $c > 0$ and limiting the analysis to the case of c close to 0, the following factorization of the fifth-order degree polynomial $p(S)$ is assumed:

$$p(S) = (S^2 - 2x_1 S + x_1^2 + x_2^2)^2(S - x_3) \tag{11}$$

Here x_3 is a real root which approaches $-\infty$ when $c \to 0$: hence, for each fixed small value of c, the ETDR is still the opposite of x_1.

By identifying the expressions (6) and (11) of $p(S)$, five nonlinear equations in the six unknowns x_1, x_2, x_3, λ, ρ and c are obtained, where κ is intended to be kept fixed. By an elimination procedure, a single nonlinear equation $f(x_1, c) = 0$ is obtained, which implicitly defines a function $x_1(c)$. The derivative $\partial x_1/\partial c$ can be obtained by a straightforward application of Dini's theorem. In particular, for $c = 0$ this derivative is given by

$$\left.\frac{\partial x_1(c)}{\partial c}\right|_{c=0} = -\frac{\kappa}{4}\frac{2\kappa^2 - 1}{\sqrt{1 + \kappa^2}} \tag{12}$$

and is plotted versus κ in Fig. 2.

It is emphasized that $\partial x_1/\partial c|_{c=0}$ is negative for $\kappa > 1/\sqrt{2}$. Hence, x_1 is a decreasing function of c around $c = 0$. As a consequence, for $\kappa > 1/\sqrt{2}$, adding a small capacitance in parallel to the inductance increases the ETDR,

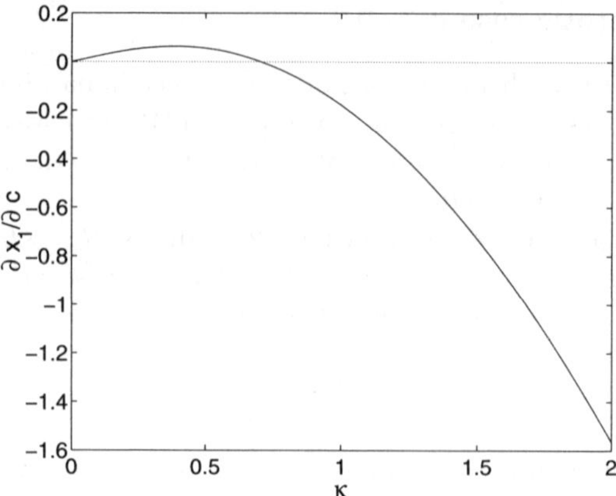

Fig. 2. Behaviour of $\partial x_1/\partial c$ versus κ for $c = 0$

proving that the proposed shunt circuit performs better than the classical RL one. It is emphasized that this conclusion is not limited by the assumed factorization (11) used in the computations. Indeed, if another factorization of $p(S)$ exist yielding a higher ETDR than the one supplied by (11), *a-fortiori* that ETDR would be better than the one provided by the RL shunt circuit.

3.2 Optimization of the $R(L||C)$ shunt circuit

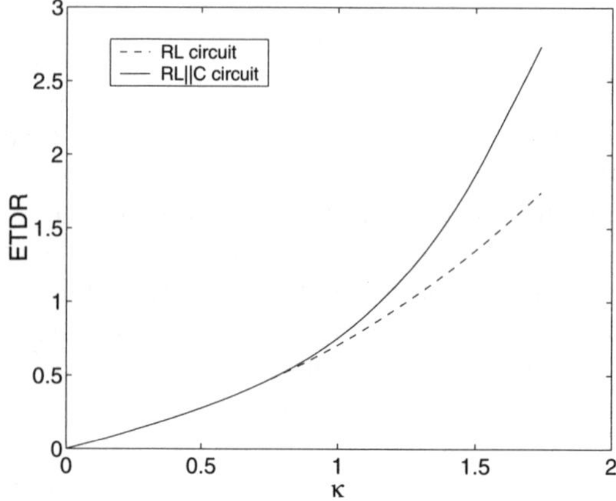

Fig. 3. Behavior of ETDR versus κ

In this section an analytical optimization of the proposed circuit is performed. In particular, the optimal values of ρ, λ and c, and the corresponding

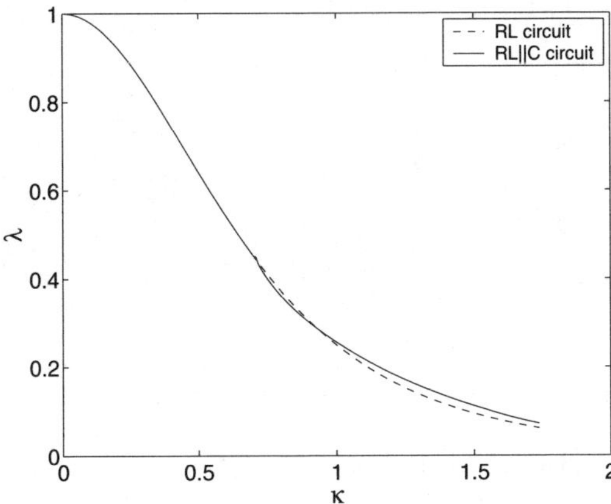

Fig. 4. Behavior of λ versus κ

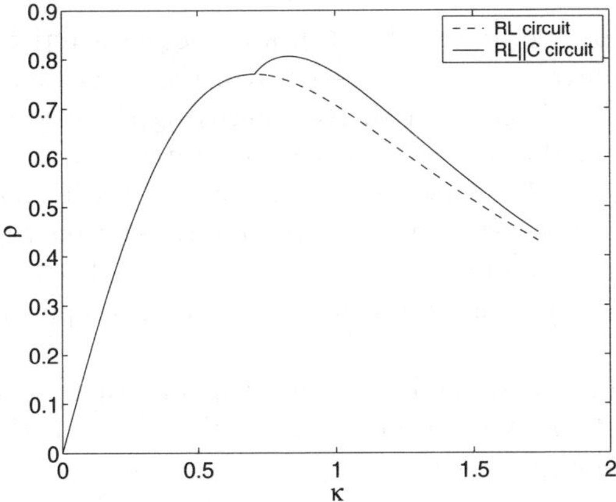

Fig. 5. Behavior of ρ versus κ

ETDR, are determined for given values of the piezoelectric coupling coefficient κ.

As a consequence of equation (12), it turns out that, for $\kappa \leq 1/\sqrt{2}$ no increase in performances can be obtained with respect to the RL circuit if a small capacitance C is added in parallel to the inductance L. Accordingly, in this range the optimal value of c, near $c = 0$, is just zero, and the optimal expressions for the other parameters are those relevant to the RL circuit, given in equation (10).

When $\kappa > 1/\sqrt{2}$, equation (12) implies that there exists an optimal value of c, strictly greater than zero, yielding the optimal ETDR. In order to apply the pole placement technique to find the optimum, a special factorization of $p(S)$ is needed. This factorization is suggested by the case of the resistive-inductive shunt circuit. In that case, two coincident couples of com-

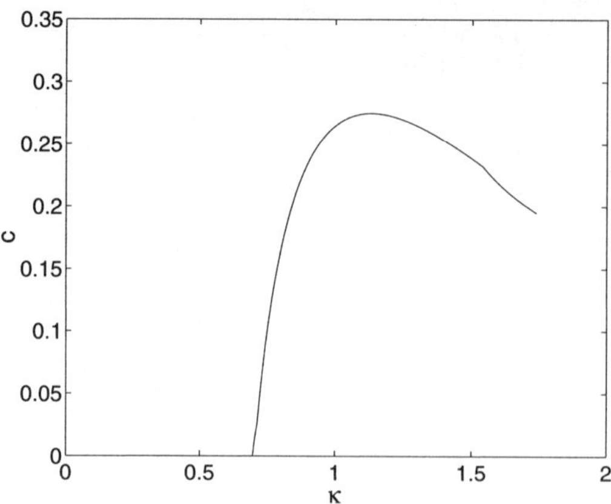

Fig. 6. Behaviour of c versus κ

plex conjugate roots can be collocated. In the $R(L\|C)$ case, the characteristic polynomial $p(S)$ is fifth-order degree. Hence, an extra real root exists and, accordingly, the factorization (11) is considered. By identifying the expressions (6) and (11) of $p(S)$, five nonlinear equations can be written involving the six unknowns x_1, x_2, x_3, ρ, λ and c. These equations define implicitly x_1, x_2, x_3, ρ and λ as functions of c. Thus, the extra condition of the stationarity of $x_1 = -\text{ETDR}$ with respect to c can be added. In the range $\kappa > 1/\sqrt{2}$ an admissible solution to this stationary condition can be found and is reported in Appendix 4.1.

For $\kappa = \overline{\kappa} \simeq 1.53$, computed in Appendix 4.1, it turns out that $x_1 = x_3$, so that all the real parts of the poles are coincident. When $\kappa > \overline{\kappa}$ the factorization (11) is no longer optimal, since x_3 would became greater than x_1. Hence, the optimal factorization enforces that the double complex poles and the real pole have the same real part:

$$p(S) = (S^2 - 2x_1 S + x_1^2 + x_2^2)^2 (S - x_1) \tag{13}$$

In this case, there are only five unknowns x_1, x_2, ρ, λ and c involved into the five nonlinear equations coming from the polynomial identification, which determine the admissible solution reported in Appendix 4.2.

Finally, for $\kappa = \tilde{\kappa} \simeq 1.75$, computed in Appendix 4.2, there exists a choice of the electrical parameters such that the polynomial $p(S)$ exhibits a fifth-order real pole, and the best possible ETDR supplied by the proposed shunt circuit is attained.

The values of the ETDR, λ, ρ and c are respectively plotted in Figs. (3)–(6) versus the piezoelectric coupling coefficient κ, and compared to the corresponding quantities relevant to the RL circuit. From Fig. 3 it turns out that, for $\kappa > 1/\sqrt{2}$, the $R(L\|C)$ circuit exhibits a greater ETDR than the classical RL circuit. Moreover the difference between the performances of the two shunt circuits increases with increasing κ. From Figs. 4 and 5 it follows that the optimal inductance and resistance relevant to the $R(L\|C)$ circuit are almost the same as the ones required by the RL shunt circuit. Finally, Fig. 6 shows that c has a maximum of the order of 0.3; hence, the capacitance C to be put in parallel to the inductance L is less than 30% of the piezoelectric capacity C_p.

The above results show that the proposed shunt circuit performs better than the RL shunt circuit for $\kappa > 1/\sqrt{2}$, and requires values of the electrical component very close to the ones required by the RL circuit.

4 Conclusions

In this paper the problem of choosing the optimal shunt circuit for single-mode passive vibration damping was discussed. It was shown that, at least for sufficiently high coupling coefficients κ, the traditional RL circuit is not the most effective. Indeed, the performance of a new shunt circuit proposed here is definitely better. It is emphasized that in modern applications very high values of κ can be achieved, for instance by using negative capacitances. In multi-modal vibration damping the basic design concepts remain the same and therefore the problem of choosing the optimal shunt circuit must be faced also in that case.

Acknowledgments

This research was developed within the framework of Lagrange Laboratory, an European research group between CNRS, CNR, University of Rome "Tor Vergata", University of Montpellier II, ENPC and LCPC.

References

1. Caruso, G. (2000) Laminati piezoelettrici, modellazione, algoritmi di calcolo ed ottimizzazione della risposta dinamica. Doctoral Thesis, Dep. of Civil Engineering, University of Rome "Tor Vergata".
2. Agnes, G. S. (1995) Development of a modal model for simultaneous active and passive piezoelectric vibration suppression. J. Intell. Mater. Syst. Struct. **6**, 482–487
3. Edberg, D. L., Bicos, A. S., and Fechter, J. S. (1991) On piezoelectric energy conversion for electronic passive damping enhancement. Proc. Damping '91 (San Diego, CA, 1991) paper GBA-1.
4. Bisegna, P., Caruso, G., Del Vescovo D., Galeani, S. and Menini, L. (2000) Semi-active control of a thin piezoactuated structure. SPIE Proc. **3989**, 300–311
5. Inman, D. J. (1997) Vibration suppression trough smart damping. Proc. of the Fifth International Congress on Sound and Vibration, Adelaide, South Australia
6. Lesieutre, G. A. (1998) Vibration damping and control using shunted piezoelectric materials. Shock vibr. digest **30**, 181–190

7. Hagood, N. W. and von Flotow, A. (1991) Damping of structural vibrations with piezoelectric materials and passive electrical networks. J. Sound Vib. **146**, 243–268

8. Wu, S. Y. (1996) Piezoelectric shunts with a parallel R-L circuit for structural damping and vibration control. SPIE Proc. **2720** 259–269

9. Park, C. H. and Inman, D. J. (1999) A uniform model for series R-L and parallel R-L shunt circuits and power consumption. SPIE Proc. **3668** 797–804

10. Park, C. H. (2003) Dynamics modelling of beams with shunted piezoelectric elements. J. Sound Vib. **268** 115–129

11. Park, C. H., Kabeya, K. and Inman, D. J. (1998) Enhanced piezoelectric shunt design. Trans. ASME **57** 149–155

12. Caruso, G. (2000) Smorzmento passivo di vibrazioni mediante materiali piezoelettrici e circuiti elettrici risonanti. Proc. XXIX National Congress AIAS, Lucca, Italy, 307–316

13. Caruso, G. (2001) A critical analysis of electric shunt circuits employed in piezoelectric passive vibration damping. Smart Mater. Struct. **10** 1059–1068

14. Tang, J. and Wang, K. W. (2001) Active-passive hybrid piezoelectric networks for vibration control: comparisons and improvement. Smart Mater. Struct. **10** 794–806.

15. Bisegna, P. and Caruso, G. (2000) Mindlin-type finite elements for piezoelectric sandwich plates J. Intell. Mater. Syst. Struct. **11** 14–25

Appendix

4.1   Optimization according to factorization (11) with stationary x_1

For $1/\sqrt{2} < \kappa < \bar{\kappa}$ the optimal value of x_1 is given by

$$x_1 = -\sqrt{\max_i\{z_i\}} \tag{14}$$

where z_i are the positive real roots of the polynomial in the unknown z

$$\begin{aligned}
q(z) = {}& 4096(\kappa^2 - 8)^2 z^8 \\
& +2048(-7\kappa^6 + 42\kappa^4 - 336\kappa^2 + 128)z^7 \\
& +512(36\kappa^8 + 91\kappa^6 + 635\kappa^4 + 448\kappa^2 + 192)z^6 \\
& +16(\kappa^{12} - 644\kappa^{10} - 5324\kappa^8 - 9720\kappa^6 - 5520\kappa^4 + 896\kappa^2 + 1024)z^5 \\
& -8(1 + \kappa^2)(7\kappa^{12} - 230\kappa^{10} - 1210\kappa^8 + 2270\kappa^6 \\
& \qquad\qquad\qquad\qquad +6238\kappa^4 + 2912\kappa^2 - 128)z^4 \\
& +\kappa^2(1 + \kappa^2)^2(73\kappa^{10} + 684\kappa^8 + 6964\kappa^6 + 14544\kappa^4 + 8096\kappa^2 - 256)z^3 \\
& -\kappa^4(1 + \kappa^2)^3(43\kappa^8 + 488\kappa^6 + 1596\kappa^4 + 1024\kappa^2 - 16)z^2 \\
& +\kappa^8(1 + \kappa^2)^4(11\kappa^4 + 76\kappa^2 + 44)z \\
& -\kappa^{12}(1 + \kappa^2)^5 \tag{15}
\end{aligned}$$

The optimal values of the electrical parameters are given by

$$\lambda = -\frac{(1 + \kappa^2 - 4x_1 x_3 - 4x_1^2 - 2y)y^2 x_3}{\Lambda} \tag{16}$$

$$c = \frac{\lambda}{(4x_1x_3 + 4x_1^2 + 2y - 1 - \kappa^2)} \tag{17}$$

$$\rho = \frac{(1+c)}{-c(x_3 + 4x_1)} \tag{18}$$

where

$$x_3 = -\frac{1}{4}\frac{y^2 - 2y - 2\kappa^2 y - 4x_1^2\kappa^2 + 2\kappa^2 - 4x_1^2 + 1 + \kappa^4}{x_1(-1 + y - \kappa^2)} \tag{19}$$

$$\Lambda = y^2 x_3 - 4x_3^2 x_1 - 20x_1^2 x_3 - 2yx_3 + x_3 + x_3\kappa^2 - 16x_1^3 - 8yx_1 + 4x_1 + 4x_1\kappa^2 \tag{20}$$

and y is given by the maximum positive real root of the following polynomial in the unknown y

$$\begin{aligned}
r(y) = {}& y^5 + (4x_1^2 - 4\kappa^2 - 5)y^4 + (10 + 16\kappa^2 - 12x_1^2\kappa^2 - 8x_1^2 + 6\kappa^4)y^3 \\
& + (32x_1^4 - 10 - 4\kappa^6 - 16x_1^4\kappa^2 + 12x_1^2\kappa^2 - 18\kappa^4 - 24\kappa^2 + 12x_1^2\kappa^4)y^2 \\
& + (5 + 8x_1^2 - 4\kappa^6 x_1^2 + 18\kappa^4 + 12x_1^2\kappa^2 - 48x_1^4 + 16\kappa^2 \\
& \hspace{4cm} + \kappa^8 - 48x_1^4\kappa^2 + 8\kappa^6)y \\
& -1/4 - 3x_1^2\kappa^2 - 3\kappa^4/2 - \kappa^6 - \kappa^2 + 4x_1^4 - 3x_1^2\kappa^4 \\
& \hspace{1cm} + 4\kappa^4 x_1^4 - \kappa^8/4 - x_1^2 + 16x_1^6\kappa^2 + 16x_1^6 + 8x_1^4\kappa^2 - \kappa^6 x_1^2 \tag{21}
\end{aligned}$$

The imaginary part of the double complex conjugate roots is $x_2 = \pm\sqrt{y - x_1^2}$.

The present factorization supplies an optimal ETDR as long as the real root x_3 is less than or equal to the real part x_1 of the double complex conjugate roots. The limit $\overline{\kappa}$ in Section 3.2 is just defined as the value of κ which makes the equality prevail. It is given by the real positive solution of the following system in the unknowns κ and x_1

$$\begin{aligned}
& A(4x_1^2\kappa^2 - 2 + \kappa^4 - \kappa^2 - 2x_1^2) \\
& \quad = 6x_1^2 + 6x_1^4 + \kappa^2 x_1^2 - 5\kappa^4 x_1^2 + 4\kappa^2 x_1^4 \\
& A(6x_1^2 + 13\kappa^2 + 9\kappa^6 - 32x_1^4 + 20\kappa^4 - 8x_1^4\kappa^2 + 2 + 16x_1^2\kappa^4 + 22x_1^2\kappa^2) \\
& \quad = -8\kappa^2 x_1^6 - 3\kappa^6 - 62\kappa^4 x_1^2 + 34\kappa^2 x_1^4 + 32x_1^6 - 26\kappa^6 x_1^2 - \kappa^8 \\
& \quad\quad + 22x_1^4 - 10x_1^2 - \kappa^2 + 12\kappa^4 x_1^4 - 3\kappa^4 - 46\kappa^2 x_1^2 \tag{22}
\end{aligned}$$

where $A = \sqrt{x_1^2(1 + x_1^2 + \kappa^2)}$.

4.2   Optimization according to factorization (13)

For $\overline{\kappa} < \kappa < \tilde{\kappa}$ the optimal value of x_1 is given by

$$x_1 = -\frac{\sqrt{2}}{2}\sqrt{\frac{2\kappa^6 - 3\kappa^2 + \kappa^4 - 2 + \sqrt{4\kappa^{12} - 8\kappa^8 + 2\kappa^{10} - 6\kappa^6}}{1 + 2\kappa^2}} \tag{23}$$

and the optimal parameters are

$$\lambda = \frac{(8x_1^2 + 2y - 1 - \kappa^2)y^2}{-y^2 + 40x_1^2 + 10y - 5 - 5\kappa^2} \tag{24}$$

$$c = \frac{\lambda}{8x_1^2 + 2y - 1 - \kappa^2} \tag{25}$$

$$\rho = \frac{1}{5}\frac{1+c}{(-x_1)c} \tag{26}$$

where

$$y = -2x_1^2 + \kappa^2 + 1 + 2\sqrt{x_1^4 + \kappa^2 x_1^2 + x_1^2} \tag{27}$$

The present factorization is admissible as long as two double complex conjugate roots exist. When $\kappa \to \tilde{\kappa}$, given by

$$\tilde{\kappa} = \sqrt{24 + 6\sqrt{5} - \frac{2}{25}\sqrt{5}(125 + 30\sqrt{5})} \tag{28}$$

the imaginary part $x_2 = \pm\sqrt{y - x_1^2}$ of those conjugate roots tends to zero, and for $\kappa = \tilde{\kappa}$ the characteristic polynomial $p(S)$ admits five coincident real roots.

Authors Index

Printing: Mercedes-Druck, Berlin
Binding: Stein+Lehmann, Berlin